Modeling Social Behavior

Modeling Social Behavior

MATHEMATICAL AND AGENT-BASED MODELS OF SOCIAL DYNAMICS AND CULTURAL EVOLUTION

Paul E. Smaldino

PRINCETON UNIVERSITY PRESS

Princeton & Oxford

Published by Princeton University Press
41 William Street, Princeton, New Jersey 08540
99 Banbury Road, Oxford OX2 6JX

press.princeton.edu

Library of Congress Cataloging-in-Publication Data

Names: Smaldino, Paul E., 1980– author.
Title: Modeling social behavior : mathematical and agent-based models of social dynamics and cultural evolution / Paul E. Smaldino.
Description: Princeton : Princeton University Press, [2023] | Includes bibliographical references and index.
Identifiers: LCCN 2023001026 (print) | LCCN 2023001027 (ebook) | ISBN 9780691224138 (cloth) | ISBN 9780691224145 (paperback) | ISBN 9780691224152 (ebook)
Subjects: LCSH: Social groups–Mathematical models. | Social interaction–Mathematical models. | Social psychology–Mathematical models. | BISAC: SOCIAL SCIENCE / Methodology | SCIENCE / Cognitive Science
Classification: LCC HM716 .S547 2023 (print) | LCC HM716 (ebook) | DDC 302.301/5118–dc23/eng/20230403
LC record available at https://lccn.loc.gov/2023001026
LC ebook record available at https://lccn.loc.gov/2023001027

British Library Cataloging-in-Publication Data is available

Editorial: Alison Kalett, Hallie Schaeffer
Jacket/Cover Design: Wanda España
Production: Jacqueline Poirier
Publicity: William Pagdatoon

Cover Image: Kandella / Alamy Stock Vector

This book has been composed in MinionPro

Printed on acid-free paper ∞

Printed in the United States of America

10 9 8 7 6 5 4 3 2 1

Brief Contents

Contents

Preface

One thing can be known only through another, all wisdom has taught.
—Piet Mondrian, *Natural Reality and Abstract Reality* (1919)

If you've ever had a serious conversation about "the way things are" or "the way things ought to be," you may have noticed in frustration that other people don't always see eye to eye with you. You may have borne witness to others arguing past each other, each failing to grasp the other's underlying assumptions. This miscommunication is a problem if we're ever going to understand and improve our world with a science of social dynamics. In order to build up such a science, we need clear assumptions that we can articulate, communicate, and work through the consequences thereof. And because we can't articulate *all* the details of the world at once, we need to simplify and focus on the most important aspects of whatever we're talking about. This simplification is called **modeling**. This is a book about modeling: about the quantitative techniques needed for modeling, about specific models that illuminate processes central to social life, and about the philosophical perspectives needed to understand previous models and design good models of your own.

The book is intended to equip those that work in the social, behavioral, and cognitive sciences with a toolkit for thinking about and studying complex social systems using mathematical and computational models. The book is therefore concerned with the study of social organisms, with a focus on what they do *together*. The examples tend to center on humans, but many of the lessons apply to non-human species as well. It is a strange historical fact that researchers studying social processes are usually divided into numerous separate disciplines whose practitioners often appear to forget that the other disciplines exist and are working on related problems. This is a real shame, as we have lots to learn from one another. Throughout this book I'll use the terms "social scientists" and "social sciences" as a shorthand for my intended audience and their fields of study, but I hope that this book is useful to anyone interested in the behavior of social creatures and in the dynamics of the social systems in which they are embedded. This includes readers with a background in psychology, sociology, anthropology, economics, political science, cognitive science, or behavioral ecology, but the book should also be of interest to researchers and scholars coming from related fields such as epidemiology, evolutionary biology, communication, computer science, applied mathematics, and philosophy.

Constraints on time and the efficiencies offered by division of labor have given rise to a minor tragedy: there is often a lack of overlap in training between those with strong quantitative skills and those with a rich understanding of social systems. There are good reasons why this is the case, but this underserved overlap is exactly where the most attention is needed. I hope that this book serves in some capacity to create a bridge between those with an appreciation of social systems who want to develop tools for modeling those systems and those with strong mathematical and/or computational skills who want to apply their abilities to the social sciences. Although many books treat agent-based modeling and mathematical modeling as completely separate techniques, this book correctly treats them as complementary approaches to the problem of modeling social behavior, and so provides instruction on both approaches. I have tried to keep the technical aspects of this book accessible to those without advanced mathematical training, in part because there are many more books on modeling written for the technically savvy than for the novice, and in part because applying old tools to new problems often benefits from a beginner's mind.

The social sciences do not have a solid, broadly agreed-upon theoretical foundation in the same way that the physical sciences and even some of the biological sciences do. This is partly because the social sciences are younger, and partly because they deal with extremely complex systems that can be described at multiple levels of organization. Yet, without something concrete to build upon, we cannot get very high off the ground. I offer this book as a starting place for a quantitative education in theory for the social sciences. An education in physics usually starts with models of mechanics that keep things very simple, like restricting systems to one or two parts, ignoring factors like friction and wind, and treating complex objects as single points. Onto such a foundation can complexity later be added. Similarly, the models presented in this book cannot be used unaltered to do predictive work in the social sciences, but I believe that they lay the foundation for such work. Along the way, they may scaffold a deeper understanding of our world and highlight important questions we need to be asking. It's a start.

How to Use This Book

Technical prerequisites

Although some models require advanced mathematical and/or computational techniques to understand and analyze, we will focus on dynamic models that are relatively simple. In terms of mathematics, proficiency with high school algebra and basic probability theory is sufficient, though knowledge of calculus, dynamical systems, and/or statistics is valuable and worth investing in. In terms of computer programming, if you have ever written any program, even a simple script, you are in decent shape. Computational modeling does, at its core, involve programming, so the more experience you have, the better. That said, if you *don't* have much programming experience, that's OK. You have to start somewhere, and this is as good a place to start as any. We'll be primarily using the NetLogo language, which, as I will discuss below, was designed to be easily learned by novices.

The results of computer simulations often need to be plotted. This book assumes that you will be able to adequately plot your data. While you can get by using mass-market software like Microsoft Excel to make your figures, I advise you to consider learning to use more advanced tools. Not only will you be able to make high-quality figures, but you will be able to re-use code and easily reproduce old figures with new data. For example, many of the graphs in this book were made using the R programming language with the ggplot package. Data visualization is an art unto itself, and we shall spend minimal time on it in this book, but it's worth investing in learning how to visualize both model dynamics and the data that result from them. Some good resources are Healy (2019), Wilke (2019), and Franconeri et al. (2021).

The book does not have a glossary, but some words are rendered in **bold type** to indicate that they are important terms worth noting. Some chapters also have shaded BOXES that contain particularly technical material or tangential asides. The casual reader may skip these.

This book involves a lot of agent-based modeling. While this is a valuable approach, I want to make clear that it is not the only way to model social dynamics. A complementary approach is to use purely equation-based mathematical modeling. We will cover models of this type as well and discuss their relationship to agent-based models.

NetLogo

The agent-based models in this book have all been coded using NetLogo. NetLogo is a useful and widely adopted software package and language for building and analyzing agent-based models. If you haven't already, you should download it at:

https://ccl.northwestern.edu/netlogo/

We will go over some of the basics of NetLogo in chapter 2, but this book does not provide extensive instruction on using NetLogo. One of the best ways to get acquainted with NetLogo is to work through the tutorials provided in the accompanying user's manual—this introduction should be sufficient for most of what is covered in this book. For those desiring further education in using NetLogo, additional resources are listed here:

https://ccl.northwestern.edu/netlogo/resources.shtml

Throughout this book, any in-line NetLogo code will be rendered in a monospaced font, so that you will know that phrases like `ask turtles [fd 1]` are computational commands. Blocks of code are indicated by shaded boxes, like this:

NetLogo code 0.1

```
ask turtles [fd 1]
```

The complete code for all the models referenced in this book is provided in the following repository, with each model located in the folder corresponding to the appropriate chapter:

https://github.com/psmaldino/modsoc/

Why This Book Uses NetLogo But You Don't Have To

Agent-based models can be programmed in nearly any programming language. It is certainly possible to employ them without ever using NetLogo. I have good pedagogical reasons for using NetLogo in this book. In particular, I have found that it is the best way to place the focus on the modeling rather than the programming. However, the savvy programmer should find many valuable lessons in these pages regardless of their preferred language.

For *any* scientist, skills in mathematics, computer programming, and data visualization are useful and worth cultivating. For the modeler, these skills are essential. Let's focus on programming. I appreciate that for some people, learning to code is not trivial. Unfortunately, there's just no getting around it. It's true that some equation-based modeling can be done entirely with pencil and paper, and we will do some of that in this book. Even then, software is helpful for plotting, for numerical simulation, and for solving particularly gnarly equations. Agent-based models are explicitly computational. To use them, you must learn to code.

This book is not a course on coding. In fact, I want to spend as much time as possible talking about modeling and as little time as possible talking about coding. If you are already an expert coder, that's great! We can get right to work. If you are a novice coder, don't worry! There are now many tools available to help you build, visualize, and analyze agent-based models that are accessible and relatively easy to learn.[1]

Most programming languages are Turing complete, which means that, in theory, anything you can do in one language you can do in another. Technically, you don't even need a computer to build an agent-based model. Thomas Schelling, who we will discuss in chapter 3, simulated his early agent-based models of segregation on a checkerboard using pennies and dimes. For any model of even modest complexity, however, computers are handy. In terms of programming languages, you have many choices. I have personally worked on agent-based models in Java, Python, R, MATLAB, Julia, and NetLogo. I have colleagues who use or have used C, C++, Pascal, and JavaScript. I once saw a paper that analyzed an agent-based model using only Microsoft Excel. The point is that you can build agent-based models using whichever language you are most comfortable with. Throughout this book, though, I will assume that we are using NetLogo. I'll tell you why.

NetLogo first appeared in 1999. It is a software package expressly designed to build and analyze agent-based models with the philosophy of "low threshold, no limit" (Wilensky, 1999). "Low threshold" means that NetLogo is easy to learn even for those with minimal coding experience. Its syntax is designed to mimic natural language whenever possible, making it relatively easy to understand what a line of code accomplishes. It has a number of drag-and-drop features, and many built-in functions common to models of social behavior. And it makes visualizing and plotting model dynamics a snap. It is for these reasons that many introductory modeling courses, this one included, use it. "No limit" means that NetLogo can be used for high-level scientific modeling. Indeed, many peer-reviewed papers are published each year describing models built and analyzed with NetLogo.

In practice, I find that there are at least two limits to NetLogo. The first concerns speed. NetLogo is itself programmed in a mix of Scala and Java, which means that its code is translated to Java and must run its wrapper software to function. This constrains the speed at which simulations run, especially compared with compiled languages like C++ or pure Java. That said, modern computers are very fast, and throughout this book we will be working with relatively simple models that will not overly tax the capabilities of NetLogo. Moreover,

[1] Perhaps not *trivially* easy, but I am confident that you are up to the challenge.

NetLogo includes tools to translate its code to a Java JAR file, which will run much faster than raw NetLogo and can be run on high performance computing clusters for added efficiency.

The second limitation concerns functionality. NetLogo comes with a slew of built-in software components that make it well suited for replicating and extending many influential agent-based models. These include features for scheduling agents, situating agents in a physical space or on a network, imbuing agents with properties that are readily accessed and transmitted to other agents, and for plotting model output and running large batches of simulations. These features are all hugely helpful. The downside is that the design also limits the modeler who requires a decomposition of reality not envisioned by NetLogo's creators. For example, NetLogo isn't great at handling some of the data structures used in complex agent-based models, like multidimensional matrices or complex network structures such as multipartite or multilayer networks.

So why use NetLogo? Let me propose an analogy to the teaching of robotics. Imagine you want to learn how to make a robot that is capable of solving simple mechanical tasks, like pushing a checker around a checkerboard. You sign up for a robotics class with Professor Wily, who hands you a circuit board, some wires, servos, an assortment of wheels and joints, and a soldering iron, and tells you to get to work. If you tackle the problem in earnest, you'll learn all the nuts and bolts of robots, but it will also be a long while before you manage to produce anything functional. Frustrated, you ditch that class and head across the hall, where Professor Light hands you a Lego Mindstorms kit. The kit comes with a more restricted set of components, and its processors are slower than the ones used in Professor Wily's class. However, you can start putting together working robots in the first week of class and focus your effort on learning how to make them do your bidding. Eventually, you get a good sense of how to program robot behavior, and you can return to Professor Wily's class with a much clearer idea of where your efforts will take you.

I think of NetLogo in a similar way. It allows us to avoid much of the pesky machinery of programming so that we can focus on the conceptual aspects of modeling. NetLogo is free and open source and comes bundled with a large library of demo code and tutorials to help get you started. There are also many resources, both free and paid, to help you to learn it. Using NetLogo yields a rapid payoff, in the sense that we can have working models with good visualizations up and running in minutes. I know of no better entry point into agent-based modeling than NetLogo. Crucially, many of the lessons you learn using NetLogo will transfer to modeling with other languages, because NetLogo allows us to focus on modeling, not programming.

So this book will use NetLogo. All that said, you don't *have* to use NetLogo. It is possible to build all of the models discussed in this book in other languages, and if you are so inclined, I encourage you to do so.

Summary of Chapters

This book was written as a course, and the chapters are intended to be read in numerical order. Some readers or instructors may find that covering the chapters in a slightly different order better suits their needs or preferences. That said, I have endeavored to present topics in a logical progression in terms of both methodological complexity and conceptual coherence.

Chapter 1 introduces the philosophical approach to modeling used throughout this book, as well as a number of key modeling concepts. Chapter 2 introduces agent-based modeling with NetLogo and walks through the creation of a simple model with embodied mobile agents. Qualitative analyses are discussed, including how to plot the output of individual simulations. Chapter 3 introduces the Schelling Segregation model, including core concepts in spatial modeling like cellular automata. In this chapter we also learn how to run simulation batches and sweep across parameter values. Chapter 4 introduces compartment models of contagion, used to study the spread of disease, information, and behavior. This chapter introduces purely mathematical approaches to complement the agent-based simulations used and will also discuss how to model counterfactuals in order to study potential interventions. Chapter 5 introduces opinion dynamics, in which individuals' opinions are represented as continuous values that can change through social influence. With these models we will explore topics like consensus, differentiation, and polarization, and how to carefully examine model assumptions.

Chapter 6 focuses on cooperation and introduces many important concepts and techniques, including game theory, evolutionary dynamics, and Hamilton's rule. In this chapter we learn how to take insights from agent-based models and then prove them mathematically. Chapter 7 extends the evolutionary modeling paradigm and discusses how coordination games can be used to study the emergence of norms, the role of group structure, and the division of labor. Chapter 8 focuses on how scientific inquiry and the norms related to its practice can be modeled. It introduces Bayes' theorem, which allows us not only to consider how individual scientists should evaluate evidence, but also to model populations of scientists responding to incentives.

Chapter 9 is an introduction to network science. Social creatures are structured in social networks that shape their interactions and information flow. In chapters 3–8, model dynamics in either well-mixed or rigidly structured populations are regularly compared, the latter using very simple network structures, usually a square lattice. In this chapter, network metrics and algorithms for more complex network architectures are explored. Chapter 10 discusses guidelines for building original models and methods for analyzing those models both as theoretical objects and in light of empirical data. Finally, Chapter 11 discusses the promise of modeling but urges caution in drawing conclusions too hastily.

Acknowledgments

I am deeply indebted to the people who gave their time to read some or even all of my early drafts of this book: Clark Barrett, Bret Beheim, John Bunce, Aaron Clauset, Tamara van der Does, Riccardo Fusaroli, Mirta Galesic, Ketika Garg, Jamie Holland Jones, Mark Lubell, Mike Makowsky, Matt Miller, Cailin O'Connor, Karthik Panchanathan, Alejandro Pérez Velilla, Matt Turner, and Matt Zefferman. Double thanks are owed to Clark Barrett, Mirta Galesic, and Karthik Panchanathan, who took the deep dive and gave me feedback on every chapter. This book is better for the contributions of all these readers, and I am humbled to be the recipient of their generosity. I thank all the students I have taught in classes and workshops on modeling who helped me refine my materials by telling me what worked and what didn't. I also thank the many friends, colleagues, and mentors with whom I have discussed models and modeling.

This book focuses mainly on models that were originally developed by others, though in many cases I have altered or reinterpreted those models for the sake of pedagogy. I am grateful to all the modelers whose work has inspired this book's chapters. They are too many to list in full, but I want to particularly acknowledge the debt owed to the work of Craig Reynolds, Thomas Schelling, Scott Page, Guillaume Deffuant, Andreas Flache, Michael Macy, William Hamilton, Robert Axelrod, Robert Boyd, Peter Richerson, Joseph Henrich, Duncan Watts, Steven Strogatz, Albert-László Barabási, Réka Albert, and Damon Centola. I do present some of my own original work where it seemed appropriate, including work done in collaboration with the inimitable Richard McElreath and Jeffrey Schank. Thanks also to Jeremy Van Cleve for help with the figure depicting fitness landscapes in chapter 10.

I am thankful to my colleagues at UC Merced who encouraged and supported this project. When I arrived at UC Merced, I was asked what I wanted to teach. Without a clear idea of what it would look like, I suggested a class on modeling social behavior. This turned into two classes, one for undergraduates and one for graduate students. I am also grateful to those who have invited me to give workshops and lectures on modeling, agent-based and otherwise. This book is the product of these courses and workshops.

I am indebted to Alison Kalett at Princeton University Press, who encouraged me to write the book, and to Hallie Schaeffer for shepherding the manuscript through the labyrinthine editorial process. I owe thanks to John and Becky Newton for the use of their house in Cambria, California, which provided me with a quiet spot near the sea in which large portions of this book were written. Finally, I thank Emily Newton for all her support throughout the lengthy process of writing this book.

1 Doing Violence to Reality

> The truth lies directly before us in the reality surrounding us. However, we cannot use it as it is. An unbroken description of reality would be simultaneously the truest and most useless thing in the world, and it would certainly not be science. If we want to make reality and therefore truth useful to science, we must do violence to reality. We must introduce the distinction, which does not exist in nature, between *essential* and *inessential*. In nature, everything is equally essential. By seeking out the relationships that seem essential to us, we order the material in a surveyable way at the same time. Then we are doing science.
>
> —Jakob von Uexküll, *Umwelt und Innenwelt der Tiere* (1909, p. 227)

The sciences of social behavior are more important than ever. These include the human social and behavioral sciences as well those branches of biology, physics, and applied mathematics that deal with social and collective behavior. Many of the most pressing questions for our time are about how groups behave and adapt, on topics ranging from disease spread and political polarization to the maintenance of cooperation, collective action, and the reliability of scientific findings.

A persistent roadblock to a focused attack on these questions is the fact that social scientists are often trained and organized in ways that impede the sorts of interdisciplinary connections needed to solve them. Researchers studying human behavior are siloed into many distinct disciplines, each with different methods, theoretical frameworks, and perspectives. This limits the sorts of questions researchers tend to ask as well as the approaches they use to answer those questions. It also means that researchers often lack frameworks or tools for dealing with complex problems at multiple scales. It would be helpful to have a bridging framework to facilitate communication between researchers and to encourage better research questions.

The past few decades have seen the emergence of several amalgamate fields that integrate key insights from across many different disciplines, helping to create new ways to understand human behavior. These fields include cultural evolution, cognitive science, network science, and complexity science. The perspectives in this book draw from all of

Figure 1.1 A murmuration of starlings.

these approaches. I believe that arming social and behavioral scientists with a basic toolkit of formalized theories and models will allow them to tackle the most important questions from multiple perspectives, and will provide a language through which richer theories of social behaviors can be developed and communicated. I also believe that when these amalgamate fields have been successful—and they have not always been so—it has often been because they relied on relatively simple models to illustrate and articulate the lessons that could be drawn from their core perspectives. In this book, we will learn about many of these models, as well as the tools needed to create and analyze them.

This book is about doing violence to reality, because that is the main undertaking of science. In the course of doing so, reality may become at least temporarily unrecognizable. This is necessary for two main reasons. First, because we have to understand simple systems before we can understand more complex systems. And second, because most of us receive precious little training in understanding social systems with much precision, particularly compared with the precision emphasized in the physical and biological sciences. There are good reasons for that—as Albert Einstein famously quipped, politics is harder than physics. The gambit of this book is that, despite this difficulty, we can still apply the sort of mathematical and computational tools that are typically used for understanding physical systems toward an understanding of social systems. Let's start with an example.

1.1 Flocking Birds and Boids

If you work through the models in this book, you'll be spending a lot of time in front of a computer. Hopefully, you also get outside from time to time to take in some fresh air, move your body, and watch the birds. Sometimes when you watch birds, you see remarkable things. An astonishing sight in some parts of the world is a murmuration of starlings (Figure 1.1), in which tens of thousands of birds gather together in midair and move as one unit. Flocks of birds, schools of fish, and herds of grazing animals all exhibit similar cohesive collective behavior.

What makes the collective behavior exhibited by the starlings so amazing, other than that it just looks cool, is that there is no central authority coordinating all the individual birds.

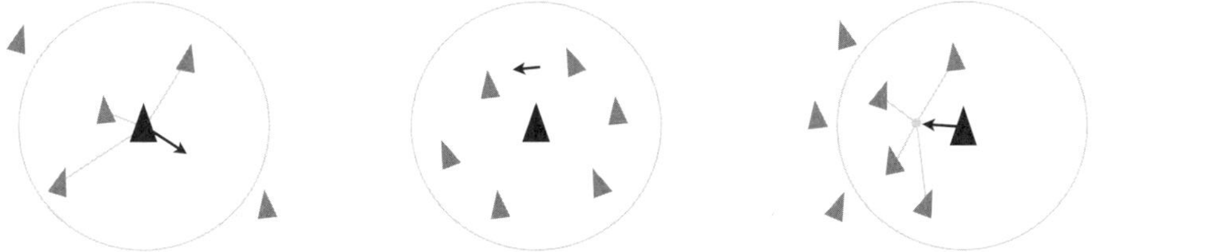

Figure 1.2 Rules of the boids model.

Each bird is aware only of its immediate surroundings, responding locally to the birds it can see, hear, and smell. And yet the collective is more than the sum of its parts—a classic example of **emergent behavior**, in which description at the level of the collective is fundamentally distinct from the behavior of the collective's individual constituents. Emergent behaviors are interesting for reasons beyond their aesthetic appeal. Flocking, schooling, and swarming are almost certainly adaptive, helping prey animals defend against predators and effectively forage in large collective units. How do they do it?

Craig Reynolds, a computer scientist working on motion picture graphics in the mid-1980s (he had been a programmer for the landmark 1982 film *Tron*), wondered how he could build a computer program to simulate realistic-looking flocks, and so he started spending hours outside watching birds. Trying to code the path of each individual bird seemed nigh impossible due to the sheer magnitude of a description characterizing the entire flock. Even if it could be done, edits to change the path of the flock would require starting the coding process all over again. There had to be a better way.

Reynolds noticed that birds seemed to be able to flock with any number of other birds, from just a handful to thousands or more. Thus, he reasoned, "the amount of 'thinking' that a bird has to do in order to flock must be largely independent of the number of birds in the flock." He considered the possibility that each bird might be using relatively simple **heuristics**, or "rules of thumb," in order to stay together with the other birds nearby, and that a whole flock of birds responding thusly—and only to local information—might lead to the appearance of a coherent collective. To investigate this idea with greater care, Reynolds created a computational model: a simulated world full of artificial creatures he named "boids" (Reynolds, 1987).

Boids are particles moving in a two- or three-dimensional space. In the simplest versions of the algorithm, they move with constant speed, varying only in their turning angle, though more complicated versions also exist that account for factors like acceleration. Each boid has only a limited field of vision so that it can perceive the location and directional heading of other nearby boids. That is, each boid is aware only of its local surroundings—no individual boid is aware of the entire flock. Boids use this information to adjust their own directional headings, using the following three rules (Figure 1.2):

- *Separation.* If there are other boids immediately in front of you, turn away from them to avoid collisions and crowding.
- *Alignment.* Turn to align with the average heading of nearby boids.
- *Cohesion.* Attempt to stay close to nearby flockmates by steering toward the average position of nearby boids.

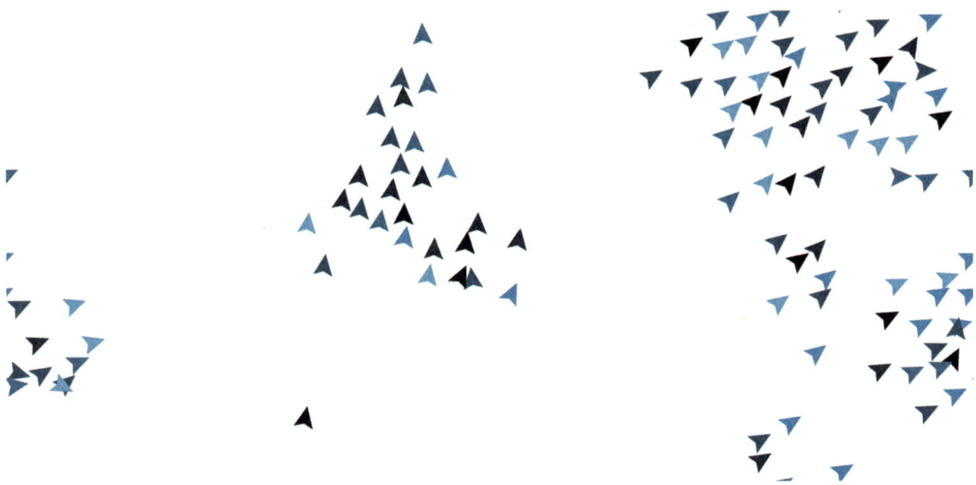

Figure 1.3 Screenshot from a simulation of the boids model.

These three rules, particularly in combination with realistic environments containing objects or agents to avoid, produce flocking behavior that is extremely realistic to the eye (Figure 1.3). Whenever you see a computer-generated flock of birds, school of fish, herd of stampeding wildebeest, or swarm of killer robots in a film, television show, or video game, their movements are almost certainly dictated by some variant of the boids algorithm. Moreover, scientists interested in the collective behavior of animals, from insects and fish to birds and human crowds, have employed computational models based on boids to understand their study systems (e.g., Sumpter, 2006; Couzin, 2009).

Boids is an example of an **agent-based model**, in which individuals are represented as computational entities (agents) that can behave and interact locally. More importantly, boids is an example of a *model*, full stop. It allows us to create a simplified representation of reality and to formally instantiate that representation to observe the consequences of our assumptions. In this case, individuals' bodies and their positions in real space are directly mapped onto the characteristics of the model agents. In this book, I will make the more general argument that models are useful for understanding all sorts of social phenomena, including phenomena for which the mappings are a bit more abstract. At this point, I think it's useful to take a little time to talk about what models *are*.

1.2 What Are Models?

The word "model" means a lot of things. To lay people, the term often refers to a fashion model. I once forgot this and typed "formal models" into Google's image search engine, hoping for images of equations or diagrams, but instead I was rewarded with images of good-looking men in formal attire. Clarification of how the term is used in this book seems in order.

For our purposes, a model is an abstract or physical structure that can potentially represent a real-world phenomenon.[1] Consider this question when conducting research: Are you

[1]This definition is also used by the philosopher Michael Weisberg (Weisberg, 2013).

Figure 1.4 U.S. Army Corps of Engineers Bay Model.

directly studying the system or phenomenon you're interested in? Sometimes the answer is yes. If you want to know the mass of a rock, you can weigh the rock. If you want to know the structure of a particular cell, you can look at the cell under a microscope. In these cases, your questions are about the thing you are directly observing. Very often, however, your questions are really about something else. Your questions might instead relate to things you can't directly and systematically observe due to constraints of time, space, budget, or ethics. Your questions might be about a general class of scenarios, rather than a particular system. In many cases like these, we use models.

Engineers make physical scale models to assess their designs. In the 1950s and '60s, the U.S. Army Corps of Engineers built a two-acre scale model of the San Francisco Bay and Sacramento-San Joaquin River Delta System (Figure 1.4; the true system is 10,000 times larger). The model, which involved a massive undertaking, was used to assess the likely impacts of building dams and rerouting various channels. Assessing this directly would involve actually building the dams and rerouting the channels, which would be extremely costly if the consequences of doing so were not well understood. The model helped convince the Army engineers not to build the dams (Weisberg, 2013). When a single failure of a system can be life-threatening, such models are invaluable.

Biomedical and behavioral scientists use animal models to make general inferences about genetics, physiology, and development. Scientists whose primary goal is to understand the Norway rat or the fruit fly surely exist, but they are just as surely in the minority among researchers studying those animals. Instead, scientists use discoveries about these and other "model organisms" with the aim of making more general claims about the biology and behavior of related animals, especially humans.

Behavioral experiments are almost always models for a larger class of behaviors and scenarios. In a widely cited psychology experiment, children are presented with a marshmallow, then told they can have a second marshmallow if they can wait to eat the first one. The children are then left alone with the first marshmallow for a length of time to see whether they succumb to the temptation to eat it. Unless they are employed by the candy industry or just like tormenting children, very few researchers really care how long children can wait for a second marshmallow. Instead, researchers have used the "marshmallow test" to model situations in which issues like willpower and trust come into play (Mischel and Ebbesen, 1970; Benjamin et al., 2020).

Finally, researchers often build **formal models**. These are mathematical or computational specifications of a system. Formal models are used widely in the more exact sciences, in which elements of a model are direct representations of measurable quantities in the world. Other times, and almost always in the social sciences, models are more abstract, intended to capture core elements of a theoretical idea without a perfect one-to-one mapping between measurement and model.

Formal models are special because they contain nothing more or less than what we put into them. Sometimes, critics of formal modeling say, "Well, you baked your result into the model, so it had to happen." They're not wrong. In fact, this is literally true of every formal model, because the model analysis is merely a series of computations based on assumptions specified by the modeler. To put it another way, a formal model is a logical engine that turns assumptions into conclusions.

It may seem curious that we should want something like this. Why do we need a model to examine our assumptions? If we know what our assumptions are, can't we just think through their consequences? In my experience, we very often cannot. Our intuitions about complex systems are frequently terrible. Good models can show us how our assumptions lead to unexpected conclusions. Moreover, we don't always know what our assumptions even *are* if we've never had to lay them all out. The process of building models often involves a lot of reflection concerning what we *are* assuming and what we *must* assume in order to produce a coherent explanation. The late physicist and complexity pioneer Murray Gell-Mann rightly called formal models "prostheses for the imagination."

1.3 The Parable of the Cubist Chicken

Communication with human language is very often imprecise, ambiguous, and indirect. For most purposes, these can actually be adaptive features of language. Ambiguity, for example, can serve to convince multiple listeners that they all agree with a speaker's big idea, which allows the speaker to vault to prominence, and it may enhance group cohesion by allowing people to cooperate toward what they perceive as a common goal.[2] In science, however, ambiguity is no good. We need to be precise. But in many fields, theories of social phenomena are still articulated in a purely verbal fashion, opening these explanations up to all the problems of ambiguity. I like to illustrate this problem with a brief anecdote I have come to call the parable of the Cubist chicken.

One evening long ago, when I was an undergraduate student, a friend and I found ourselves waiting in the basement of a theater for a third friend, an actor about to finish his

[2] There is a fascinating literature on the adaptive and strategic uses for ambiguity. See Eisenberg (1984); Sperber and Wilson (1995); Aragonès and Neeman (2000); Flamson and Bryant (2013); Smaldino and Turner (2022).

Figure 1.5 An artist's interpretation of the Cubist chicken parable. Drawing by Nicky Case.

play rehearsal. There was a large collection of Legos in the room, and we, being of a jaunty disposition and not entirely sober, amused ourselves by playing with the blocks. My friend idly constructed an assembly of red, white, black, and yellow blocks and declared, "Look! It's a Cubist chicken!" (Precisely what it looked like has also been lost to time, but an artist's interpretation is presented in Figure 1.5.) I laughed and heartily agreed that it most definitely looked like a Cubist chicken. We were extremely satisfied with ourselves, not only because it was very silly, but also because if in fact we *both* understood the design to be a Cubist chicken, then it surely was one. We had identified something *true* about our little masterpiece and had therefore, inadvertently perhaps, created *art*. This is how liberal arts students amuse themselves.

Our conversation moved on to other topics, but we continued to occasionally comment on the Cubist chicken. After some time had passed, the rehearsal ended, and our actor friend entered the room. "Check it out!" we exclaimed. "A Cubist chicken!" Our friend smiled bemusedly and, with a raised eyebrow, asked us to explain exactly how the seemingly random constellation of Legos represented a chicken. "Well," I said, pointing to various parts of the assemblage, "here is the head. And here is the body and the legs, and here is the tail." If you squinted, I thought with some satisfaction, it sort of looked chicken-ish. But my satisfaction was short-lived. "No!" cried my co-conspirator. "That's all wrong. The whole thing is just the head. Here are the eyes, and the beak, and here is the crest," for my friend had envisioned our chicken as a rooster. And thus, the illusion of our shared reality was shattered. We thought we had been talking about the same thing. But when more precision was demanded, we discovered we had not.

As many a late-night dorm room conversation can attest, humans are capable of very elaborate theories about the nature of reality. The problem is that, as scientists, we need to clearly communicate our theories so that we can use them to make testable predictions. In the social and behavioral sciences, the search for clarity can present a problem for verbal models and can lead to a depressing recursive avalanche of definitions. For example, many researchers are interested in preferences. But before we ask about the sorts of things that influence preferences or are influenced by preferences, we should first ask: What *is* a preference? Perhaps a preference is a tendency to choose certain behaviors over others. This leads to a new question: What are the available behaviors? Of course, the set of possible behaviors depends on the social and environmental context. What are these contexts, and what determines them? This can go on for a while.

Formal models provide a means of escape from the recursive abyss. By restricting our discussion to the model system, we can clearly articulate what we are—and, just as importantly, aren't—talking about. This generally leaves us with something that, on the surface, might seem pretty stupid. The statistician George Box famously noted that "all models are wrong, some are useful." I would add that not only are all models wrong, they are *obviously* wrong. They often appear to be gross oversimplifications that leave out seemingly important details. However, the apparent stupidity of a model can be a strength. By focusing only on some key aspects of a real-world system (i.e., those aspects instantiated in the model), we can investigate how such a system would work if, in principle, we really *could* ignore everything we are ignoring. This only sounds absurd until one recognizes that, in our theorizing about the nature of reality—both as scientists and as mere humans hopelessly entangled in a complex and confusing world—*we ignore things all the time.* We can't function without ignoring most of the facts of the world. Our selective attention ignores most of the sensory input that innervates our neurons (consider the well-known "cocktail party effect," in which you ignore the jumble of noises in a crowded party until your attention is suddenly drawn to a mention of your name). This ignorance is fundamentally adaptive; the bounds to our rationality are severe, and dedication of cognitive resources entails balancing benefits and costs.

By ignoring all but the most relevant information, we are able to impose a modicum of order upon the world. Modeling helps us avoid some of the problems that arise when we try to verbally communicate our systems for ordering the world. Each of us has likely focused our attention on a slightly different notion of the world, highlighting some aspects and ignoring others. We might use the same words but still talk past one another. Left unchecked, this sort of ambiguity renders a science of social behavior all but impossible. Formal models help solve the problem by systematizing our stupidity, and ensuring that, at the very least, we are all talking about the same thing.

1.4 Decomposition

Most of us are taught a version of science that goes something like this: a scientist observes the world and constructs a hypothesis. The scientist is presumed to enter the scene as a passive observer, taking it all in with calm, quiet contemplation. They then form a hypothesis, which compels them to action. The scientist devises a test, the results of which will help to confirm or refute the hypothesis. Much of the literature on scientific methodology focuses on rigor in hypothesis testing—statistical power, multiple comparisons, p-values, Bayesian something or other, etc. This stuff is all important, but there is also a problem with this view of science as hypothesis testing, which is that it glosses over the question of *where hypotheses come from* in the first place.

A hypothesis is usually a proposal that the parts of a system are organized in some way and/or that *because* the parts are organized in a particular way, some phenomena and not others occur. But what are these parts? If we want to understand some aspect of a system and form hypotheses and theories[3] about it, we first have to articulate the parts of the system, a process I'll call **decomposition** after Herbert Simon, who used the term similarly.[4] We must answer the following questions: What are the parts of the system we are interested in? What are their properties? What are the relationships between the parts and their properties? How do those properties and relationships change? Decomposition consists of usable answers to these questions.

What is the right decomposition? Dauntingly, there is no one right way to decompose a system. An economist interested in supply chains might model a local economy as a set of firms defined by their assets, sectors, and dependencies. The individual humans actually making the decisions in those firms and consuming their goods and services would be ignored, as would the weather, the position of the moon, and the migration of butterflies. A cognitive scientist interested in the same system might instead focus on the decisions made by the individual stakeholders, and thus their model would include the perceived costs, benefits, and affordances of those individuals. How you decompose a system depends on the questions you are trying to answer with your model and on the granularity required for answers to those questions. The value of a model largely depends on how well its decomposition usefully answers the questions the modeler is asking.

No idea is theory-free. We parse the world based on categories and schemas—the mental set comprising and organizing our world. We build models to help us understand them. All models are ultimately unrealistic. But reality cannot be fully captured by our minds, and so we come up with explanations that work for us, that help us to see the world in a meaningful and useful way. Models, then, are reflections of how we parse the world. The key point I want to make here is that there are *many* ways to parse the world, and how we parse it determines the questions we are able to ask about it.

As a graduate student in the summer of 2008, I attended a workshop on computational modeling run by John Miller and Scott Page at the Santa Fe Institute. On the first day, the attendees were given the following assignment:

> People enter and leave an elevator as it travels up and down. Model, using whatever techniques you wish, the above scenario. Explicitly state your model and key assumptions.

We split into small groups of two or three and spent a day working on the project. My group considered the perspective of an individual rider making the decision to wait for the elevator or take the stairs. The decision calculus was based on the distance required to travel, the time of day, and the number of people currently in the elevator. The relevant parts were the individual decision makers (specifically their locations and intended destinations) and the elevator (specifically its location and occupancy level). We programmed our model and explored the relationships between building occupancy, distribution of destinations, and the number of stories in the resulting patterns of elevator usage. When we reconvened with the other workshop attendees to share our results, we discovered a wide range of perspectives we hadn't considered at all. One that stands out in my memory was a group that modeled optimal ways for agents to arrange themselves *within* an elevator so that they avoided crowding while also minimizing the likelihood of being blocked in when the elevator reached their floor. This was a completely different model, using a completely different decomposition of the system! In this model, the layout of the elevator itself was a key component, and the

[3] See also Box 1.1: Hypotheses, Theories, and Theoretical Frameworks.

[4] This section is strongly influenced by Simon (1963) and Kauffman (1971). The 1960s and '70s saw a boom in what is sometimes called *systems thinking*.

decision calculi of the individuals, including how their decisions influenced each other, bore little resemblance to those in our model.

The reason such vastly different model designs could emerge from the same prompt is that although the elevator scenario is a reasonably well-defined system, the prompt provides no specific questions to be addressed. This point is worth repeating: *it is the question that determines the relevant parts of the system*. When it eventually comes time to build your own models (which we will discuss in chapter 10), you will need to think carefully about the parts you will include and the parts you will ignore. What questions does your theory address? What parts do you need to include to answer those questions? Is your model a satisfying representation of your theory? If not, why not? There are lots of ways to represent any particular system, and these representations matter. At least some great scientific advances occur because new decompositions are introduced that allow us to ask better questions or to explain more empirical phenomena in a coherent framework. Good models also serve as fuel for analogical reasoning, whereby the parts of the model can be mapped onto the relevant parts of reality (Brand et al., 2021).

In this book we will examine a number of models. Each decomposes a system in a particular way and is well suited to answering certain questions and not others. Exploring some models means excluding others, and there are natural constraints on the questions any set of models can address. That being said, I have tried to be deliberate in my choice of models. There is value in having a set of well-known or even (if I may be so bold) canonical models. If such models are widely known throughout the social sciences, they can help drive theory forward (by focusing attention on a set of well-understood questions) and also encourage communication and collaboration (because the model formalisms remove ambiguity and ensure that researchers are using concepts in similar ways).

BOX 1.1: Hypotheses, Theories, and Theoretical Frameworks

Throughout this book, I will be talking about models and their connections to both hypotheses and theories, so it is worth clarifying how I will be using these terms. I will also distinguish between individual theories and overarching theoretical frameworks. This breakdown draws somewhat from a lovely paper by Muthukrishna and Henrich (2019), though my definitions differ slightly from theirs. The following definitions are nonstandard, but I think they make sense in terms of the modeling philosophy used here.

A **hypothesis** is a prediction that if a particular set of assumptions are met, a particular set of consequences will follow. In practice, this is a prediction that either (1) the parts of a system are organized in a particular way—in other words, that a particular decomposition carries explanatory power for some observed phenomena—or that (2) *because* the parts of a system are organized in a particular way, certain phenomena and not others will occur. Good hypotheses allow us to exclude and distinguish between competing theories.

A **theory** is a set of assumptions upon which hypotheses derived from that theory *must* depend. Strong theories allow us to generate clear and falsifiable hypotheses.

A **theoretical framework** is a broad collection of related theories that all share a common set of core assumptions. An example of a theoretical framework is Darwinian evolution by natural selection, from which many subordinate theories have been derived.

1.5 Formal Theory in the Inexact Sciences

From the above discussion, it should be clear that one of the perennial challenges involved in doing science is pinning down exactly *what we are talking about.* Many people have an intuition that there is something fundamentally different between the "soft" social sciences like sociology, anthropology, behavioral ecology, or social psychology, and the "hard" sciences like physics, chemistry, or geology. The philosopher Karl Popper (1963) suggested that a key feature of what one might call "hard science" involves **falsifiability**. Popper proposed that scientific theories should make clear predictions, so that if an empirical result contradicts the prediction of a theory, the theory has been falsified. An unscientific theory, in contrast, is not specified clearly enough to delineate whether a result does or does not falsify it. "Soft science" presumably straddles the line between science and non-science. I've always found the soft-hard distinction somewhat unsatisfying, however, and so I would like to offer up a slightly different way of characterizing the sciences.

In the decades since Popper, a great deal of commentary has been made on his doctrine of falsifiability (reviewed in Oreskes, 2019), grappling with the reality that much of what we regularly call "science" involves neither predictions nor measurements that are precise enough to determine whether and when falsification has occurred. In order to make sense of this, I propose a distinction between exact and inexact sciences, with the understanding that exactness is more like a continuous variable than a binary characteristic. In the *exact sciences*, theories involve direct mappings between measurable constructs and model predictions—the terms in their fundamental equations all have universally-accepted units. In classical physics, for example, theories concern quantities like mass and velocity, charge and voltage. In other words, the theories are exact specifications of the relationships between quantities that can be measured with high levels of precision.

The *inexact sciences* are those in which the mappings between measurements and theories are imprecise. This creates a challenge for theory in the inexact sciences. Formal theories of social behavior, which involve mathematical or computational models, are themselves quantitative and exact, but the model parameters typically won't align precisely with empirical measures in the way that theories do in the exact sciences. Indeed, the quantities being measured are usually mere proxies for the concepts at the heart of theories in the inexact sciences. This imperfect mapping may be why there are more widespread preferences in these fields for empirical, heuristic, or verbal models rather than formal models.

The social sciences are almost always inexact. Social scientists are interested in theories about things like communication, cooperation, norms, identity, contentment, and prosperity. You can measure these things, but you will almost always encounter arguments for why your measurements don't capture key aspects of the concepts dealt with in your theory.[5] In what units do we measure cooperation or contentment? This inexactness sometimes leads social scientists to dismiss formal models as failing to adequately capture the phenomena of interest, over being useless oversimplifications. Models of social phenomena *do* usually involve extreme simplification of complex systems—this simplification is exactly the point of models. If we're not clear on how our theories work, how can we agree on the value of the data we collect to address those theories?

Social systems are gonna be modeled, because scientists keep coming up with theories about those systems that benefit from formalization. As such, it's important to ensure that

[5]There are many things that social scientists *can* determine exactly, such as the identity of the UK prime minister or the location of the U.S.-Canada border. However, it is much more difficult to articulate *general* theories about prime ministers or borders.

at least some of the people modeling social systems are social scientists, if not by training then at least by inclination. By this, I mean that while you don't necessarily need to have a degree in social science to do social science, you *do* need to take the questions posed and the research done by social scientists seriously. There's a lot of poor modeling of social systems done by people who can do some cool math or programming but haven't taken the time to understand many of the finer aspects of human behavior,[6] which is the sea in which social scientists swim.

Due to the inexactness of the social sciences, the mapping between models and data is complicated and often more analogical than precise. Because of this, we will mostly ignore the question of fitting theoretical models to empirical data until the book's penultimate chapter, and instead focus on how to articulate models of social phenomena and gain intuition about the dynamics that emerge from them.

1.6 Why Model?

Despite the inexactness of the social sciences, formal models of social systems still get you quite a lot. If you ask most people why scientists build models, they'll probably respond with something about prediction. We want to predict what will happen. Prediction is often valuable, and even occasionally achievable. In the inexact social sciences, our predictions will rarely be precise, and when they are precise, they will rarely be accurate. However, qualitative predictions are still valuable. It can be of great use to know that when some variable X increases under condition Y, variable Z is likely to increase in turn. Relatedly, researchers often use statistical models to identify predictive correlations in their data. Such models are extremely valuable but can also fail catastrophically if the conditions that generated the associations change. This book focuses on generative or mechanistic models, whereby the processes that generate particular conditions are modeled explicitly.

Many models do not make even qualitative predictions about the real world. Certainly, the models we will explore in this book will rarely be useful for making specific predictions about the future, at least in their unaltered forms. Nevertheless, even non-predictive models have a lot of value, which I will condense into three categories: precision, tractability, and insight.[7]

Precision

There are many, *many* ways to parse the world. We could not go about our day without creating simplified descriptions of how the world works in various contexts—what cognitive scientists call **mental models**. In this sense, we are all modelers, but only some of us (and only some of the time) can write our models down. In the social sciences, many theories are expressed as **verbal models**: descriptions of the assumptions required to purportedly explain some phenomenon, written in plain language. Verbal models are very important, and are often the first step toward articulating a theory that can later support formalization. The most successful verbal model—from a scientific point of view—may be Darwin's theory of evolution by natural selection, which provides the foundations for explaining much of life as we know it. Darwin's writings contain no mathematical formalisms, only richly described

[6]Or even the coarser aspects, for that matter.

[7]Lengthier explorations of uses for models beyond prediction can be found in Wimsatt (1987), Bedau (1999), Epstein (2008), Smaldino (2017), and Page (2018).

ideas. Most verbal models, however, also contain at least some ambiguity. This is because, as noted earlier, language is inherently ambiguous. Words and phrases often afford multiple interpretations despite our best efforts to be clear. This flexibility can be useful when we are first developing our ideas because it can delay committing to a particular path before sufficient clarity is obtained, but ultimately a path must be chosen. In contrast to verbal models, formal models lend themselves to severely limited, if not wholly unique, interpretation because they describe relationships and processes exactly and unambiguously.[8] This approach can facilitate important theoretical advances. For example, Darwinian natural selection was for many years presumed to be incompatible with Mendelian genetics until formal models were developed to illustrate how the two approaches could be reconciled (Plutynski, 2009).

Formal models require us to articulate exactly what is and is not included in our theory. That is, the model makes explicit all the assumptions required to generate the consequences that it implies. This precision creates several benefits. I'll highlight two. First, formal models can provide a clear **scope**: an indication of the constraints required for the theory to apply. Among other things, this helps us to know when an empirical finding does or does not affect the validation of a theory. And second, precision aids in communication by avoiding the ambiguity inherent in verbal models. We sidestep the problem of the Cubist chicken, because the formalism tells us exactly what the parts are and how they fit together. Relatedly, models can provide communities of researchers with a common language to talk precisely about their systems.

Tractability

It is the precision inherent in formal models that makes them tractable. Formal models act as logical engines that turn assumptions into conclusions. Stating our assumptions precisely allows us to know what outcomes necessarily follow from those assumptions, which in turn can help us to identify potential gaps in our explanations. The real world is extremely messy, and constraints on time, resources, causality, and ethics all serve to limit what we can learn about it directly. But by studying models, we can simulate dynamics on time scales that would be impossible to test empirically, on spatial or organizational scales that would be impossible to study practically. We can explore counterfactual scenarios that would otherwise be prohibited either by ethical concerns or by the fact that (as far as we know) we cannot actually go back in time to see how things might have played out under different circumstances.

Insight

Studying models can provide insight. To me, this seems like the most obvious benefit of working with models, but it is often overlooked, perhaps because insight is difficult to quantify. A model can be a playground to explore the dynamics of complex systems with different features. This sort of exploration tends to be cheap—running a simulation is usually much less costly in terms of both time and resources than collecting sufficient empirical data to build and test dynamical theories—which means one is free to do a lot of it.

[8]Some ambiguity may remain about the mapping between model and reality. This is not always easily resolved, but the model formalism at least lays out constraints that must be met for a mapping to apply.

The world is complicated. Learning about models provides us with a cognitive arsenal for understanding complex systems—and not only those systems we have directly modeled! When you know that a system in question involves features and constraints similar to those in models you have seen before, insight into how they operate can follow.

1.7 Some Models of Note

To understand what models can do for us, it will help to describe some of the major insights from a few example models. Compiling a list of all the interesting and useful models in the sciences is a fool's errand. Let it suffice to say that such a list would be vast. Instead, I want to merely illustrate via a few pointed examples how simple models can be not only useful, but fundamental to good science. I will briefly describe two well-known examples of models that changed our understanding of basic concepts in the physical and biological sciences. I have chosen these examples from other disciplines to highlight how models contribute in domains that are more exact than those usually tackled by social scientists. In the section that follows, I will explore a model of disease transmission—among the more exact domains in the social sciences—in greater detail.

1.7.1 Newton's Model of Gravity

In seventeenth-century Europe, astronomers faced a great challenge. Following the pioneering work of Copernicus, and building on the meticulously collected data of Tycho Brahe, Johannes Kepler had not only confirmed that Earth and the other planets revolve around the Sun, but had also shown definitively that their orbital paths describe ellipses rather than perfect circles. It was a great mystery why this should be. Enter Isaac Newton.[9] Newton was not the first person to propose that the heavenly bodies might be attracted to one another with a force that varied with the inverse square of the distance between them, but he was the first to build a model based on that proposition (Gleick, 2004). His model was startlingly simple, consisting of only two objects: the Sun and Earth (Figure 1.6). The model ignored the Moon as well as the five other known solar planets, not to mention all the celestial bodies that were unknown in Newton's time. The size and topology of the Sun and Earth were also ignored; they were modeled as points identified only by their mass, position, and velocity. Nevertheless, the model's strength lies in its simplicity. By restricting the analysis to only two bodies, the resulting planetary orbit was mathematically tractable. Using a simple rule stating that the force of gravitation was proportional to the product of the objects' masses and inversely proportional to the square of the distance between them, Newton was able to show that the resulting orbits would always take the form of conic sections, including the elliptical orbits observed by Kepler. And because he could show that the same law explained the motion of falling objects on Earth, Newton provided the first scientific unification of the Terrestrial with the Celestial.

Although Newton's model helps explain a great deal of observed phenomena, there are nevertheless inconsistencies—observations about planetary orbits that do not align with what would be predicted by Newtonian gravitation. Newton's model also simply asserts that the gravitational force exists and does not provide a mechanism for why it should do so.

[9] It should be acknowledged that Isaac Newton was a super weird dude. He and Edmund Halley once dissected a dolphin in a coffee shop and he notoriously shoved a bodkin into his eye socket to compress his eye's lens in order to observe its effect on color perception.

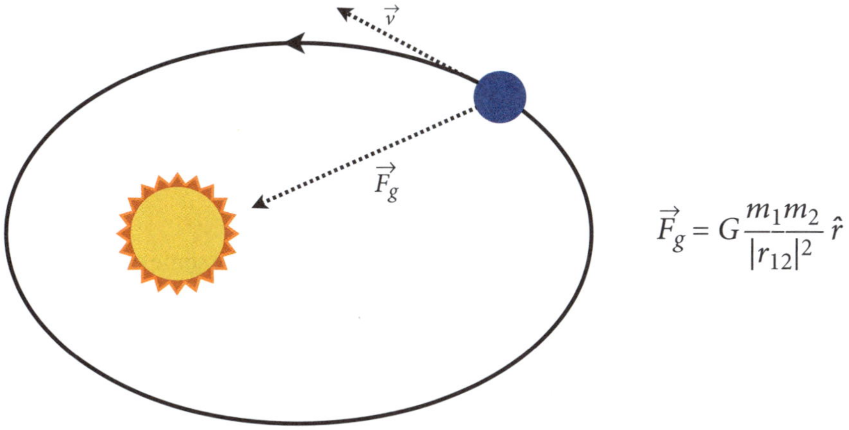

Figure 1.6 Newton's model of planetary gravitation. Earth has a forward velocity v, which is continuously altered by the gravitational attraction of the Sun, F_g, resulting in an elliptical orbit. In reality, the model is even simpler than implied here, because the Sun and Earth are represented as point masses rather than spheres.

Modern astrophysicists now prefer Einstein's theory of general relativity, in which masses create curvature in space-time. Nevertheless, by modeling the consequences of the inverse-square law of attraction, Newton showed how the motions of the planets could be explained, and the model provides exceptionally good approximations of gravitational forces—so good that NASA's Moon missions have relied upon them.

1.7.2 The Lotka-Volterra Predator-Prey Model

For many years, fur trapping organizations like the Hudson's Bay Company in Canada kept meticulous records on the pelt-producing animals in the regions where they trapped. These records illustrated that linked predator and prey species, like the Canada lynx and the snowshoe hare, tended to have cyclical population levels whose dynamics were tightly correlated (Figure 1.7). How to explain this? In the early twentieth century, Alfred Lotka and Vito Volterra, working independently, applied ideas from the chemistry of autocatalytic reactions to generate a simple model of two interrelated populations, which can be instantiated as a pair of coupled differential equations.

This model specifies two animal species: a prey species with a positive rate of growth in the absence of predators, and a predator species with a negative growth rate in the absence of prey. The number of predators negatively influences the number of prey, and the number of prey animals positively influences the number of predators. The model can produce correlated oscillations in the two populations that bear a striking resemblance to data from many predator–prey systems. The model also identifies conditions under which the two growth rates can instead give rise to more stable equilibria or yield complete population collapse—phenomena that have been empirically observed under conditions consistent with the model. Of course, the model is also extremely simplistic. It assumes perfect mixing, so the probability of a prey animal encountering a predator is simply the relative frequency of predators in the population. It ignores seasonality, circadian cycles, migration, density dependence in the growth rate of the prey species, development, and interactions with other species. Thus, in cases where these features matter, the model may fail to align with empirical

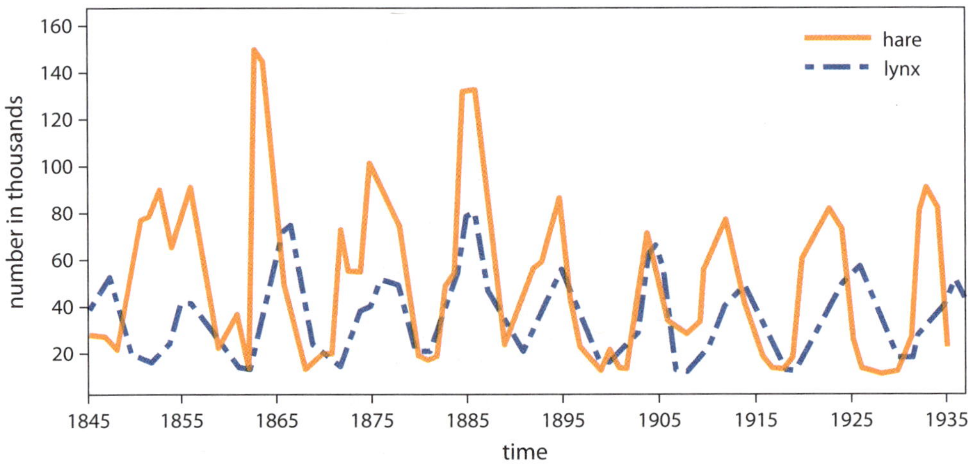

Figure 1.7 Changes in the abundance of the Canada lynx and snowshoe hare, as indicated by the number of pelts received by the Hudson's Bay Company.

fact. Nevertheless, the core assumptions of the model yield insights into the emergence of cyclical dynamics and provide opportunities for extensions and refinements of the model when additional features cannot be ignored. The model has even been extended to help understand cycles of war and peace in human societies (Turchin, 2003, 2016). The Lotka-Volterra model remains one of the core tools for understanding the relationship between predator and prey populations.

1.8 Equation-Based Models and Agent-Based Models

Both Newton's model of gravitation and the Lotka-Volterra predator-prey model are typically expressed as differential equations, which describe mathematically how quantities change over time. This differs from the boids model of flocking behavior that we discussed earlier in this chapter, in which each member of a rather large population was explicitly represented in computer simulation. How do these two modeling approaches relate to one another?

One can distinguish these approaches as equation-based and agent-based models, as some authors do (e.g., Smith and Conrey, 2007). **Equation-based models** involve writing down, well, *equations* that specify the key relationships between the parts of a system, such as the dynamics of how a population changes. In population models, classes of objects or individuals are treated as aggregates for the sake of tractability. Equation-based models can provide quite a bit of precision as well as a certain kind of mathematical elegance. Exploration of parameters is generally quite easy, since we can simply plug new numbers into the equations, and we can often derive the exact conditions under which particular outcomes will or will not occur. Even when closed-form solutions[10] are not possible, equation-based models can be explored through numerical simulation—a technique that will be used in several chapters in this book, beginning with chapter 4.

[10]Having a closed-form solution means that outcome measures can be described as equations involving known or measurable parameters.

Equation-based models are limited primarily in their ability to deal with heterogeneity. For example, we may want to explore the spatial or network structure of a population, or keep track of how individual differences in traits or behaviors are distributed; such things are challenging with equations only. In cases where additional complexity is desired and analytical tractability is not feasible (or is beyond the mathematical ability of the modeler), computational modeling can provide a useful alternative.

Agent-based models (ABMs) are a particular class of computational models in which individual agents are simulated as explicit computational entities. Agents often represent people or other animals, but agents can also represent anything from biological cells to economic firms to political municipalities. In addition to allowing for greater heterogeneity, agent-based models have other attractive features. One is that learning to code ABMs often represents a lower bar to entry than learning the requisite mathematics for analyzing equation-based models, especially for those with less formal mathematical training. A related advantage is that ABMs can sometimes provide the sort of intuition for the behavior of complex systems that typically comes only from direct observation. For those without strong mathematical training, equations can be opaque or cryptic. Observing the behavior of agents in a visualized simulation can also help accomplish something that is often difficult to do with equations: confer understanding upon non-modelers. A point I want to emphasize is that these are complementary rather than competing approaches. Throughout this book, whenever possible we will explore *both* equation-based and agent-based models for each system we encounter.

In the interest of getting you more deeply into a modeling frame of mind, I am going to walk through equation- and agent-based versions of a simple epidemiological model of disease transmission. We will revisit this and related models in greater detail in chapter 4. Consider the scenario where an infectious disease has broken out and is spreading through the population.[11] We can characterize individuals as either *susceptible* to the disease (S), *infected* (I), or *recovered* (R) and immune (or, alternatively, *removed* from the population in some versions). This is the well-known SIR model, the dynamics of which can be expressed as three coupled differential equations:

$$\frac{dS}{dt} = -\beta SI$$

$$\frac{dI}{dt} = \beta SI - \gamma I$$

$$\frac{dR}{dt} = \gamma I$$

If you are unfamiliar with this sort of equation, don't get scared! Familiarity with differential equations is not required for any of the exercises in this book, although it is useful knowledge to have more generally for understanding dynamical systems. I will focus on discrete-time dynamics throughout this book. At times when discussion of differential equations is particularly valuable, such information will be relegated to text boxes for the interested reader. Learning differential equations is useful, but there's a lot you can still do without them. Discrete-time versions of the SIR model equations can be represented in a way more familiar to some social science readers like this:

[11] This should not be difficult for anyone who lived through 2020 or the few years thereafter.

$$S(t+1) = S(t) - \beta S(t)I(t)$$
$$I(t+1) = I(t) + \beta S(t)I(t) - \gamma I(t)$$
$$R(t+1) = R(t) + \gamma I(t)$$

These equations define how the relative numbers of susceptible, infected, and recovered individuals change over time, and represent two propositions about disease contagion. First, that susceptible individuals become infected via contact with infected individuals, at a rate that is proportional to the expected number of interactions between susceptible and infected individuals, tempered by the transmissibility of the infection, β. Second, that infected individuals recover at a constant rate, γ. An implicit assumption is that the rate of interactions[12] between individuals in different states is exactly proportional to the frequencies of those states in the population—that is, that the population is **well-mixed**. This model is simple but powerful. It can be used to estimate the time course of an epidemic, the maximum number of individuals who will be infected at a given time, and the number of individuals requiring immunity (such as through vaccination) needed to prevent an outbreak from becoming an epidemic, thereby providing "herd immunity." Variations on the model have considered a number of other factors, including non-contagious periods after exposure, age-structured populations, non-random assortment, and even simultaneous "behavioral contagions" that could alter transmission rates.

Beginning in the first months of the COVID-19 pandemic in 2020, before vaccines were available, authorities urged people to maintain physical distance from one another in order to "flatten the curve" of the epidemic. Reducing physical contact would decrease the effective transmission rate of the disease and, critically, reduce the number of individuals infected at any given time. This can be illustrated by numerically simulating the differential equations in the SIR model above for different values of β (Figure 1.8). Although articles explaining this curve-flattening process proliferated, my personal experience was that many people did not find it intuitive that individual behaviors could translate to reduced transmissibility on a large scale.

To provide an alternative framing, we can build a simple agent-based model of SIR dynamics, in which agents are situated on a two-dimensional surface and move around using a random walk.[13] Any time a susceptible individual is sufficiently close to an infected individual, they become infected with some probability (the transmission rate). An infected individual then recovers with a probability dictated by the disease's recovery rate (Figure 1.9A). Rather than modifying the disease transmission rate directly, as in the equation-based model, I modified the size of the step taken by agents during their random walks, so that they took either large steps (thereby rapidly traversing the space and interacting with many different individuals) or small steps (thereby staying close to where they started and interacting with a smaller number of distinct individuals; Figure 1.9B). This is meant to represent differences in physical distancing. Comparing the dynamics of the infected populations in these two movement conditions produces a plot that is comparable to that produced with the equation-based model[14] but illustrates more directly how a

[12]Because the parameters β and γ are *rates*, their values in the discrete-time model are only equivalent to those in the continuous-time model in the limit as $\Delta t \to 0$. Otherwise they must be calibrated to the discrete time unit assumed by the modeler.

[13]A similar model will be explored more extensively in chapter 4.

[14]I made no attempt to keep the transmission and recovery rates the same between the two models—my purpose here is to illustrate how both models can produce the same qualitative patterns.

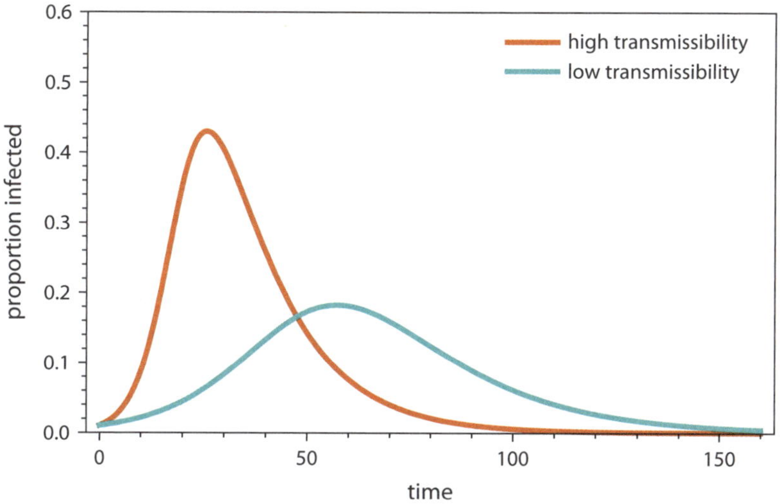

Figure 1.8 Temporal dynamics of infected individuals in the equation-based SIR model with a recovery rate of $\gamma = 0.07$. This compares populations under either high transmissibility ($\beta = 0.3$) or low transmissibility ($\beta = 0.15$).

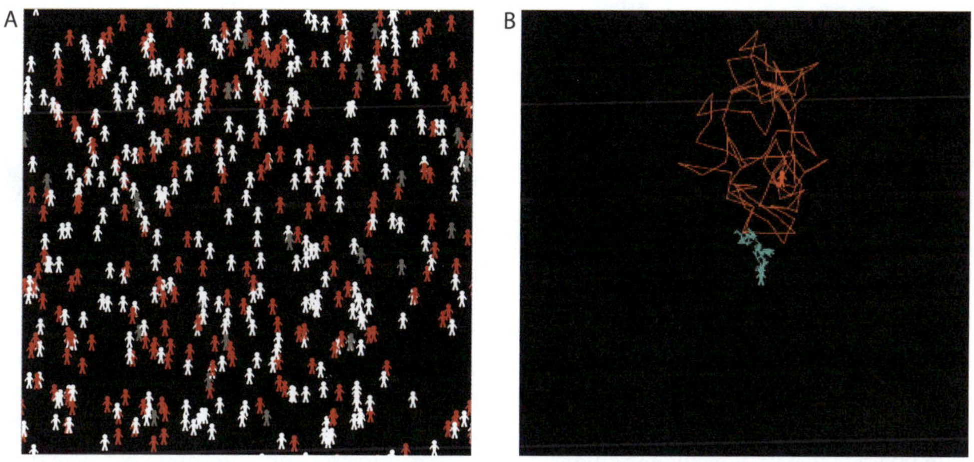

Figure 1.9 (A) Visualization of a spatial agent-based SIR model. There are 500 agents, which can be either susceptible (white), infected (red), or recovered (grey). (B) Example random walk trajectories over 100 steps for agents taking either large (orange) or small (blue) steps.

reduction in social contact flattens the curve (Figure 1.10). It also reveals that "transmissibility" in the equation-based SIR model is an aggregate variable that incorporates properties of both the disease and its hosts.

I want to make it clear that both agent-based models and purely equation-based models are valuable, and attempts to paint them as competing techniques are misguided.[15] Both

[15]Indeed, the distinction is more heuristic than technical. Equation-based models can be explored computationally, and even complex agent-based models can, at least in theory, be reduced to a set of recursive mathematical functions (Epstein, 1999; North, 2014).

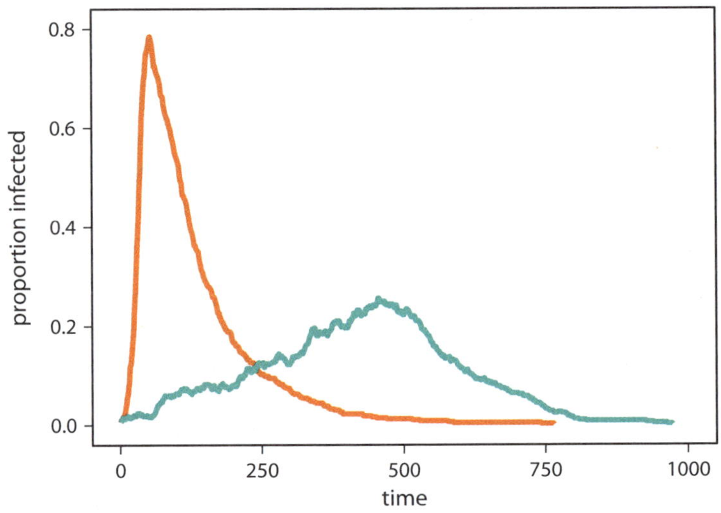

Figure 1.10 Temporal dynamics of infected individuals in the agent-based SIR model. Agents move using a random walk with either a large (orange) or small (blue) step size.

techniques are part of the modeler's repertoire, and one method is not inherently superior to the other. In many cases, it is valuable to combine both analytical equation-based models and agent-based simulations to provide richer coverage of the model system.

1.9 Fine-Grained and Coarse-Grained Models

Another distinction worth noting is one between fine-grained and coarse-grained models. Fine-grained means that there are data in the world that can be used to precisely parameterize and test the models. Many models in physics are like this; the model parameters are precisely measurable quantities like mass, pressure, or voltage. In epidemiology, some agent-based models are calibrated using high-precision data on demographics, geography, schools, travel matrices, and so forth, with the goal of predicting the time course of an epidemic. In neuroscience, biophysical models might exactly predict the dynamics of action potentials or motor behaviors.

Coarse-grained models focus on broad, qualitative patterns in the data, not on reproducing exact measurements. In the social and behavioral sciences, most models are fairly coarse-grained (the SIR model just described is an example). Measurement in the social sciences is often very difficult. Processes related to cognition, behavior, and social organization involve interacting parts at many levels of organization and time scale. While the simpler sciences have focused their study on things that are readily measured like mass and motion, the social sciences are concerned with emergent phenomena like emotions, perceptions, and norms. The utility of these concepts in lay thought and communication is indisputable, but it is less obvious how they should be measured for scientific study. Even when a concept is precisely defined, measurement is often made difficult by constraints of time, resources, or ethics.

Complex systems are by their very nature difficult to model with great precision. This is partly because they involve many interdependent components that interact in nonlinear ways. It is also because the mapping between the constructs we are interested in and the measurements we are able to make are rarely one-to-one. For this reason, modeling social behavior usually begins with quite abstract representations of phenomena. Exploring these models equips us with a toolkit for ever more nuanced approaches to studying complex social systems.

Coarse-grained models are valuable despite their inability to generate precise quantitative predictions. As far as I'm concerned, the only alternative to using a formal model to articulate a theory or hypothesis is to use a verbal model, or worse, an unspoken mental model. In those cases, it is much more difficult to identify implicit assumptions or show how the explicit assumptions lead to particular consequences, and therefore it is much easier to enter into the territory of unscientific vagueness. To repeat, everyone is using some model, but it is hard to know how good that model is without writing it down. We will begin learning how to write down our models in the next chapter.

1.10 The Journey Begins

Our understanding of human mind and behavior is in many ways barely out of the Dark Ages. We have some theories, and some of them are kind of decent. But we are only at the beginning of real understanding, and we often are stymied by norms and inertia and the capture of our institutions by people who aren't really interested in understanding anything. I charge you with doing better. Sometimes working with these models might seem tedious or difficult. And sometimes it *is* difficult. But let me offer you this: If you can't be bothered to understand how a simple model system works and how its assumptions generate the resulting dynamics and population-level patterns, how are you going to be able to understand a real-world system, which is way more complex and for which there are way more variables you can't observe?

In the social sciences, there is a great need for formal theory based on mathematical and computational models. Yet there is still a paucity of training programs for students and researchers interested in learning how to build and analyze those models. I designed this book with the following question in mind: If I were training a collaborator with whom I could work on models of social behavior, what would I want them to know? The book is intended to equip social, behavioral, and cognitive scientists with a toolkit for thinking about and studying complex social systems using mathematical and computational models. The approach marries two traditions: complex systems-style modeling, which focuses on mathematical and computational tools for studying emergent phenomena, and a more theoretically informed approach from work in human behavioral ecology and cultural evolution. Rather than focusing on general prescriptions for modeling, we will work through a variety of modeling topics in detail, each exemplified by one or more archetypal models. With some luck, we will thereby build up strong theoretical foundations for understanding social behavior.

Particles

2

> We have the tools, we have the talent!
>
> —Winston Zeddemore, *Ghostbusters* (1984)

Before we can build something, we need to be conversant with our tools. This chapter introduces the fundamentals of programming agent-based models with NetLogo, after which we will explore a simple toy model called "Particle World."

In this chapter and the next, we'll stick to agent-based models and leave aside purely mathematical formulations. For many people, simulating explicit agents provides greater intuition than mathematical models based on seemingly sterile equations. There's something visceral about seeing embodied agents moving around on your screen, which may explain some of the popularity of agent-based modeling. Many people also find the task of writing down models as equations intimidating. This is a hurdle we will have to vault, but not just yet. Hopefully, the intuitions gained from simulating social systems will provide a scaffold for understanding the equations. If you're raring to start writing down equations, sit tight and enjoy the ride for now. We'll get there soon enough.

This book is predicated on the idea of learning by doing. Therefore, I will not spend too much time on general principles of coding or modeling. Rather, I will try to give you just enough information to get going right away with coding and analyzing some simple models, and then, later, some more complicated models. There will be lots to learn from the models we study, and I want to get us started as soon as possible. In the next two sections, I will introduce some of the basics of NetLogo and some general principles of computer programming. If you are already broadly familiar with both NetLogo and programming, feel free to skip ahead to section 2.3.

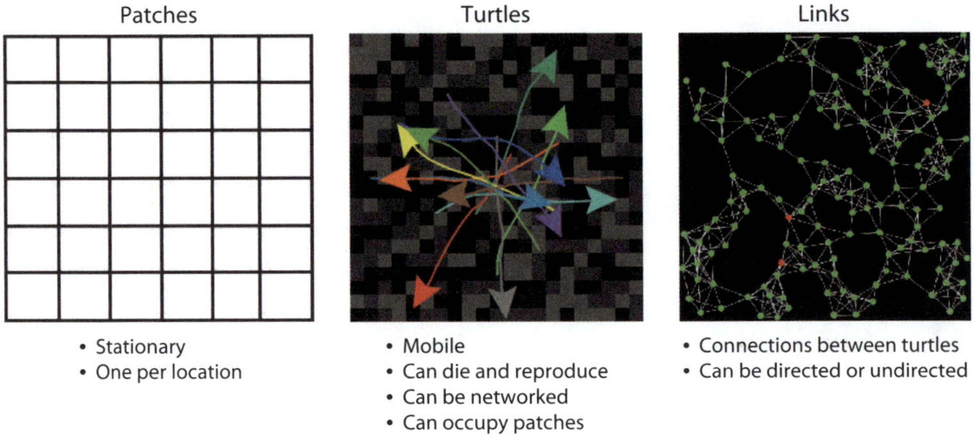

Figure 2.1 The building blocks of a NetLogo model: patches, turtles, and links.

2.1 NetLogo Basics

If you have not done so already, you should download NetLogo onto your machine, which you can do for free here:

https://ccl.northwestern.edu/netlogo/

As noted, this is not a book on how to use NetLogo. You should make sure to spend sufficient time on your own getting familiar with its workings.[1] First and foremost, you should work through the tutorials included in the NetLogo User's Manual. Doing this will give you a sense of how NetLogo works. In addition, the NetLogo Models Library comes loaded with lots of well-documented sample models that will give you a good sense of the program's possibilities. Finally, the NetLogo Dictionary at the end of the User's Manual contains information about all the NetLogo **primitives**—the built-in procedures that make coding models in NetLogo so efficient. This dictionary is a valuable resource. When I work in NetLogo, I refer to it often.

2.1.1 Patches, Turtles, and Links

NetLogo has three important building blocks for agent-based models that have many useful built-in properties: patches, turtles, and links (Figure 2.1). **Patches** are the default spatial organization in NetLogo: a set of discrete square cells laid out in a two-dimensional rectangular grid. Each patch has a fixed, unique location, but otherwise can be assigned arbitrary properties. Indeed, patches can be used as agents themselves in a model, for example when there are a fixed number of agents in stable locations.[2] More typically, patches are used to represent properties of the agents' environment.

When there is need for agents with a more versatile set of properties than those afforded by patches—such as the ability to move, die, reproduce, or form nodes in a network—NetLogo has a useful object class called **turtles**. Turtles will be used to represent agents

[1] If you are using a language other than NetLogo, with or without an explicit modeling library, make sure you are comfortable with it and able to create a simulated world with simple agents in it.

[2] Patches can easily be used to simulate one- and two-dimensional cellular automata.

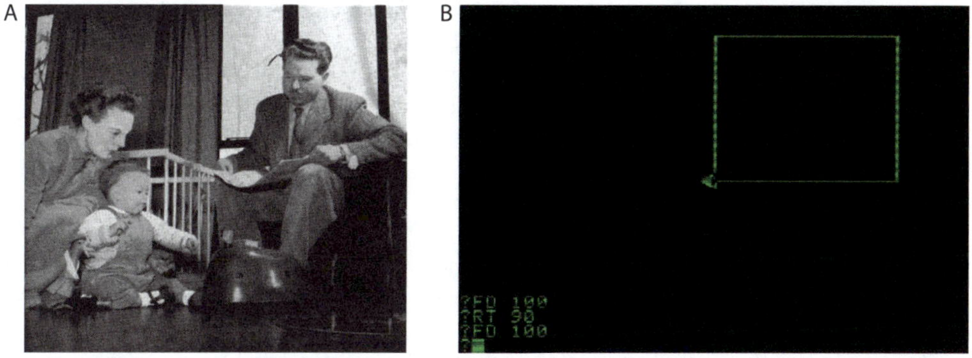

Figure 2.2 (A) W. Grey Walter introduces his cybernetic tortoise to a mother and her child. (B) Logo commands to draw a rectangle with the turtle.

in most of the simulations explored in this book. Turtles are spatially embodied and have a position in continuous space (which overlaps with the discrete space on which the patches exist) as well as a directional heading, though one may also ignore these pieces of locational information if they are not important to a model's dynamics.

Why are these objects called *turtles*? The term is an homage to the neuroscientist and roboticist W. Grey Walter,[3] who created a simple, programmable, mobile robot in the 1950s called a "tortoise." This idea was inspirational to the creators of the Logo programming language—developed in the 1960s—which allowed a user to issue commands to move and draw with a small triangular figure on a computer monitor.[4] The commands were similar to those given to Walter's tortoise, and the green triangle may have also given the impression of a turtle (many screens in those days were monochromatic and produced only green on black; see Figure 2.2). NetLogo uses many of the same commands to move its turtles as did the original Logo language; its agents are named in Logo's honor.

Finally, NetLogo conveniently allows turtles to be connected with the class of objects called **links**. More will be said about the use of links when we discuss network models in chapter 9.

2.1.2 NetLogo Tabs

The NetLogo window has three tabs: Interface, Info, and Code (see Figure 2.5). A NetLogo file contains information about all three tabs, and their union is properly considered the model code. The **Interface** tab displays the model visualizations and graphs, buttons that execute commands for initializing and running the model, and controllers (sliders and switches) for any variables that will be varied in the analysis of the model. You should make sure you are familiar with how to create, delete, and manipulate these aspects. The Interface tab allows for real-time inspection of the model output as well as rapid parameter

[3]Walter worked within the then-trendy discipline of cybernetics, which combined insights from psychology, neuroscience, and the nascent field of computer science. Cybernetics can be viewed as a precursor to the modern fields of artificial intelligence and cognitive science.

[4]Biographical note: Logo was the first language I ever programmed in, as a grade-schooler back in the optimistic heyday of the 1980s. Perhaps this helps to explain my affinity for NetLogo. For a history of Logo, see Solomon et al. (2020).

manipulation. The tab is also where the **Settings** window can be accessed, which allows the user to update the size and shape of the patch grid.

The **Info** tab is essentially a template for extensive documentation of the model and associated code. Nothing in this tab directly affects the run of the model itself. Rather, it is a place to describe the model for the benefit of others who may be using the model (including your future self, who will likely benefit from a refresher). In general, I will ignore this tab throughout the book. However, the library of example models included in the NetLogo download is very helpfully documented using the Info tab.

Finally, the **Code** tab is where the instructions for the model's operation are coded. All the code that will be presented throughout this book is entered in this tab. One thing worth noting is that calling this the "code" tab is somewhat misleading, because the interface is really also part of the code. For example, we will often instantiate global variables in the Interface tab using sliders and switches. How this works will become clear with practice.

2.2 Programming Basics

Once upon a time, a person could obtain an advanced degree in the social or biological sciences without ever typing a line of computer code. Those days are over, and rightly so. There are many skills a scientist can hone, and no one can do everything (at least not well). We create stronger sciences if specialization is allowed to flourish. Nevertheless, programming is simply too valuable a skill to leave entirely to specialists—gaining at least some programming experience is essential for the modern scientist. There are currently many, many resources available to help you learn to code, and it is likely you have already been exposed to some of these. As such, I will generally assume some familiarity with the conceptual language of computer programming. However, for novice programmers, it's also important to make sure that some of the foundational concepts are understood. For this reason, I present below a very short primer on some key programming concepts with examples of how they can be instantiated in NetLogo. Readers comfortable with programming can probably skip ahead to the next section.

Variables

Variables are the secret weapons that give mathematics its power—the ability to perform computations on an object without knowing the value of that object. In computing, variables often represent numbers, but they can also represent character strings, Boolean (true/false) values, or agents, as well as lists or sets of these things. In some languages, like Java or C++, when declaring a new variable you are required to specify what type of object will be represented, be it an integer, a floating point number, or a Boolean array. In other languages, including interpreted languages like Python or NetLogo, you are not required to do this, because the interpreter will automatically figure out what sort of variable it is dealing with.

When we write mathematical equations, we usually represent variables with Latin or Greek letters, like this:

$$y = 0.1x - 1$$

In this case, the variable y is a linear function of the variable x. Perhaps x represents the time you invest in the Skee-Ball game at your local penny arcade, and y represents the value of the knickknacks you can purchase with the tickets you win (the negative intercept is due to the entry cost for the arcade). We represent each value with a single character, because only philistines use multiple letters to represent a variable (excluding subscripts) in purely mathematical formulations. However, in programming we can use almost any set of characters we want, excluding spaces. And we should. We want our variable names to be both succinct and easily identifiable. Code can be confusing, so let's lessen the confusion by using names that help us figure out what the variable refers to when we show our code to others or to our future selves.

NetLogo uses the equals sign (=) exclusively as a logical test (to check if quantities are equal), and instead uses the function `set` to change the value of a variable. So, in NetLogo, the code to assign the current value of y would look like this:

NetLogo code 2.1

```
set value-win (0.1 * money-invested) - 1
```

Variables are the heart and soul of any model. They represent the quantities we need to describe our study systems, and they represent the features of those systems whose dynamics we are interested in understanding. In an agent-based model, we often distinguish between several categories of variable: global variables, agent variables, and local variables. **Global variables** have a single value regardless of how they are accessed. These can be fixed for the duration of a simulation (such as the initial number of agents) or they can change over time (such as the current number of agents in a model with birth and death events), but their values apply globally in the simulation. **Agent variables** are values held by each member of a class of agents, with values that can vary between agents. For example, each agent might keep track of its group identity and current resource level. Global and agent variables are sometimes collectively described as a model's **parameters**. Within a block of code, it is also sometimes useful to define a **local variable**. This is a temporary marker that is used only within a particular block of code and then automatically deleted.

Functions

A function, also called a **procedure**, is a label for a set of commands that are run together whenever the function is called. A function may also take **arguments**, which are values that are assigned to variables within the function. For example, an addition function called `add(x, y)` might take the two arguments `x` and `y` and add them together, so that the output of `add(2, 3)` would be 5. Functions need not return any values, but they can, and when they do the output of a function can in turn be assigned to some other variable. In NetLogo, functions that return a value are also called **reporters**.

NetLogo has a number of built-in values and functions, called *primitives*. These are part of what makes NetLogo so powerful, because they make it easy to program the computations that make the dynamics of a model possible. For example, the primitive reporter `color` returns a turtle's color, while the primitive function `create-turtles` takes as arguments a number and a code block. The function then creates the given number of turtles, each of which executes the commands in the code block upon being created. There are many useful NetLogo primitives, all of which are described in the dictionary at the end of the NetLogo User's Manual.[5] We will be using many functions, both custom and primitive, throughout this book.

[5]For a useful introduction to NetLogo primitives, see here: https://ccl.northwestern.edu/netlogo/bind/article/what-is-a-primitive.html.

Loops

Looping is one of the major capabilities that make computers so amazing: the ability to perform the same computation over and over and over at great speed. Loops can take several forms, but are often of one of two varieties: **for-loops** perform a computation over a fixed set of values, and **while-loops** perform a computation for as long as some condition is met. For example, a while-loop in NetLogo can be instantiated like this:

NetLogo code 2.2

```
while [any? other turtles-here]
    [forward 1]
]
```

The code block above contains several NetLogo primitives—in fact, every word is a primitive except for the number 1. When a turtle executes this code block, it calls the function `any?`, which takes as an argument an agentset (a set of agents) and returns false if the set is empty and true if the set contains at least one agent.[6] The agentset is defined by the primitive function `turtles-here`, which returns the set of all agents on the same patch as the **caller**—the code object that initiates the function call (generally either an agent or the global observer). The function `other` takes an agentset and returns a copy of the same agentset excluding the caller. Thus, the set of brackets on the top line returns true if there are any other turtles on the same patch as the turtle executing the code, and false otherwise. If true, the turtle will move forward one unit (the width of a patch) in the direction it is currently heading. It will then check the conditional again to see if there are still any other turtles on its current patch, and it will keep moving forward until there are not. In this way, the turtle moves away from others, behaving in a somewhat antisocial manner.

NetLogo additionally has a very special sort of looping command that allows the user to loop over a set of agents in a random order. This is done with the primitive procedure `ask`. The randomization is key, as we usually don't want all of our agents to be scheduled in the same order every time step, because this could create undesirable artifacts in at least some cases. Most agent-based modeling libraries allow for a similar random scheduling of agents' behaviors. NetLogo is a very polite language, so we don't tell agents what to do, we ask them. The `ask` procedure allows us to create a list of instructions that each agent will then execute one by one. For example, suppose we had a population of agents located at various positions on a two-dimensional space and facing a particular direction, and we wanted each agent to turn 45 degrees to the right and then move forward one unit. We could do this with the following code:

NetLogo code 2.3

```
ask agents [
    right 45
    forward 1
]
```

[6]There is a convention among NetLogo coders that Boolean variables and functions that return Booleans have names that end with a question mark. This is not required, as NetLogo treats the question mark as a regular character, but it is often useful.

If agents all start at the center of the space and this command is their only behavior, looping over this ask command will yield our population of agents spiraling outward.

Conditional statements

Conditional statements are another major capability that makes computers awesome. Contingent rules allow for code to be executed only when specific conditions are met. NetLogo provides the function `if`, which takes a Boolean argument and executes a series of commands if and only if the argument is true. If the argument is false, nothing happens. The command `ifelse` is similar, but it involves two sets of commands, executing the first if the argument is true and the second if it is false. For example, consider a scenario in which agents have an energy reserve, stored as a variable called `energy`, and we want the agents to turn slightly to the right and move forward only if they have more than one unit of energy, otherwise they run some custom function called `forage` that we can imagine involves some other meaningful behavior. We can achieve this with the following code:

NetLogo code 2.4

```
ask agents [
   ifelse energy > 1
   [
      right 45
      forward 1
   ]
   [
      forage
   ]
]
```

More recent versions of NetLogo also allow `ifelse` to operate as a "switch" function, which allows the user to delineate an arbitrary number of commands with each to be run under some condition. Building models of behavior very often involves constructing sets of decision rules for how agents will behave under different circumstances. Conditional statements are a key part of making this happen.

Object-oriented programming

Early programs were simple lists of commands, performing sequential operations on stored variables one line at a time. And indeed, many programs today are still of this type. In these programs, variables typically refer to values or sets of values (such as a matrix of integers). **Object-oriented programming** allows for something more nuanced. The user can define an arbitrarily large number of object classes, each of which characterizes new *types* of variables for the program to work with. Each class can store its own variables and procedures, and all instantiations of that class (that is, the objects) will possess these variables and procedures.

This type of programming lends itself very well to agent-based modeling, because it is easy to define a class of agents with the desired properties. NetLogo uses a special type of object-oriented programming called **agent-oriented programming**, which differs primarily in that commands can be written as instructions directly to the agent. This allows for what many consider a more natural style of programming for agent-based models. We saw this above in our use of the `ask` command. A common problem faced by novice NetLogo programmers is a "confusion of levels," in which a set of code is written at one level—for example, at the level of the observer (the perspective of you, the coder)—but is placed in the code in such a manner that it will be interpreted at another level—e.g., by a turtle. This usually generates an error, but if you're on the lookout for it, it is easily handled.

Through practice, the elements of programming—both those common to most languages and those specific to NetLogo—will gradually become more and more like second nature.

2.3 Particle World

Now that we've introduced our tools, let's get on with seeing how we can put them to work. We are going to build a very simple agent-based model. The goal is to get you comfortable with the tools of model building, so for now we won't worry about how well the model represents reality. Instead, we are going to start with a toy model of a make-believe world: Particle World.

This ignoring of reality may appear to run counter to what I said in the previous chapter about how models, by analogy, help us to understand the real world. However, we must learn to stand before we can run. To do this, we need to engage in a bit of play. Throughout the animal kingdom, young animals (humans certainly included) play with simplified versions of the objects or scenarios they will eventually need to take seriously as they grow into adults. Juvenile meerkats are given injured scorpions with their stingers removed by their mothers in order to become familiarized with the tasty but deadly prey. Young children in the West might play "family" or "cops and robbers" to become familiar with social roles in a low-risk context. Similarly, our first "model" is a sort of playpen, in which agent dynamics can be explored entirely within the world of the model without us worrying about how the model maps onto anything in the real world. These agents will be simple dots moving around on your computer screen. Of course, humans are natural pattern finders and storytellers, so you may not be able to stop yourself from imposing a narrative onto our model. I certainly couldn't.

2.3.1 The Story of Particle World

Particle World is a magical land inhabited by strange but simple creatures called particles, who require no sleep or food, only space to move around. And move they do, *constantly*. The space they live on is a **torus**, a donut-like surface, so a lone particle could happily move around forever. The torus (Figure 2.3) is a commonly used surface in agent-based modeling. We will represent space as a toroidal grid, sometimes referred to as a grid with **periodic boundaries**. This means that if you move past the rightmost edge, you end up on the left side, and if you move past the topmost edge, you end up at the bottom. If you've ever played a game of Pac-Man you will be familiar with this concept—when Pac-Man exits the maze through an opening on one side of the maze, he emerges on the opposite side. The grid is *toroidal* rather than a true torus because each cell or patch is presumed to be a perfect square with an area equal to that of all the other cells.

Particles vary in their propensity to wander around in new directions. Particle World psychologists have determined that the various cultural groups of Particle World exhibit varying levels of the personality trait they call "whimsy," which is the extent to which individuals fail to stick to their current directional heading (Figure 2.4). The opposite of whimsy is stubbornness. A stubborn particle heading due north at one moment in time will still be heading north in the following moment, barring a collision. A more whimsical particle, on

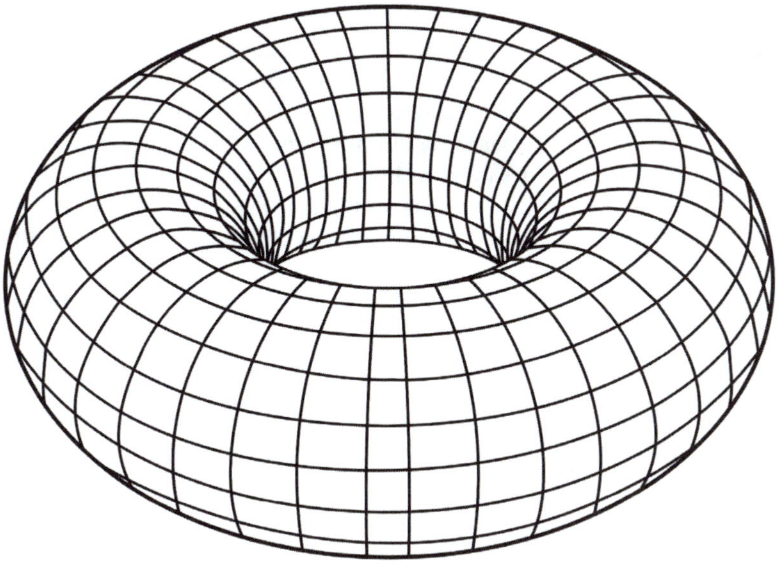

Figure 2.3 A torus.

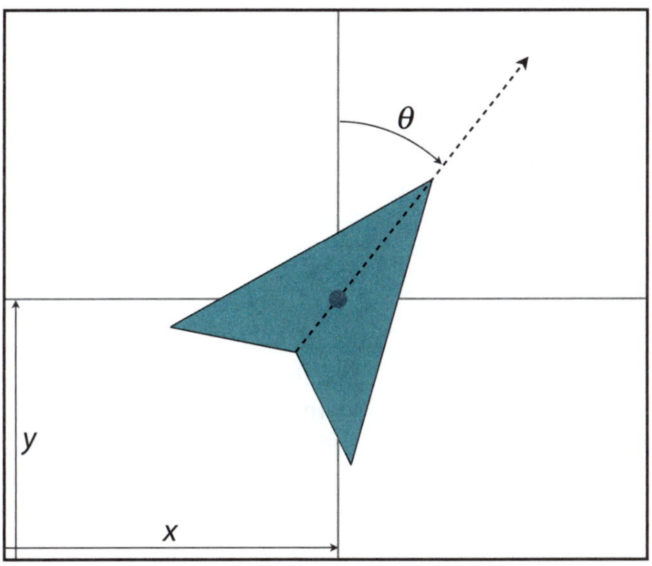

Figure 2.4 A depiction of a particle. The particle has position, (x, y), and a directional heading, θ. Its heading and position are updated on every time step.

the other hand, may have drifted some during the same time interval, so that in the moment following its northerly heading, it might be heading in a more easterly or westerly direction.

Things get hairy if a particle bumps into another particle. Keeping track of where they are heading takes a lot of concentration, which is broken by the collision. After they collide, both particles get confused and forget in which direction they were heading, and so each sets off in a new, randomly chosen direction.

Particle World physicians have become concerned about the long-term health consequences of these collisions. As such, they are particularly interested in how often the collisions occur. Some regions of Particle World are more densely populated than others, and they suspect that higher densities lead to more collisions. But how *many* more? Particle World scientists also know that individuals in some areas tend to be very whimsical, while individuals in other areas are more stubborn. They wonder: How does the whimsy of a region affect the number of collisions observed in that region?

We can make Particle World a reality and get to work answering these questions. Let's get coding!

2.4 Coding the Model

Now that we've got a good idea of how the model should work, let's get to coding it in NetLogo. For many of us, this is the fun part. I've provided a working version of the code in the repository,[7] titled **particles.nlogo**, and you may wish to simply examine that. However, there is also value in coding something up from scratch. For most of this book, I will not work through the NetLogo code in detail, but will instead focus on the algorithms and organizational aspects of the model code (this is also in service of readers who may wish to use other programming languages). This is our first time working through an agent-based model, however, and I want to make sure you understand what goes into coding up a new agent-based model.

Open up a new NetLogo model. You will be faced with a blank Interface window (Figure 2.5). We'll start by adding a few buttons and sliders for some of the commands and variables we know we'll need (you should be familiar with how to do this after going through the NetLogo tutorials). We'll need `setup` and `go` buttons, which will each correspond to procedures in our code. We'll also add sliders for the following global parameters: `num-particles`, `whimsy`, and `speed`. Before we head over to the Code window, we'll increase the size of our grid by clicking on the *Settings* button. The default grid in NetLogo is a 33×33 square. We'll increase the size to 101×101 to give our agents lots of space in which to move around (Figure 2.6).

With that accomplished, we can head over to the Code tab to start coding the model. If you look through my code, you'll notice throughout that I have added comments on many lines using semicolons. NetLogo ignores anything on a line that comes after a semicolon (two semicolons are often used as a convention, but one will work just as well). In general, documenting your code is a good habit to get into. It helps other people read and make sense of your code. Don't care about other people being able to understand your code? Well, consider one specific other person: your future self. You may know exactly how your code works now, but many a programmer has returned to code they haven't seen in months or years and found themselves baffled. Comment!

Moving on, NetLogo has the interesting feature that variables introduced in the Interface tab are treated as declared global variables, and so they do not need to be reintroduced in the Code window. In fact, there's only one other global variable we need to declare, and that is our outcome variable, `collisions`. Global variables are traditionally declared at the very top of the code, and so we can write the following:

[7] See the Preface for the repository URL.

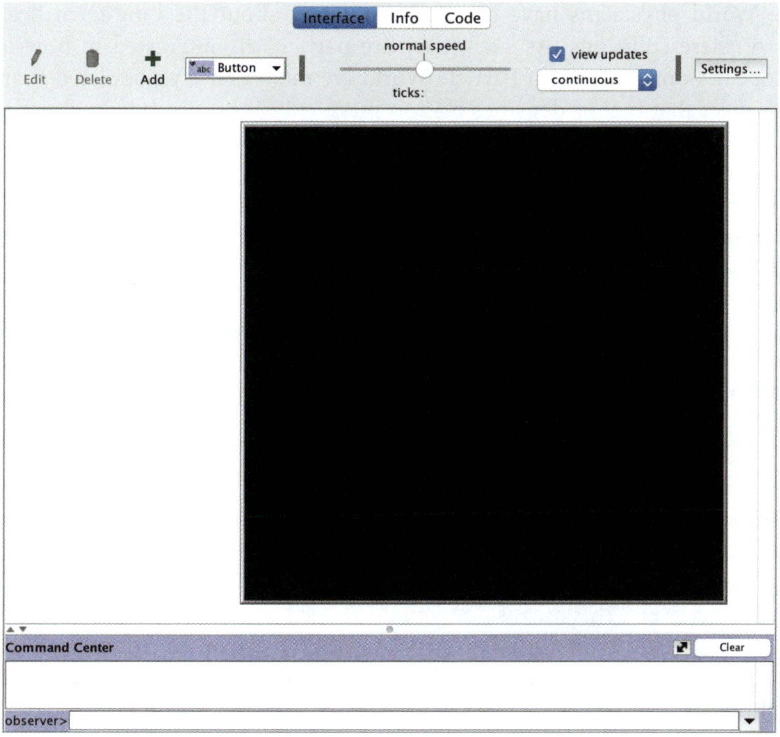

Figure 2.5 Blank Interface window in NetLogo.

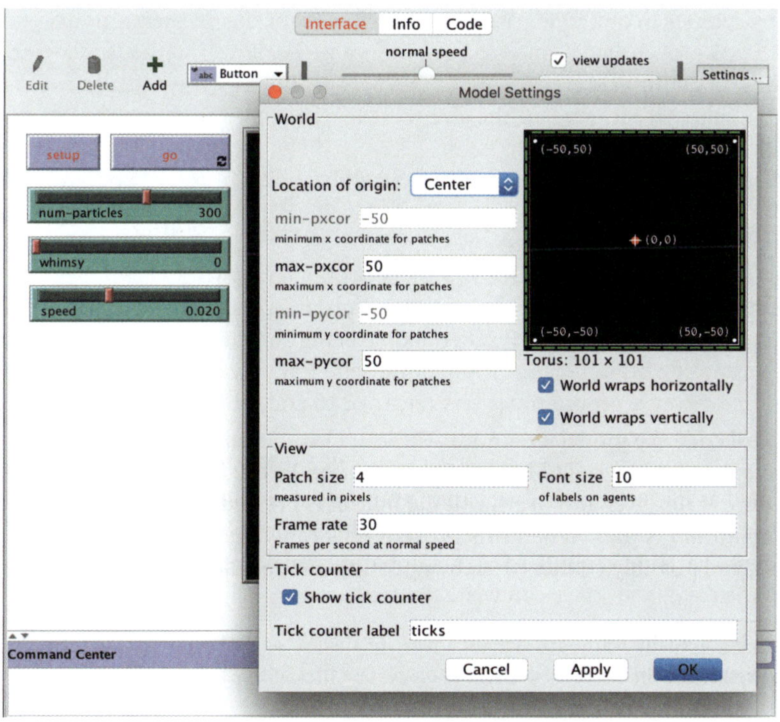

Figure 2.6 Interface window setting up for the Particle World model.

NetLogo code 2.5

```
globals [
  collisions
]
```

Next, we'll write the code for the function to initialize the model, the `setup` procedure, using the primitive function `to` to introduce the definition of a new procedure. The `setup` procedure must accomplish a few things. The first and last commands are standard and will be part of most NetLogo programs. `clear-all` resets all variables to their default initial values and removes all turtles and links from the simulation. `reset-ticks` resets the tick counter to zero and initializes all plots. These two commands are NetLogo primitives and are described in the NetLogo User's Manual. The first command after `clear-all` is aesthetic: we'll ensure that our agents take on the default shape of an arrow, which will allow us to easily see their directional headings.[8] We can then create the agents (as turtles). We make each agent green (or whatever color you like), make it larger to help us see the agents more easily, set it in a random location, and give it a random heading. The primitive reporters `random-xcor` and `random-ycor` are extremely useful, because each returns a random floating point number from the allowable range of spatial coordinates along the *x* or *y* axis, respectively. This means that you don't need to update the code if the size of the world is changed. After creating the agents, we will set our collisions counter to zero. After this procedure is run, all of our agents will exist in our simulated world, but they won't have done anything yet.

NetLogo code 2.6

```
to setup
  clear-all ;;clear/reset all variables
  ;;make the turtles look like arrows
  set-default-shape turtles "default"
  ;;make a bunch of green turtles
  create-turtles num-particles [
    set color green
    set size 2 ;;make them easier to see
    setxy random-xcor random-ycor ;;give them a random location
    set heading random 360 ;;give them a random heading
  ]
  set collisions 0 ;;initialize collisions at zero
  reset-ticks ;;set the clock to zero
end
```

Next, we'll code the model dynamics. These are laid out in the `go` procedure, which you should have set up to be called by a "forever" button in the Interface tab. This means that the procedure is run over and over until the button is pressed again or until some specified stopping condition is met. When this procedure is called, it will cause all of the agents to turn, move, and possibly collide (thereby updating our collision counter). We do this using

[8] The arrow is actually NetLogo's default shape for turtles, so declaring this isn't strictly necessary. However, doing this helps us to be conscious of the fact that other shapes are possible.

the command `ask turtles`, which takes as its argument a block of code that each turtle, in random order, will execute. First, each agent will turn to the right and then to the left, with the angle of each turn consisting of an integer randomly drawn from a uniform distribution between zero and (`whimsy`−1).[9] The agent will then move forward a distance specified by `speed`. I personally like the speed to be quite slow (around 0.02 or less) so that collisions are easy to visually observe and agents' behavior approximates smooth movement through space.

After an agent moves, we need to determine if it has collided with another agent. NetLogo has some neat tools for making this happen. The conditional statement here that determines a collision is `if count turtles in-radius 1 > 1`. The agent checks whether there are any other agents whose distance to the **focal agent**—the agent that we're focusing on at the moment—is less than or equal to one unit, indicating that they are touching. (For more information about any of the primitive commands used here and elsewhere in this book, see the NetLogo Dictionary.) If this condition is met, all of those agents with which the focal agent has now collided, as well as the focal agent itself, are asked to set their headings to a new random direction. Each of these agents then moves forward a small amount. Why do we do this? It's worth experimenting with removing this command (you can "comment it out" by adding semicolons at the front of the line) to see what happens. Agents will continuously jostle around, because they will remain near one another and are just as likely to move closer together as to move farther apart. Having each move a bit forward substantially reduces this possibility, making for a much more "natural" collision. It is debatable whether this sort of hack is reasonable in a model of behavior, and as a modeler that is the sort of decision you may often be faced with. The procedure ends with the command `tick`, which advances the simulation clock and updates the plots.

NetLogo code 2.7

```
to go
  ask turtles [
    ;;turn a random amount
    right random whimsy
    left random whimsy
    forward speed ;;move forward

    ;; if at least 1 other turtle near, set new headings
    if count turtles in-radius 1 > 1 [
      ask turtles in-radius 1 [
        set heading random 360
        fd 0.1
      ]
      set collisions (collisions + 1)
    ]
  ]
  tick
end
```

[9] Note that this will emphatically *not* produce a uniform distribution between negative and positive `whimsy`, but rather a binomial distribution bounded in ±`whimsy`. See Box 2.2: Correlated Random Walks and the Central Limit Theorem.

That is all the code you need to make the model work. Play around with it and see what kind of behavior you observe, fiddling with the various sliders in the Interface tab. We'll talk more about how to analyze the model below. Before we move on, however, it's worth saying something about code **modularity**. We have put all the model's code in just two procedures: `setup` and `go`. It may seem economical to reduce the number of procedures in a program, but in fact the opposite is often preferable. By having many small procedures, code becomes more modular. Modular code is usually easier to read and debug, and code snippets are more likely to be directly transferable to another piece of software. Below, the `go` procedure is rewritten to be more modular, supported by newly created supporting procedures: `move` and `bounce-turtle`.

NetLogo code 2.8

```
to go
  ask turtles [
    move
    bounce-turtle
  ]
  tick
end

to move
  right random whimsy
  left random whimsy
  forward speed
end

to bounce-turtle
  if count turtles in-radius 1 > 1 [
    ask turtles in-radius 1 [
      set heading random 360
      fd 0.1
    ]
    set collisions (collisions + 1)
  ]
end
```

Now that the model dynamics are coded, you can observe the simulation unfold in the Interface tab, watching the agents move around and collide. Wheeee! Visualization is useful when studying dynamic models, as watching those dynamics unfold under different parameter assumptions can give you intuition that is otherwise difficult to come by merely from reading the algorithmic model description or inspecting summary statistics. But summarize we must if we are to produce more than merely qualitative reports of the model behavior. We'll start very simply, by plotting the number of collisions as a function of time. This sort of plot is very easy to set up in NetLogo, which is yet another reason it's a nice platform for studying simulation models.

Back in the Interface tab, use the plus button or right-click to create a new plot. This is, by default, a plot of how one or more values change over time. Since we have already defined a variable to record the number of collisions, you can simply tell NetLogo to plot this variable. Filling in the window as shown in Figure 2.7 should do the trick. In addition to creating a plot, it may be useful to add a Monitor that reports the exact number of collisions at each time step. You can do this by means very similar to those you used to add the plot. You

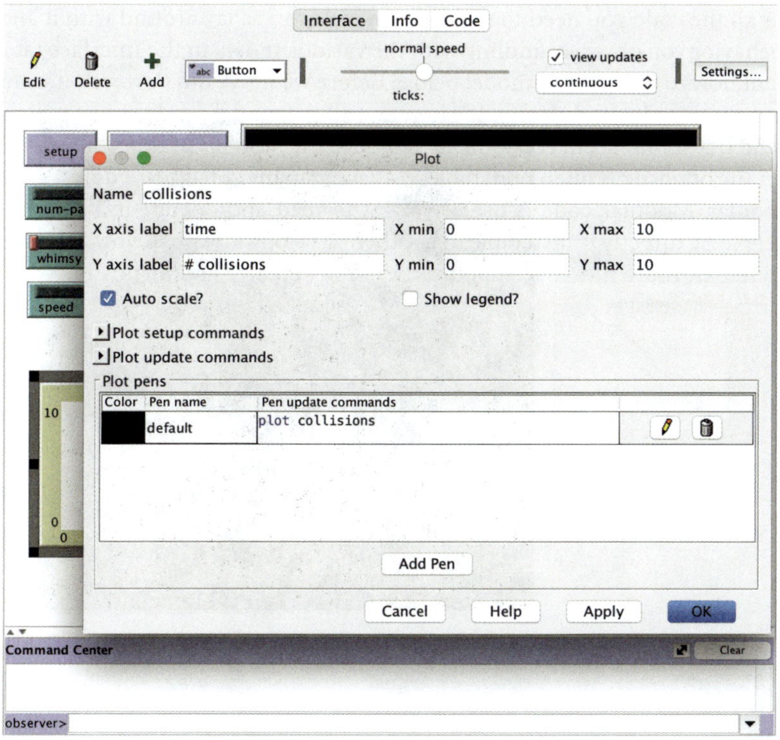

Figure 2.7 Adding a plot of the collisions as a function of time.

should end up with something that looks similar to what is depicted in Figure 2.8. You will observe that the number of collisions tends to increase linearly with time. Thus, you can use the *rate* of collisions per unit time to compare different scenarios of whimsy and population density—doing this is left as an exercise.

You now have a simulation model to take to the scientists of Particle World to help them study how the different population densities and cultural amounts of whimsy influence the rate of collisions each group experiences. You might even be able to use your model to suggest some interventions! For example, you might observe that while both population density and whimsy influence the number of collisions, the effect of density tends to dominate (Figure 2.9). This suggests that restricting density might be more practical than trying to change the (presumably strongly culturally and/or genetically imbued) degree of whimsy in each group.

The Particle World model is very simple and perhaps a little silly. However, the basic skills involved in putting it together constitute much of what will be needed to build models we can use to ask deeper questions about social behavior. I recommend you spend some time playing with the model parameters and seeing what kinds of outcomes you can get. Think about some other assumptions you could make about your agents, and try to code them. Get creative. Playing around to see what you can make your model do is a valuable habit to cultivate when designing and building models. I encourage you to play as much as you like, and I have provided some suggestions in the Exploration section of this chapter. In the next chapter, we'll study a slightly more serious model as well as a more serious approach to analyzing it.

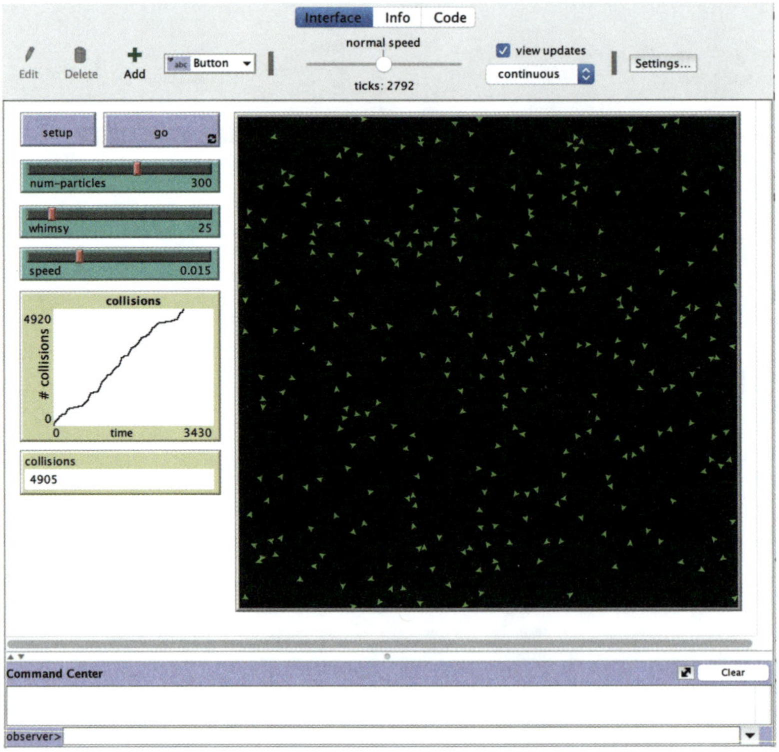

Figure 2.8 The Interface of the Particle World model in NetLogo.

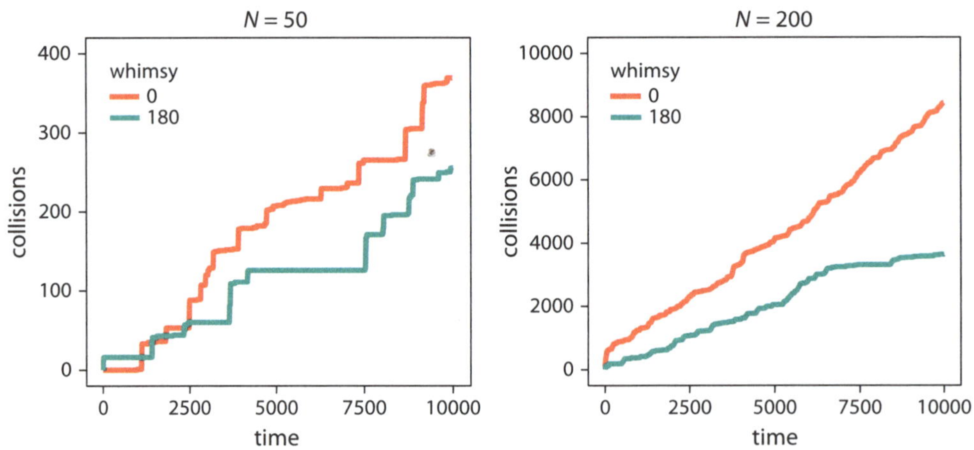

Figure 2.9 The cumulative number of collisions for several runs of the Particle World model over 10,000 time steps, for different values of the population size (N) and `whimsy`. In all cases, `speed` = 0.02.

You may notice that I have been a little vague about exactly how Particle World works. I could give the written description above to a few modelers to code up, and, without access to the NetLogo code, we might see slight or even prominent differences between their interpretations. This is a problem. If we are going to do serious science with models, we need to be able to talk with precision about how models work and to communicate that precision to others so that they can understand and reproduce our work. We therefore need to know how to talk about the components of a model—any model—and how to effectively write up a model description. In the next section, I'll discuss how to describe and communicate a model design, using Particle World as a case study.

2.5 The Components of a Model

Models are representations of a slice of reality, of some system of interest. A model decomposes the system into a simplified set of parts, properties, and relationships. As discussed in the previous chapter, there is no one right way to decompose a system. Rather, the value of a decomposition depends on how useful it is in explaining or illuminating some phenomenon or set of phenomena. Throughout this book, we will consider many systems involving social behavior and explore some representative decompositions. In doing so, we should always keep in mind this question: What are we assuming, and what are we excluding from those assumptions?

When it comes to presenting a model to others, our task is to help them to answer this question as easily and precisely as possible. Model descriptions should therefore be clear and transparent. In particular, they should include clear statements about (1) what the parts and properties of the model are, (2) how the model is instantiated and initialized, (3) how the dynamics of the model are scheduled, and (4) how the outcomes of the model simulations are measured or computed.

2.5.1 Parts and Properties

We need to lay out all the components of a model when describing it. What is being represented, and what is the nature of that representation? If we are talking about an agent-based model, what are the agents like? What properties do they have? What behaviors do they exhibit? If the agents interact, how are those interactions structured? What is the nature of their environment, and what are its properties?

The Particle World model consists of agents and their rather sparse environment. Let's start with the environment. The "real" Particle World might contain rocks and trees, lakes and rivers. However, we have made the simplification (which for now I will assert is reasonable) that particles move on large stretches of fairly flat land. So we'll ignore those geographic features and assume a fairly featureless landscape, and we'll model the topology of Particle World as a continuous square space with periodic boundaries. Let's unpack what that means. The world is square. For a spatially explicit simulation, a square is one of the simplest landscapes to program, analyze, and visualize.[10] As such, a square is one of the most commonly used spatial arrangements, particularly when a modeler doesn't have specific domain knowledge that would make the square a poor choice. The space is continuous, which means that

[10]A line is even simpler. However, when considering agents moving around and colliding, a line makes collisions unavoidable, while a 3-D space is unnecessarily complicated (not to mention more difficult to visualize). A 2-D space is best for our purposes.

agents can travel arbitrary distances in arbitrary directions. A commonly used alternative is a discrete grid, in which cells or patches are arranged in a regular pattern in the space and an agent's location is defined by its cell. Finally, and perhaps most notably for the novice modeler, the boundaries of Particle World are periodic, or toroidal (Figure 2.3). An agent that moves down past the lower bound of the space will emerge at the top, and an agent that moves right past the rightmost bound of the space will emerge at the left edge. Periodic or toroidal boundaries are commonly used in spatially explicit agent-based models. This may seem strange, given that almost no earthly environments are toruses. However, toroidal surfaces help us to avoid tricky artifacts like agents getting stuck in the corners of the grid, or asymmetries where agents near the edges have fewer spatial neighbors. Finally, we'll have to specify the size of our square, denoting the length of one side as L in arbitrary units. The spatial length unit is arbitrary, but it serves to determine the meaning of things like speed and distance in the model, and in some models this unit may correspond to real lengths or distances. In NetLogo, this is the width of a patch.

Onto this space we'll place our agents, whose number we will specify with one of our global parameters. The population size will also determine the population *density* if we hold the size of the space constant. Each agent has only two intrinsic properties: its location and its directional heading. That is, each agent will keep track of where it is and where it is heading. These need not be the only agent properties. For example, we discussed the importance of the speed at which the particles move. However, for our analysis, we are interested in speed as a property of the cultural group, so we will code speed as a property of the environment that governs the movements of all the agents in the population. We similarly discussed whimsy as a property that varied primarily between cultural groups, so we will likewise code whimsy as a global variable (that is, a property of the entire simulation) rather than as a property of the individual agents.

Finally, let's briefly talk about visualization. For reasons that will hopefully become clear, it is often useful to visualize the model dynamics in detail. Dynamic visualization isn't always practical for some model designs or some programming language choices—the ease with which dynamic visualizations are possible is a strength of NetLogo. When it is possible, it can help the modeler gain intuition about the system they are studying, and even observe outcomes they otherwise wouldn't think to test for. When making a visualization, we'll have to make some arbitrary choices concerning the aesthetic appearance of our agents and their world, including agents' color and shape. These aspects are often purely cosmetic, because they do not affect the model's behavior or its outcomes, and such cosmetic aspects are typically not reported in formal descriptions for simulation models. Indeed, these aspects of the code are often bypassed when running batches of simulations to reduce computing time. For the Particle World model, I have chosen to represent agents as green circles on a black background, but you can choose any appearance you like. Now that we know the components of the model, we need to specify how they are arranged.

2.5.2 Initialization

You cannot tell someone how to get somewhere if you don't know where they are starting from. Similarly, you cannot fully understand the dynamics or end states of a model if you don't know what things were like at the start. What's going on when a model simulation begins? How many agents are there, and what are their properties? Where are they in their environment and in relation to each other? What does the environment look like? Research on nonlinear dynamics has shown that in complex, interconnected systems, long-term dynamics can be highly sensitive to initial conditions. Although many social systems are also quite resilient, the social systems we are interested in are often complex and subject to feedback processes that can amplify small variation. It is therefore vital to answer these questions about initialization thoroughly.

At the beginning of a Particle World simulation, the spatial environment is established and the agents are placed upon it. Recall that agents have only two uniquely held properties—location and heading—and so each agent will keep track of its own values for these variables (NetLogo does this automatically for turtles). How should they be initially assigned? There are lots of possibilities. For example, all the agents might start in the same location, or be evenly spaced in a grid pattern, or be arranged in a circle, or in the shape of a T-rex, or be placed on the grid in locations representing actual places in the real world. Similar concerns apply to their directional headings. If we don't have good reasons to choose one of these, or if we have a good reason to think it doesn't matter, we might as well draw both locations and headings at random from a uniform distribution of choices, and this is what we have done. That is, `num-particles` agents are created and placed at random x and y locations on the square grid. Our agents are ready to go!

2.5.3 Dynamics

Once we know what the parts of the model are and how they are initialized, we need to know how they change. In other words, how does the state of the model system update from one moment to the next? We usually think of time as progressing continuously, and this sort of continuous change *can* be modeled to some extent by mathematical formulations using infinitesimal time increments, as with coupled differential equations. In this book, we will generally stick to modeling time as advancing in discrete increments, though these increments can be arbitrarily small. This assumption also fits the nature of computational representation, which is naturally discrete. In NetLogo, discrete temporal increments are usually called "ticks," and I will use this term interchangeably with the more widely used phrase "time steps." Time steps can represent short units of time like a second or a day, or longer units like a year or a reproductive generation. The description of a model's dynamics typically involves what happens in one time step of a model simulation.

You should carefully consider the specifics of what happens during a time step and in what order those things occur. This ordering of the computations performed during each time step is called **scheduling**. Choices about scheduling can be consequential. For example, all agents might calculate their next move before any actions are taken, so that each agent is responding to the exact same environment. In this case, the ordering in which agents make their decisions probably doesn't matter. Alternatively, each agent could calculate and execute their move in one fell swoop, so that each agent potentially responds to the actions of the agents who are scheduled before it. In this case, the order in which agents are scheduled can matter, and it is usually wise to randomize the order in which agents are "stepped" at each tick[11] (NetLogo's `ask` procedure does this automatically). Whether agents respond **synchronously** (all responding to the same environment) or **asynchronously** (one at a time) can affect how the dynamics of the model unfold, though the qualitative results of most models are usually robust to both styles of scheduling.[12] A related issue concerns

[11]For example, Turner and Smaldino (2018) studied how stochastic decisions in initialization and scheduling could dramatically affect the dynamics of a single simulation run in a model of opinion dynamics.

[12]An interesting debate on this issue can be observed by reading Nowak and May (1992), Huberman and Glance (1993), and Nowak et al. (1994) in sequence, which deal with a spatial model for

the scheduling of multiple actions within a single time step. Should all the agents complete one action before any agent can move to the next action? Or should each agent complete all actions before the next agent is scheduled? There is no right answer to this question without knowing the details of the system being modeled, but the modeler should be mindful that different answers can lead to different outcomes. Finally, specification of a model's dynamics includes any stopping conditions for the model—that is, when a simulation is considered finished. Sometimes this will be after a fixed number of time steps, while in other cases it may be whenever a particular system state is reached (such as an equilibrium). In the former case, the number of time steps should be justified. For example, it may reflect sufficient time such that all observed simulations have reached a roughly stable state.

For each time step of the Particle World model, each agent, in a random order, turns, moves, and potentially collides. If the agent is stubborn, it will not deviate much from its previous heading when it moves. If it is more whimsical, it will turn quite a bit more. More precisely, the variable `whimsy` denotes an agent's maximum turning angle. Each agent will adjust its heading at each time step by first turning to the right an amount randomly chosen from a uniform distribution between zero and `whimsy` degrees. It then draws another number at random from the same distribution and turns that many degrees to the left. Thus, more whimsy translates into more frequent wide turns and less correlated random walks, while less whimsy translates into smaller turning angles and more highly correlated random walks, such that at the limit of zero whimsy, each agent simply travels in a straight line. After turning, the agent will move forward `speed` units in the direction of its current heading (ending up on the other side of the space if it moves across the edge of the grid). If no other agents are located where the agent has moved to, the agent is finished for the current time step. However, if another agent is sufficiently close to the agent in question (within one spatial unit), then a collision occurs. Recall that when an agent collides with another, it gets confused and heads off in a new direction. So, when two agents collide, they both receive new directional headings chosen at random from a uniform distribution between 0 and 359 degrees. A time step is completed when all agents have performed these actions.

2.5.4 Outcomes

The design of a model is driven by the questions we are asking about our system, and the process of modeling must therefore include deciding how the model's outcomes will be quantified and how the model's behavior under different conditions will be characterized for comparison. In other words, we need to know about our outcome measures. Sometimes this is as easy as counting the proportion of agents in the population exhibiting some trait. Other times more computation will be necessary, as when we calculate the structural properties of an evolving network.

In the Particle World model, we are concerned with the number of collisions that occur over the course of the simulation, as a function of both `whimsy` and `num-particles`. That is, our outcome is the number of collisions that occur over some standardized length of time. We measured this by creating a variable that is initialized to zero at the beginning of a simulation and is then incremented by one every time a collision occurs.

---

the evolution of cooperation that initially used synchronous updating. Some results were robust to a change to asynchronous updating, others were not.

2.6 Describing a Model

If you are sharing your model with others, as in a scientific paper or even in a blog post, you should describe it well. Make sure you've included details about the parts and properties, how the model is initialized, how the dynamics work, and any outcome measures you are collecting. A careful reader who is a competent coder and familiar with models—but not necessarily familiar with *your* model—should be able to replicate your model in a programming language of their choice, based solely on your written description. That is, the model description should be clear and complete, and should minimize ambiguity.[13] This is really important. The lessons one can draw from a model come directly from understanding how the model's assumptions lead to the consequences highlighted by the modeler. If a model is described poorly, the reader won't be able to discern exactly how it works, and any results of the model's analysis border on useless.

Writing up a clear description of a complicated model is a skill that requires practice to hone. There are many good suggestions in the **ODD protocol**, widely used in ecology, for describing agent-based models (Grimm et al., 2010, 2020). The protocol suggests a three-stage strategy of model description: the Overview (the "story" of the model), the Design (the computations involved in the model's dynamics), and the Details (all of the model's algorithmic details, sometimes relegated to an appendix). I am hesitant to recommend elaborate all-purpose protocols for describing research conducted across disciplines, but I quite like the general approach of an iterated description with increasing detail given at each stage. Nascent modelers often forgo the Overview stage, preferring to plunge ahead with the formal mathematical or computational details of the model. This is usually a mistake for all but the simplest models. A model is a representation of something else. When we make assumptions about a model system, we are mapping them onto our representation of the corresponding real-world system. However, as noted, a model simplifies—it omits details, and it even introduces falsehoods in the service of simplicity (for example, the falsehood that there are only two types of people). So, in order to make sense of how the model details map onto the real-world system, it helps to explain the model system verbally. The subsequently presented formal details can then be used to clarify how the model works without requiring the reader to simultaneously figure out the underlying analogy to the real world.

Because my intent in this book is pedagogical, I will often introduce models in a more narrative style than I would use were I to describe them in a scholarly article written for experts. In order to demonstrate more clearly what I am talking about, however, I have presented a complete formal description of the Particle World model in Box 2.1. This information is redundant with what is described above, but it illustrates how a model might be described in a publication.[14] The box also uses the convention of labeling parameters using single letters rather than the names actually used in the computer code. I find this approach easier to read, and it offers continuity between mathematical and agent-based modeling.

---

[13] For very complicated models, such as those used in systems science or artificial life, it may not be practical for every presentation of the model to include a full description. Even in these cases, however, that description should be accessible *somewhere*, and that somewhere should be directly referenced in all write-ups.

[14] For recent examples of how I recommend describing agent-based models of much greater complexity, see Smaldino and Turner (2022), Smaldino et al. (2019a), and Smaldino et al. (2019c).

BOX 2.1: Particle World Description

This model features a population of spatially embodied agents moving through continuous space, each using a correlated walk. When agents collide, they become confused, and each sets off in a new direction. We run each model simulation for 10,000 time steps and compare the number of collisions during that time.

Initialization

A population of N agents is initialized with each agent placed at a random real-valued locations on an $L \times L$ grid with periodic boundaries. Each agent i has a direction heading θ_i, which is initially chosen at random from a uniform distribution of integers [0, 359]. Each agent is fully defined by its location and directional heading. Other model parameters are the whimsy, w, which determines the turning angle agents use on each time step, and the speed, s, which determines the size of the step they take when moving. Finally, we keep track of the cumulative number of collisions over time, $C(t)$.

Dynamics

At each time step, each agent, in a random order, turns, moves, and collides. An agent first *turns* by adding to its direction heading an integer value that is randomly drawn from a uniform distribution in $[0, w]$. The agent then subtracts a newly drawn value from the same distribution from its directional heading. In other words, its new directional heading is $\theta_i + \epsilon$, where ϵ is randomly drawn from a binomial distribution bounded in $[-w, w]$. The agent then moves s units forward. If there are any other agents with a position within one unit of the focal agent's new location (defined by the Euclidean distance between the centers of each agent), a *collision* occurs between the focal agent and all of these other nearby agents. In this case, all of the agents involved in the collision update their heading to a new value randomly selected from the uniform distribution [0, 359]. Each of the involved agents then moves forward 0.1 spatial units in order to move away from the site of the collision and avoid cycles of perpetual collision. If a collision occurs, the cumulative collision counter $C(t)$ is incremented by one.

BOX 2.2: Correlated Random Walks and the Central Limit Theorem

In our Particle World model, the agents use a kind of **correlated random walk**, which just means that an agent's directional heading at time $t+1$ is correlated with its heading at time t. We implemented this by having agents turn a random amount to the right and then a random amount to the left. Both of these turning angles are drawn from a uniform distribution bounded between 0 and `whimsy`. So, the resulting distribution of agent turning angles (their right turn minus their left turn) should be uniform as well, right?

Actually, it's not. To see that it's not, we can plot the distribution of the resulting turning angles. I wrote a simple script that repeatedly generates two random numbers between 0 and 90, and then subtracts the second number from the first, which is

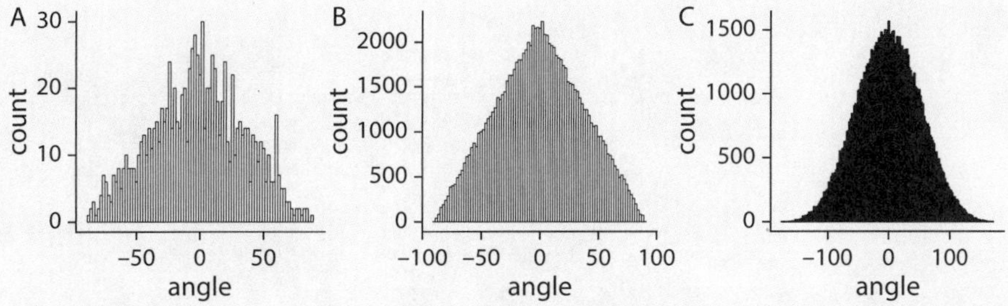

Figure 2.10 (A) Distribution of 1,000 turning angles generated by subtracting one random value from the other, where both random values are drawn from $U(0, 90)$. (B) The same, but with 10^5 turning angles generated. (C) Summing more numbers better approximates a normal distribution. Here 10^5 turning angles were generated by adding two positive random values together and then subtracting two positive random values from that sum, with each value drawn from $U(0, 90)$.

exactly what our agents do if we assume `whimsy` $= 90°$. I repeated this random process a thousand times and then plotted the distribution of the sums generated, shown in Figure 2.10A. There seem to be a lot more values close to zero than at the extremes. If we repeat the process a few more times and get more data points, we can see that the distribution of turning angles is essentially triangular (Figure 2.10B). This definitely isn't a uniform distribution. Why not?

It turns out that the distribution of the sum of draws from a uniform distribution is not itself a uniform distribution. This is because there are simply more ways to get values near the center of the range than values at the extremes. Consider a simpler case: rolling two standard six-sided dice. Even if you are not a professional gambler, some familiarity with dice or board games has likely given you the impression that rolling a seven is more common than rolling a two or a twelve (Figure 2.11). This is more than an intuition, of course, it's a mathematically necessary fact of probability. Consider a case where you want to roll a two: snake eyes. Your first die must come up one, and your second die must also come up one. This is the *only* way to roll a two. Next, consider a case where you want to roll a seven. If the first roll is a one, the second must be a six. But the first roll could also be a two, in which case the second must be a five. For *any* roll of the first die, there is a one in six chance that the second roll will yield a total of seven. There are thirty-six possible outcomes when you roll a pair of dice ($6 \times 6 = 36$), so one in every six rolls will come up seven, while only one in every thirty-six rolls will produce snake eyes.

When we repeatedly sum integers drawn at random from uniform distributions, the distribution of those sums will be given by the **binomial distribution**. For a large enough number of integers being summed, the binomial distribution is well-approximated by the normal distribution. And indeed, the continuous version of this scenario is the **central limit theorem**: the sum of a large number of uniformly distributed quantities will yield a normal distribution. Technically, the variables being summed need not be uniformly distributed, but merely *identically distributed* and independent. That is, each number must be drawn from the same probability distribution and be independent of all other draws.

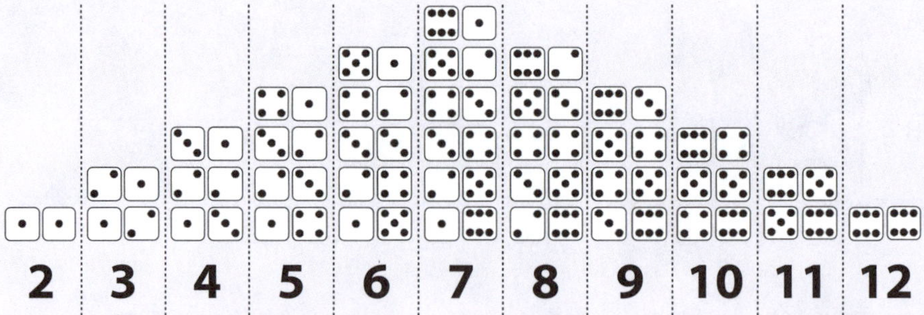

Figure 2.11 The many ways to roll two dice. Outcomes near the middle of the range of possibilities are more probable.

The purpose of this divergence is to illustrate the importance of examining the consequences of seemingly trivial modeling decisions. The distribution of turning angles produced by summing two uniformly distributed turns is not itself uniformly distributed, but highly concentrated so that small turns are always more likely than large turns even for highly "whimsical" agents. If turning angles were uniformly distributed, large turns would be much more common for those agents, which might dramatically change the relationship between whimsy and the resulting number of collisions. In general, it is wise to invest time considering the consequences of even the most minor or seemingly trivial assumptions when modeling.

2.7 Flocking

In our basic Particle World model, the agents just move on their own and crash into one another willy-nilly. Each agent is just moving along without any heed for what others are doing. As a result, there are tons of collisions. Indeed, the only social behavior we have modeled is collisions, and this behavior is confrontational to say the least. But now let's make things more interesting and imagine that our agents are a bit more civic-minded. They go with the flow. In practice, that means that they go the same way that others go.

One of the big ideas behind agent-based modeling is that coherent population-level phenomena can emerge from the aggregation of local, decentralized behaviors. In the last chapter, we briefly discussed the boids model, which illustrated how realistic flocking behavior could emerge from three simple rules for mobile agents: separation, alignment, and cohesion. Here we will explore the idea of emergence by adding just one of these rules to our Particle World model: alignment. The particles will observe their nearby neighbors and adjust their headings to match.

We'll implement flocking by adding two new global variables to the Particle World model: a Boolean variable called `flock?`, which will allow us to toggle between the original Particle World model and the version with flocking, and a parameter to control the size of the local neighborhood that agents can observe, `vision-radius`. In general, when extending a model it is often desirable to set up the code so that you can recover previous or alternative incarnations of the model. The code for this model in the repository is titled **particlesflock.nlogo**.

In our go procedure, we'll have the agents call a new procedure called flock, which, if flocking is turned on, will be executed before the agents whimsically turn and move forward. The go procedure now looks like this:

NetLogo code 2.9

```
to go
  ask turtles [
    if flock? [flock]
    move
    bounce-turtle
  ]
  tick
end
```

Note that flock is only called when flock? is true, so that our code allows us to directly compare conditions with and without flocking. When the flock procedure is called, the agent scans an area described by a circle with radius vision-radius, centered on itself. The agent first checks whether there are any other agents in that circle. If there aren't, the agent does nothing. If there are, it records the directional headings for each of the other turtles and averages them. It then adjusts its own heading to match that average. The NetLogo code for this procedure is below. Note that we define a local variable called mean-heading, which is the average of the headings of the other agents in the focal agent's vision radius.

NetLogo code 2.10

```
to flock
  if any? other turtles in-radius vision-radius [
    let mean-heading (mean [heading] of other turtles
     in-radius vision-radius)
    set heading mean-heading
  ]
end
```

What happens when flocking is turned on? Even when agents respond only to very close spatial neighbors, cohesive flocks encompassing most or even all of the agents in the simulation emerge, with large groups of agents all traveling in the same direction (Figure 2.12, top). If we compare the simulation before and after flocking is enabled, we also see that flocking allows the agents to avoid all but a few rare collisions (Figure 2.12, bottom).

The flocking algorithm we have implemented is a highly simplified version of the boids algorithm introduced by Craig Reynolds in 1987. This algorithm is also based on simple particles moving at a constant speed, but the full version requires not one but three observational radii. Within the smallest circle, an agent turns to avoid collisions. If no collisions are imminent, the agent looks within the second radius, and turns to align its heading with its neighbors. It is this aspect—alignment—that we have added to the Particle World simulation. Finally, if there are no nearby agents with which to align, the agent looks in a wider radius and heads toward the center of mass of any agents observed, thereby maintaining social cohesion. Having all three rules generates collective behaviors that are a bit more realistic than what we observe in our version. This simple flocking model forms the basis of more detailed explanations of collective behavior in a wide variety of species, including schooling fish, flocking birds, and swarming humans (Sumpter, 2010). It is also the basis

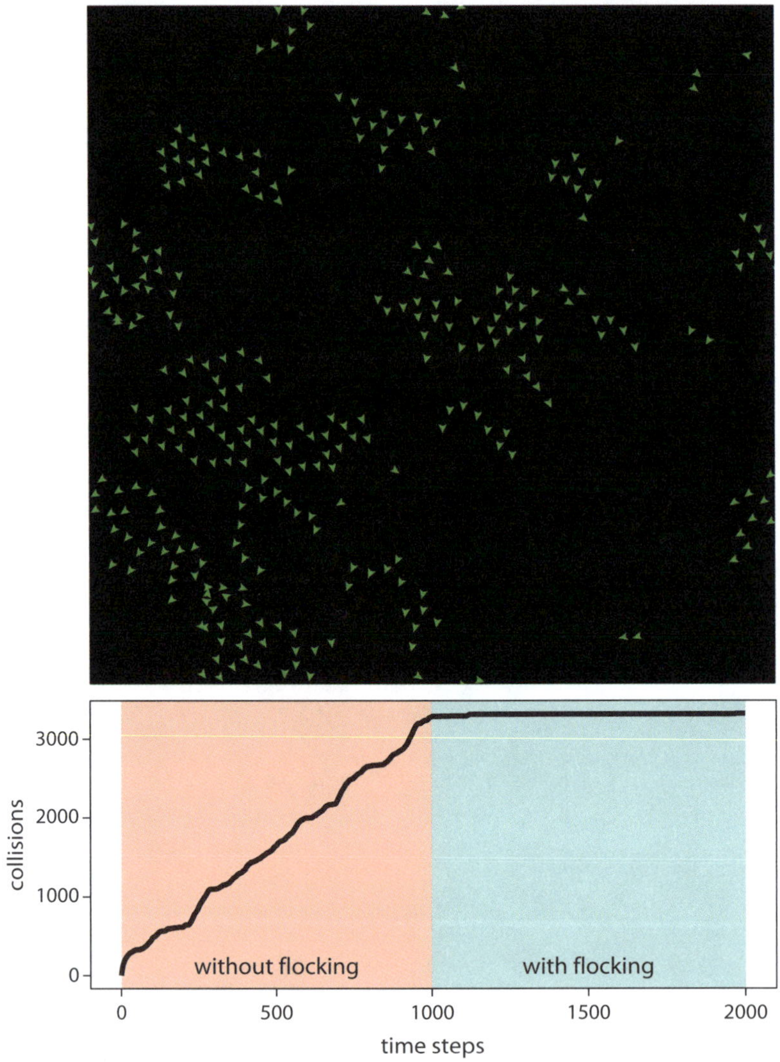

Figure 2.12 Top: The emergence of coherent flocking when agents align with those nearby. Bottom: Number of collisions over time. Simulation starts out without flocking, but flocking is turned on (that is, `flock?` is set to true) at $t = 1000$. Flocking effectively eliminates collisions. For this simulation, $N = 300$, `vision-radius` $= 3$, `speed` $= 0.03$, and `whimsy` $= 10$.

of almost all CGI (computer-generated) schools, flocks, and swarms in movies and video games (Gerdelan, 2010). Play around with the simple model we have constructed and see what behaviors you can produce by altering parameters and modifying rules.

2.8 Reflections

Throughout this chapter I have emphasized the importance of play. It is hard to overstate its value for becoming a competent modeler. Play allows you to test the failure modes of your code and provides opportunities for solving novel problems you might not otherwise encounter until the consequences are more serious. Spending time exploring in a domain without severe consequences provides opportunities to build up competence and confidence in navigating within that domain.[15] We can learn so much when we give ourselves license to simply try things for the sake of trying them. Play is important even when coding artificial worlds, partly because play helps you to develop stronger competence in turning your ideas into reality. This facility can, in turn, help you to be less constrained by the details of any particular model or software package, and instead allows you to be constrained only by what you can imagine. Play also helps to cultivate your imagination. For any system of interest, there are many ways to model it. Even once you have specified your parts, properties, and relationships, there are still details to be decided upon, and these details sometimes matter. Gaining some familiarity with how seemingly small decisions affect the behavior of a model is a valuable muscle to train.

2.9 Going Deeper

In this book we will use NetLogo to code and analyze agent-based models. However, there are also other languages and libraries you might take advantage of. Of these, the most complete and best supported is probably the MASON library (Luke et al., 2005), which is a Java library for agent-based modeling that has a rich collection of demo code and a broad community of users. Several newer libraries are available in Python, notably Mesa (Kazil et al., 2020) and the more recent AgentPy (Foramitti, 2021). The Agents.jl library, written for Julia, is also promising (Datseris et al., 2022). At the time of this writing, these Python- and Julia-based packages are still not as well developed or as widely used as MASON, but due to the popularity of these languages among social scientists and data scientists, they are growing their user bases along with all the trappings that come with that growth. NetLogo, MASON, and Mesa can all be integrated with geographic information systems (GIS), allowing agent-based models to be mapped onto real-world landscapes, including cities, roads, and ecosystems. Of course, you don't actually need a specific software library to do agent-based modeling (though it often helps, particularly with visualization and scheduling). A competent programmer should be able to write a working model in any language they choose. For example, Acerbi et al. (2022) have recently provided an open access textbook on coding simple agent-based models of cultural evolution using R. For a deep conceptual discussion on modeling complex social systems, see Miller and Page (2007).

There is a very rich literature on modeling the behaviors and patterns that emerge from the movement behavior of embodied agents. To the extent that we will model mobile agents in this book, their movements will generally be quite abstract and represent movement through social space rather than physical space. But there are large and important literatures on using models to understand the movement of collectives, including the behaviors and patterns that emerge from various movement strategies. These include flocking and schooling, the dynamics of crowds, and the decisions of social foragers. For a deeper look into these topics, see Sumpter (2010) and Ball (2004).

[15] This is supported by a wide range of work in child development (Gopnik et al., 2015), animal behavior (Smaldino et al., 2019b), and even robotics (Cully et al., 2015).

2.10 Exploration

1. **I'm walkin' here.** Create a new NetLogo model in which a user-defined number of agents are created in initially random locations and then walk around randomly.

   (a) Create a new NetLogo model with a `setup` procedure that creates turtles.
   (b) Create a slider for a parameter called `num-turtles` that controls the number of turtles created.
   (c) Write a `go` procedure that makes the turtles wander around the screen randomly. To do this, have each turtle turn a random angle and then walk forward one spatial unit.

2. **Pen down.** Observing the pathways of collisions in the Particle World model is not so easy when all the agents look the same and you can't see their trajectories over time. This is easily fixed, however. NetLogo has primitives you can use called `pen-down` and `pen-up`, which trace the movement trajectory of an agent using the same color as the agent itself. To implement this tracking, change the `setup` procedure so that all the agents are assigned a random color. Then add a Boolean switch to the Interface and corresponding code in the Code tab that allows you to toggle whether or not the agents leave colored trails as they move. Report your code and some screen captures of your output. You're a regular Jackson Pollack.

3. **Collision analysis.** Characterize how density (`num-particles`) and `whimsy` influence the number of collisions in 1000 ticks. Collect the data for at least three values of each of the two variables, and report them in a plot and/or a table. Characterize the results verbally. What did you learn? What are some of the limitations of your ability to draw conclusions from this sort of analysis, and how might they be improved?

4. **Exploding collisions.** Create a new version of your Particle World model called Particle Smash. In this version, you will let the space be bounded rather toroidal, so that agents will bounce off the walls. Moreover, crashing into other agents will now have more dire consequences. To keep things relatively simple, restrict agent movement to straight lines in the absence of collisions (`whimsy` = 0).

   (a) Set the boundaries to fixed instead of toroidal. Change the dynamics so that the turtles "bounce" off the walls if they contact the edge of the world. Recall from physics that the angle of incidence equals the angle of reflection. For example, when a moving turtle hits the wall on a shallow angle, it should bounce off at a similarly shallow angle.
   (b) Alter the code so that every time an agent collides with either a wall or another agent, it changes color. Report your code.
   (c) Update the code so that the first time an agent collides with another agent, each agent splits into two smaller agents heading in random directions. It may be useful to know that the NetLogo primitives `hatch` and `die` cause an agent to spawn new agents and to be removed from the simulation, respectively, and that the primitive `size` controls an agent's size. When smaller agents collide, they should remain the same size.

5. **Bust a move.** Study the properties of random walks. How far will one agent tend to move from its initial location using a correlated random walk as a function of the maximum turning angle, θ (this is whatever parameter controls the turning angle)? Create

a model in which a single turtle is initialized in the center of the grid. Each time step, the agent should turn first to the left and then the right, where each turning angle is a random draw from a uniform distribution between zero and θ. The agent will then move forward some fixed distance. Ensure the grid is sufficiently large, relative to the step length, that the agent cannot reach an edge in 1000 time steps.

(a) Code this model.
(b) Create trajectory plots for several values of $\theta = \{0, 15, 30, 60, 180\}$.
(c) Create a plot or monitor that displays the agent's Euclidean distance from the origin. You may wish to use the NetLogo primitive `distancexy`.
(d) What do you conclude about the relationship between turning angle and distance traveled for correlated random walks?

6. **Color spread.** This is a more challenging task to test your ability to translate a verbal description into a working simulation. Make sure to document your code and describe how each aspect works. You will create a new NetLogo model called Color Spread. When the model is initialized, all patches will start as white or black (your choice) except a single patch of some other color either at the center of the space or in one of the corners. Your first task will be to allow the color to spread to adjacent patches.

   (a) Build this model. At each time step, the color should spread from colored patches to any adjacent noncolored patches (there are many possible ways to do this). Create a button to launch the procedure. Describe verbally how the color spreads, and include screenshots of the process.
   (b) Add a plot that graphs the number of colored patches as a function of time.
   (c) Create a chooser to allow the user to select which color will spread. NetLogo has primitives for several colors, but it can also represent colors both as single numbers from 0 to 139 and as RGB triplets.

7. **Rainbow spiral.** Here's the tricky one. Create a NetLogo model that produces a rainbow spiral. When the model is initialized, all patches will start as white or black (your choice) except a single patch of some other color either at the center of the space or in one of the corners. When the model runs, it should produce a spiral (either inward, starting in a corner, or outward, starting in the center) of spreading colors in which patches change color one by one. Each colored patch should have its color chosen at random. In this version, only one new patch should become colored at each time step, and the model should stop running when the spiral is complete. For the ambitious: you can choose to leave "layers" of black patches so that the spiral pattern is easier to discern, as in Figure 2.13.

8. **Flock of seagulls.** Let's do some experiments with our flocking model.

   (a) Consider the parameter `vision-radius`, which controls how close agents need to be to each other in order to influence each other's heading. Fix the population at $N = 300$, `speed` $= 0.03$, and `whimsy` $= 10$. Turn on flocking and initialize the model with different values of `vision-radius`: $\{0, 1, 2, 3, 4, 5\}$. Plot the number of collisions over 5000 time steps for each of these values, and then plot the rate of average collisions per time step against the value of `vision-radius`. Is the effect that flocking has on collisions linear with the radius of vision? If not, why not?
   (b) Now consider how a propensity for random turning interacts with flocking. Consider the following values for `whimsy`: $\{0, 15, 30, 45, 60, 75, 90\}$, and how each of

Figure 2.13 A run of the rainbow spiral model in a 101 × 101 grid at $t = 46$.

these values interacts with the following values of `vision-radius`: {2, 4}. From visual inspection, what sort of relation do you see between the parameters and the emergent behavior of the agents? Next, run simulations out to 5000 time steps, and plot rate of average collisions per time step against `whimsy` for each value of `vision-radius`.

(c) Relate the quantitative patterns you saw in your plots to the dynamic visual patterns you observed from simply watching the model run. How well do the graphs accurately capture the relationship between your observed patterns and the parameters used? What sort of important information about agent behavior or emergent patterns is not captured by these graphs? Can you think of other metrics that might capture this information?

3 The Schelling Chapter

> It is inevitable that you are drawn back into human drama.
> —Caterine Vauban, *I Heart Huckabees* (2004)

In this chapter we will start thinking more concretely about questions of social importance and about how models can clarify our understanding of relevant real-world systems. We'll also begin to consider more carefully the assumptions we make in modeling, how to turn an abstract idea into a concrete model, and how a model should be analyzed in depth.

Our topic is, ostensibly, segregation, and we tackle this topic using a model introduced in the late 1960s by the economist and game theorist Thomas Schelling. Schelling's segregation model is arguably the most well-known model in the social sciences, and is almost certainly the most influential agent-based model of human social behavior. Variants of this model continue to be used to study segregation and other problems of social sorting. Schelling's 1978 book *Micromotives and Macrobehavior* remains a classic and is recommended reading for almost anyone reading this book. That said, our charge in this chapter is not only to study Schelling's model, but also to learn how to turn an idea into a model and how to then analyze that model. As we proceed, it's worth paying attention to the sorts of assumptions that go into building the models we study. In particular, consider how we break down the world into components and relationships, and how we examine the consequences of our assumptions.

3.1 The Puzzle of Segregation

Cities, neighborhoods, and countries are often segregated. This means that individuals of different races or ethnicities tend to be geospatially clustered. The city of Baltimore, for example, is approximately 64% Black, but that does not mean that two-thirds of each neighborhood's residents are Black. Instead, it is easy to find neighborhoods where 95% of residents are Black, and other neighborhoods that are overwhelmingly white. Patterns like this are hardly unique to Baltimore—Figure 3.1 shows the intense ethnic segregation

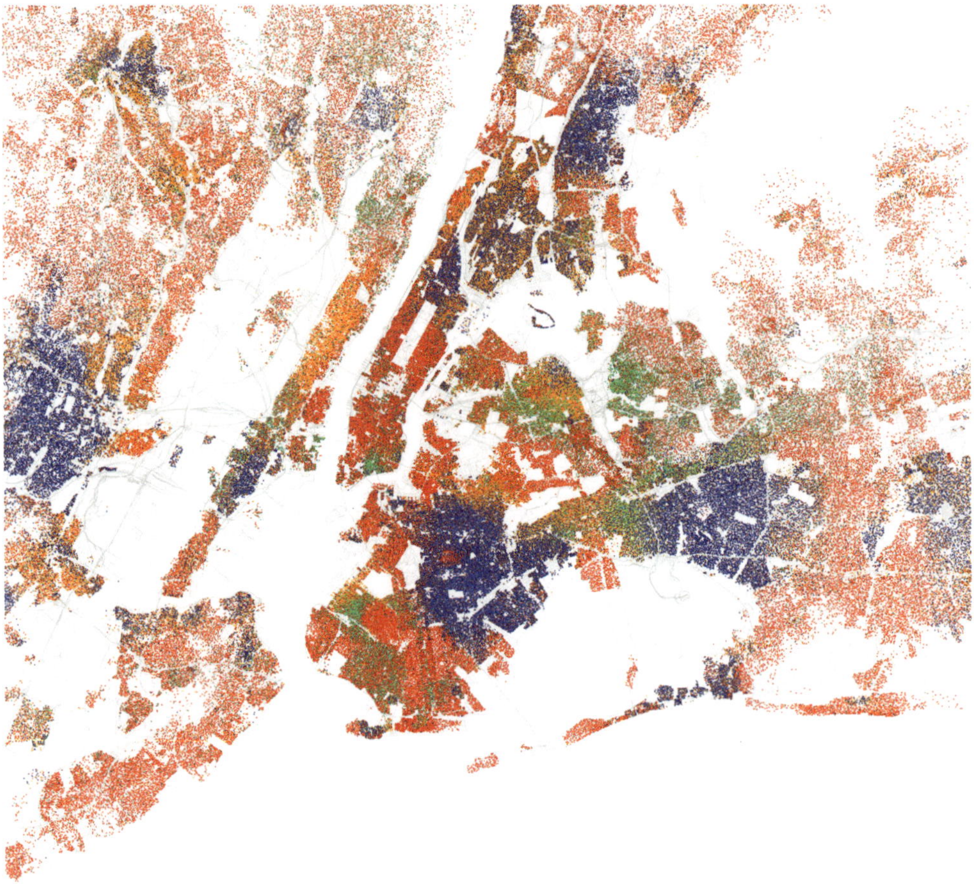

Figure 3.1 Map of racial and ethnic divisions in the New York City metropolitan area. Red is White, Blue is Black, Green is Asian, Orange is Hispanic, Yellow is Other, and each dot is 25 residents. Data is from the 2010 U.S. Census.

in New York City with data from the 2010 U.S. Census (the high population density of New York makes the segregation particularly visually stark). Racial and ethnic segregation can be found in most multicultural cities. What explains this severe segregation?

In the late 1960s, while the civil rights movement in the United States was going full force and racial segregation was a particularly hot-button issue, Thomas Schelling set out in search of an explanation of how neighborhoods come to be segregated. He considered several possible explanations:

1. *Organized action*, whereby only certain classes of individual are permitted to live in certain neighborhoods. This action may be "legal or illegal, coercive or merely exclusionary, subtle or flagrant, open or covert, kindly or malicious, moralistic or pragmatic" (Schelling, 1971, p. 144). In other words, there may be laws prohibiting certain groups from living in certain neighborhoods, or merely such systemic racism that members of certain groups are effectively kept out of particular areas.

2. *Socioeconomic filters*, by which race or ethnicity, once associated with different socioeconomic classes, retain those associations through structural differences in opportunities and incentives for advancement and mobility. In other words, the richer get richer and the poor stay poor. If race or ethnicity correlates with wealth, education, and mobility, then neighborhoods will be more homogeneous than expected under an assumption of fluid mobility. Nowadays we might consider these phenomena through the lens of social networks.
3. *Individual preferences*, by which individuals may choose to live near others who are similar. In other words, people may simply prefer to be around others who are like them.

Schelling believed that the first two explanations were probably of greater concern than the third, and I agree.[1] Still, Schelling argued, "in a matter as important as racial segregation. . ., even third place deserves attention" (1971, p. 145). Social scientists often talk about the importance of **homophily**, which is the tendency for similar individuals to preferentially assort and interact, and more recent research indicates that individual preferences probably *do* play a substantial role in where people end up living (Clark and Fossett, 2008). Even so, if it's overshadowed by other, more nefarious factors, why is it important to consider this third explanation? One reason is that if individual preferences have a measurable effect on population structure, then such effects should be accounted for. Only then can the true effects of organized actions and socioeconomic filters be accurately estimated and, perhaps, neutralized. Another more general reason is to study the relationship between individuals' preferences and the population-level patterns that emerge as a result.

To study the potential for individual preferences to create segregation, Schelling introduced a simple model that allows us to examine how homophilic inclinations on the part of individuals translates to segregated spatial distributions at the population level. Specifically, he asked: In the absence of any other factors, to what extent do individual preferences to live near similar neighbors influence the population-level pattern of segregation?

3.2 A Model of Segregation

The world is complicated. Part of the art of modeling is to propose a drastic simplification that captures the essence of the system under consideration in light of the central problem being investigated. The system under consideration here is an urban population in which each individual is defined by a home location and an observable group identity (e.g., race). Individuals have preferences, which can vary in strength, related to the proportion of their neighbors they can tolerate belonging to a different group. If their neighborhood falls below their preference threshold, they move.

A funny thing about building models is that a model almost always contains models within it. A model of segregation needs a model of geographical space and models of individuals' ethnicity and behavior. It's models all the way down. We are interested in how individuals belonging to multiple groups become spatially segregated. So we need our model to contain individuals who belong to one of some number of groups. How many groups? We can probably imagine a population with arbitrarily many groups, but in order to study group preferences, we'll need at least two. Individuals affiliated with either group must then

[1] It's also likely that these three explanations do not capture the full extent of reasons for ethnic segregation.

be located within some minimal model of geographical space and have some mechanism for changing their location. And our individuals must have preferences related to the group affiliations of their neighbors, which they will use to make decisions regarding whether to stay in their current location or to move. In other words, we must lay out our assumptions about the space in which agents live and about the agents themselves. Let's take these one at a time.

Space is the place

Urban segregation is intrinsically spatial. Neighbors are usually defined by proximity, and segregation occurs when neighborhoods have less diversity than the larger populations in which they are embedded. We will consider how individuals make decisions about where to live in a spatially structured world based on the ethnicity of their neighbors. So, minimally, our model must allow for some way to define neighborhoods, and a good way to do this is with an explicit spatial structure. There are many ways to model structure, and we will encounter many of them in this book. For now, we will use a structure we will be seeing quite a lot throughout this book: a discrete, square grid (Figure 3.2, left). This structure is also called a square lattice; lattices and other network structures are discussed in more detail in chapter 9.

The grid is discrete rather than continuous (as it was in the Particle World model) because dwellings in most urban areas are discrete. Homes tend to have fixed locations, and you cannot move your home arbitrarily close to another. The square grid assumes a fixed number of evenly spaced locations. This structure has three important properties. First, it is intuitive to visualize. If one looks at colored agents on a square grid, it is easy to identify agents and their neighbors, as well as neighborhood structures formed by collections of different types of agents. The value of good visualization should not be underestimated. Second, it is symmetrical. On a lattice of infinite size in which all cells are occupied, all agents will have an equal number of neighbors. In network terms, all nodes have the same degree. Of course, for most agent-based models, we cannot and should not assume a grid of infinite size. Instead, there will be boundaries along the edges of the square. This can lead to boundary effects, because agents along the boundaries will have fewer neighbors, and mobile agents will be restricted in their movement. On occasion the ability to account for such effects can be desirable, but often we wish to avoid such boundary effects, so we will use toroidal boundaries. In addition, agents must have the ability to move, which requires that some locations be empty if we assume that multiple agents cannot share a location (if you prefer, you can imagine agents as families rather than individuals). If the grid is not fully occupied, this eliminates perfect symmetry. However, because the space is itself symmetrical, we still eliminate any effects of a preferred axis of orientation, such as would be found if, for example, some areas of the space were intrinsically more desirable or had more or less capacity for residency. Finally, a discrete square grid is easy to code. No matter what language you use, a simple square matrix can be used to represent agent positions. In NetLogo, a square grid of patches is the default spatial organization.

Of course, the square grid ignores many features common to cities in the real world. It ignores variation in the density of housing, as well as natural features like parks and rivers. Indeed, we will ignore all variation in the desirability or feasibility of living in any particular location based on factors other than the ethnic makeup of the neighborhood. You might find yourself a bit uncomfortable with this oversimplification. This discomfort is healthy—hold on to it. It is important to keep in mind the simplifying assumptions one makes

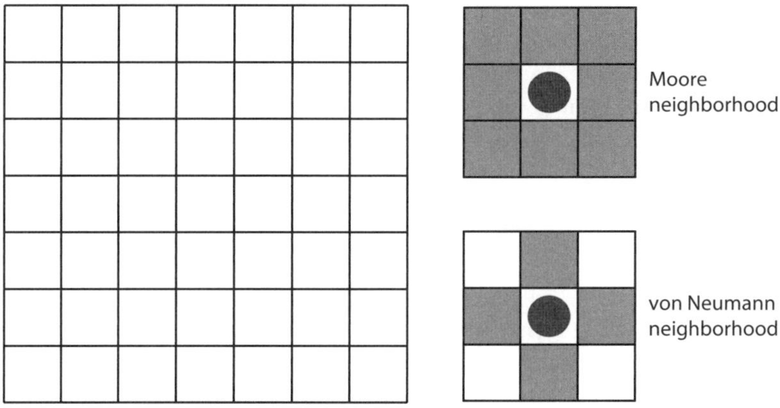

Figure 3.2 Left: a square lattice. Right: Moore and von Neumann neighborhoods of the focal agent (the grey circle).

when modeling, because the results of our simulations are direct consequences of those assumptions. Full understanding of a model often requires exploring alternate assumptions. Nevertheless, we must always do violence to reality to make any sense of our world, and we have to start somewhere. The grid captures the essential need of our decomposition: a structured environment in which our agents can reside. Meaningful variations can and should be explored when they matter, but establishing a baseline is important. Now, let's talk more about those agents.

Agent properties

Schelling's question was how individuals' preferences to live near others of the same ethnicity might influence patterns of assortment and segregation at the population level. Our agents are spatially embodied, each occupying a single cell of the grid. We also need to give our agents an observable ethnicity to be used in decision-making. We probably don't need to give them *real* ethnicities, with all the cultural and psychological baggage that ethnic identity entails. Our minimal approach requires the minimal aspects of ethnicity: an identification that connotes membership in one group and not others. We will therefore give each individual an arbitrary trait that others can observe, and we can call that trait anything we like—ethnicity, race, religion, type, hair color, sports fandom, political affiliation, whatever. It is worth pausing here to consider at least two potential objections. First, not all identities are necessarily observable by outsiders. And second, individuals often have a multiplicity of identities—racial, religious, professional, political, etc.—and these may interact in complex ways. Once again, our model here is a simple baseline, upon which additional nuance may be added later.

Since our model is spatial and will be easy to visualize, we can represent our trait by assigning agents different colors. While we use colors to represent variation among the agents, it is worth emphasizing that these are not directly related to race but instead could represent any discernible trait. Recall also that colors can be important for visualization and communication of the model dynamics but are not strictly necessary for model construction and simulation—we could just as easily code the same trait using numbers or letters. It's clear that we won't get any interesting dynamics if everyone is the same color; we need at least two. And since that's the smallest useful number, and because the simplest models are often the easiest to code, analyze, and extend, we'll stick with two colors for now.

So, each of our agents will be defined by a location and a trait (represented by a color). The agents will use this information—their own color and the color of others in their immediate vicinity—to update their locations. The idea we wish to model is that individuals prefer to live near others who are similar to themselves. There are many ways one could model this. For example, individuals could choose a location with a probability proportional to the fraction of neighbors who are similar to themselves. Probably the simplest possible rule is to use a strict threshold: stay if the fraction of similar neighbors is sufficiently high, and move otherwise. This is what Schelling did in his initial model, and it is what we will replicate. However, it is worth noting that other formal decision rules consistent with the idea that "individuals prefer to live near similar neighbors" are possible, and that using these may somewhat change the model outcomes (Bruch and Mare, 2006).

To make a decision based on how many of their neighbors are similar to them, each agent must have a definition of a neighborhood and be able to accurately assess the colors of the other agents in that neighborhood. There are, as always, several ways one could accomplish this algorithmically. We will begin with a fairly intuitive operationalization of a neighborhood on a discrete square grid: the nearest eight cells to the focal agent. A focal agent considers all the other agents on the cells defining the smallest square of cells that is centered on the focal agent's cell and includes other cells. On a square lattice, this is sometimes called a **Moore neighborhood**. A closely related definition is the neighborhood of the four closest cells, often called the **von Neumann neighborhood**. Both neighborhoods are named after the mathematicians who were among the first to use them in computer simulations, and both are often used in spatial models of social dynamics (see Box 3.1: Cellular Automata and the Game of Life).

Sometimes it may be desirable to extend the Moore or von Neumann neighborhoods to capture larger neighborhoods of more cells (Figure 3.3). The Moore neighborhood is easily extended by simply making squares of larger size, so that the area of a square with radius r is $(2r+1)^2$. Excluding the focal cell in the neighborhood's center means that the focal cell will have $4r(r+1)$ neighbors. Extending a von Neumann neighborhood is usually accomplished by creating a diamond shape, so that a cell with a von Neumann neighborhood of radius r will have $2r(r+1)$ neighbors. These neighborhoods can often come in handy when you want agents to assess or otherwise interact with spatial neighbors on a square lattice. These larger neighborhoods are also easily processed in toroidal spaces. In bounded finite spaces, the neighborhoods will become truncated, leaving agents near the edges of the space with fewer neighbors than those nearer the middle. For most of the analyses in this book using these neighborhoods, we will assume $r=1$.

Dynamics

Now that we've got a world and agents to reside in that world, let's talk about how the world changes. After all, nearly everything interesting is related to change. Each time step, every agent takes one of two actions. It can do nothing and remain where it is, or it can move. Our assumption is that agents prefer to live near others of the same ethnicity. An agent doesn't necessarily mind being a minority in its neighborhood, but it wants to have at least some similar neighbors.

Each agent can perceive itself and the other agents in its neighborhood, and so it can easily calculate the proportion of its neighbors who are the same color as itself. Agents will tolerate some threshold proportion of their neighbors to be drawn from the outgroup,

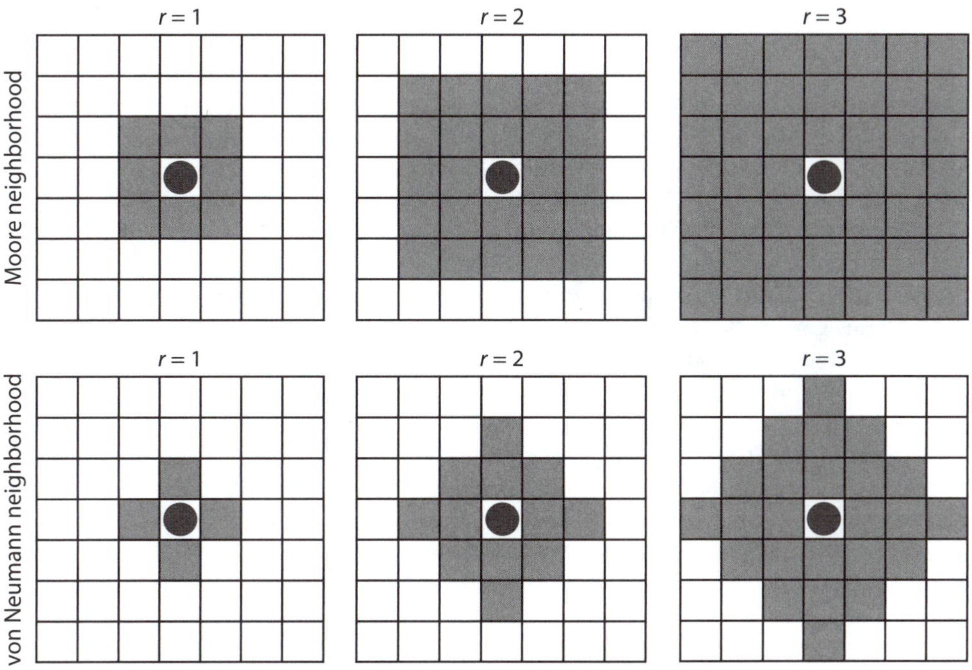

Figure 3.3 Moore (top) and von Neumann (bottom) neighborhoods of varying radius, *r*.

but no more. The agents will therefore use a very simple rule to decide whether to move or stay put: *If the proportion of my neighbors who are the same color as I am is below my tolerance threshold, move. Otherwise, stay put.* There is no strategic calculation in this decision. Rather, an agent that moves will search at random until it finds a new empty cell in which to reside. Each agent executes this rule, one at a time, until no agents move anymore, at which point the system will be at **equilibrium**—a stable state that will not change further if the system is not perturbed. This model will eventually settle into an equilibrium under most conditions, though as we will see there are some interesting exceptions.

3.2.1 Outcomes

What do we want to learn from this model? Once the system is at equilibrium, we would like to know just how segregated the population is. We therefore need a measure of segregation, and there are several possibilities. A simple metric is to have each agent count the proportion of its neighbors that are the same color as itself, and then average this proportion over all agents. We'll call this the population's *average similarity*. In a world with two colors equally represented in the population and no segregation (e.g., when agents' colors are randomly assigned at initialization), the expected average similarity will be 0.5. We will consider alternative measures in the Explorations section.

For clarity, I'll also include a more formal, technical description of the model in the following section, such as you might find in a scientific paper. Pay attention to how the model works, and also how the description is implemented in the accompanying code (**segregation.nlogo**).

BOX 3.1: Cellular Automata and the Game of Life

Some of the earliest computational modeling experiments took the form of **cellular automata**, which were introduced in the 1940s by the physicist Stanislaw Ulam and the mathematician John von Neumann. The basic framework is very simple: a grid of cells, each of which is in one of a finite number of states (such as on or off) at any given time. The states are updated by each cell considering its own state and the states of the cells in its neighborhood. For a square lattice, common neighborhoods are the von Neumann and Moore neighborhoods (Figure 3.2). The von Neumann neighborhood is slightly easier to code and is definitely easier to approximate mathematically, but the Moore neighborhood is often deemed more realistic, and thus is somewhat more commonly used in agent-based models of social behavior. The grid need not be a square. Many analyses have been conducted on the surprisingly complex behaviors that can emerge from some simple one-dimensional cellular automata, in which cells are arranged on a line. Each cell then updates its state based on the current states of itself and its two neighbors according to fixed transition rules. Some of these rules can yield complex or even chaotic behavior (Wolfram, 1984).

The Schelling segregation model is not technically a cellular automata model, because of the need for agents to move randomly. However, it bears many similarities to cellular automata (CA), and concepts related to CAs are worth being aware of. One of the best-known CA models is **Conway's Game of Life**, introduced by the mathematician John Conway in 1970 (Gardner, 1970). Not strictly a model of anything (in the same sense as our Particle World model), it helped launch an interest in what became known as **artificial life**—a field in which computer simulations and robotics are used to explore mechanistic ideas about how living systems work. The Game of Life takes place on a square grid with toroidal or effectively infinite boundaries. Each cell is either "live" or "empty." Cells update synchronously each time step by considering their own state and the state of the cells in their Moore neighborhood. The rules are as follows:

1. *Survival.* Each live cell with two or three live neighbors survives for the next generation.
2. *Death.* Each live cell with four or more live neighbors dies, as if from overpopulation. Each live cell with one or zero live neighbors dies, as if from isolation.
3. *Birth.* Each empty cell adjacent to exactly three neighbors—no more, no fewer—becomes alive, as if by reproduction.

The Game of Life is famous because of its ability to generate persistent and even self-replicating dynamic patterns, yielding tremendous complexity from very simple rules. For example, the well-known "glider" pattern will appear to move diagonally across the grid until it collides with other live cells (Figure 3.4). Turing complete computers have even been coded in the Game of Life. The science communicator Alan Zucconi has recently produced a wonderful short documentary on the game, available here: https://youtu.be/Kk2MH9O4pXY. NetLogo's Model Library also includes a version of the Game of Life. Playing around with this is recommended.

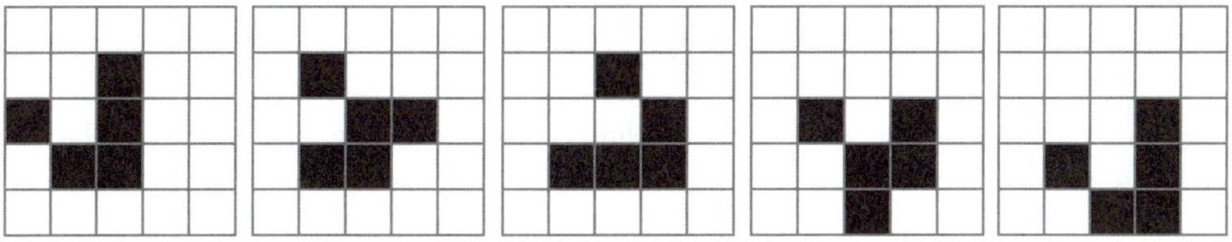

Figure 3.4 A glider in Conway's Game of Life. Black cells are live, white cells are empty. The pattern of the glider moves down and to the right every four time steps.

3.3 A Formal Description of the Model

As you read through this formal description, notice how the previous section makes this section easier to interpret than it would have been otherwise. You know how the model is *supposed* to work, so the explanation of how it actually *does* work is easier to process. The Schelling segregation model is quite simple, and so can be fully described in just a few paragraphs. More complicated models may sometimes require you to partition your model description, focusing on the important computations in the main text of your write-up while relegating the implementational details of your algorithms to appendices. This is often advisable for readability, and is (in my opinion) fine as long as the full description of the model can be readily accessed somewhere. Notice also that I often use variables rather than specific values to describe the model (for example, using L rather than specifying a specific grid size). This allows the model to be described generally and should also serve to remind you that you must make choices about which parameter values you will include in your analyses and which you will exclude.

3.3.1 Initialization

Consider an $L \times L$ square grid with toroidal boundaries. At initialization, one agent is placed upon each cell with a probability p, which characterizes the population density relative to the available space ($0 < p < 1$, so that each agent can relocate to an empty location). The expected population size is therefore $N = pL^2$. The population is divided into G groups, such that each agent i is randomly assigned a fixed group identity $g_i \in \{1, 2, \ldots, G\}$, each chosen with equal probability. For all of our analyses, we will use $L = 51$ and $G = 2$. The population is also defined by a similarity threshold, S, which defines the minimum proportion of an agent's neighbors that must be similar for it to refrain from moving.

Note that the use of probabilistic assignments means that, even holding p and G constant, there will be some variation between simulation runs in terms of exactly how large the population is and how many agents belong to each group. This stochasticity is often seen as a positive because it allows us to assess how robust the model is to minor fluctuations. However, one could also impose stricter requirements, so that, for example, the population size was always the nearest integer value of pL^2.

3.3.2 Dynamics

At any given time, each agent is either "happy," in which case its neighbors are sufficiently similar and it will not move, or "unhappy," in which case it will move because its neighborhood does not contain enough similar neighbors. At the beginning of each time step, each agent i first determines whether or not it is happy. The agent considers the other agents in its neighborhood (defined as a Moore neighborhood with radius $r = 1$) and determines the proportion of its neighbors that share its group identity, s_i. If $s_i < S$, the agent is unhappy, otherwise it is happy. After each agent has made this assessment, each unhappy agent, in random order, moves to a random empty cell (happy agents remain in their current locations). These dynamics repeat until no agents are unhappy. And they all lived happily ever after.

3.4 Thinking about Consequences

Before we get to coding, let's think carefully about the dynamics that can ensue from a model such as this. This sort of careful thought is always a valuable exercise. Part of a modeler's job is to explain why their assumptions lead to the observed outcomes, so being able to think through the model dynamics (with or without assistance from pen and paper or their digital equivalents) is an important skill that should improve with practice.

Let's consider a relatively small grid of $L = 7$, with a population of $N = 35$ (a possible outcome if, say, $p = 0.7$) and a similarity threshold of $S = 0.33$, so that an agent is unhappy only if its outgroup neighbors outnumber its ingroup neighbors by more than two to one. In other words, all of our agents are perfectly comfortable being in the minority. We assume that each group is represented in roughly equal numbers, and that agents are initially distributed randomly on the grid. As such, each agent should expect roughly half of its neighbors to be of either type, and therefore it should expect to be happy. However, random variation means that some agents will end up being surrounded by more outgroup neighbors than they are comfortable with, and will therefore be unhappy and need to move. This can then cascade to produce quite stark patterns at the population level.

I've worked out this example in Figure 3.5, with the two types of agent depicted as blue circles and yellow triangles.[2] The initial distribution of agents is shown in the top left. To make this example easier on our imaginations, let's ignore the toroidal nature of the space for now (so that, for example, patches on the far left will *not* count as neighbors any patches on the far right). On the right side of the grid there are two blue agents. The bottom one is perfectly happy with one blue neighbor and two yellow neighbors. However, the top blue agent has one blue neighbor and *three* yellow neighbors. Their unhappiness necessitates a move to a neighborhood with more blue agents. This move also causes the remaining blue agent on the far right to become unhappy, because it now has zero blue neighbors, and so it moves as well (Figure 3.5, top center). This dynamic repeats on the left side of the grid, as the two yellow triangles find themselves unhappy and move. In the lower left of Figure 3.5, I have identified all the remaining unhappy agents. I then moved each unhappy agent to a location where they were no longer unhappy, ending up with the configuration on the lower right.

Even though all of the agents are perfectly happy to be in an integrated neighborhood, and even to be in the minority as long as they are not outnumbered by more than two to one, we end up with a scenario that appears highly segregated! This can occur when the movement of one agent changes the happiness of other agents, triggering a **cascade** of relocations that leads to segregation.

[2]The color scheme is an homage to the excellent interactive tutorial on the Schelling model, "Parable of the Polygons," by Vi Hart and Nicky Case, available at https://ncase.me/polygons/.

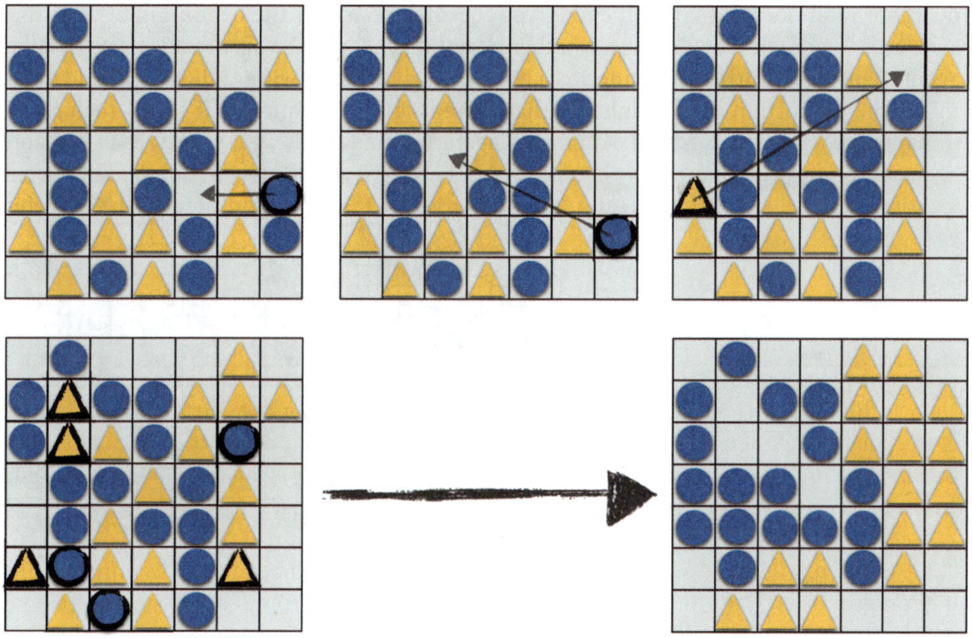

Figure 3.5 Example dynamics of the Schelling model, with $S = 0.33$. Top row: in each consecutive plot, an unhappy agent (indicated by a black outline) moves to a new location where it is happy. Bottom left: all remaining unhappy agents are outlined. Bottom right: an equilibrium is achieved once all unhappy agents have relocated or have otherwise ceased to be unhappy.

Demonstrating this phenomenon slowly by hand can aid our understanding of the system, but it is time consuming, especially if we want to play around with our assumptions (for instance, by changing the population density p or the similarity threshold S). It will be more useful to have a computational model with which we can explore these variations both quickly and systematically. Let's get coding.

3.5 Coding the Model

Now that we know how our model should work, we can code it. As with all the models featured in this book, I have provided NetLogo code for this model in the online repository. The code for the Schelling model is titled **segregation.nlogo**. My intention throughout this book is to focus on the process of modeling, and leave it to the reader to figure out how the model assumptions can be translated into working code. For those working in NetLogo, this can be worked out directly by looking over the provided code. I also wish to separate skills with any particular programming language from more general skills in building and analyzing models, which can be done with any language. As such, I often don't repeat all of the provided NetLogo code in the pages of this book. However, there are times when it will be useful to provide certain computational algorithms in greater detail. For these, I will use NetLogo code, which has the advantage of being explicitly designed to resemble pseudocode. I also recommend working through the provided NetLogo code as you read through this section. Because this is the first model for which we will perform detailed analyses, I will include more code here than in subsequent chapters.

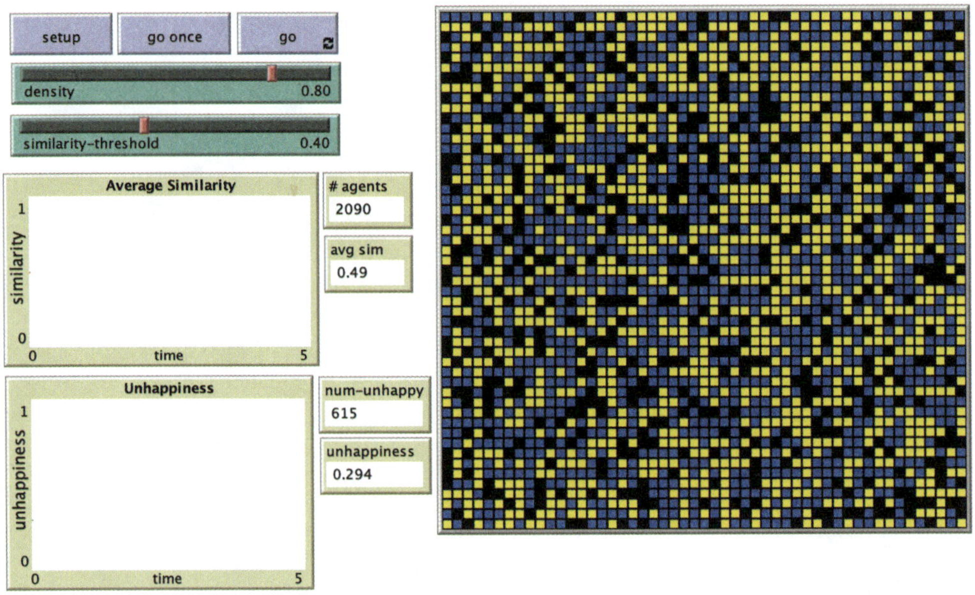

Figure 3.6 Interface tab for the NetLogo Schelling model.

For the Schelling model, we must first establish the square grid upon which our agents will reside. In NetLogo, this is readily accomplished by simply using the default grid of patches. In the Interface tab, you can click on Settings to change the grid dimensions to be 51×51 patches and ensure that the boundaries are toroidal. In order to initialize our model, we should first establish two global parameters: the population density, p, and the similarity threshold, S. We could make the model more complicated or extendable by adding more parameters, but these two will suffice for now. In NetLogo, I have titled these parameters `density` and `similarity-threshold`, respectively, and mapped them to sliders so that their values can be easily changed on the fly. The Interface tab of the completed NetLogo model is shown in Figure 3.6. All of its features will be explained by the end of this section.

3.5.1 Initialization

Our model is initialized by creating agents and locating them on our grid. Because we have assumed (for now) that all groups will be of approximately the same size and that all agents will use the same similarity threshold, our agents are in fact fully defined by their group identity and location. In most languages, these agent-level variables will have to be formally declared in the code. In NetLogo, we can use the fact that turtles automatically keep track of their locations and colors. Each agent will also need to assess its happiness in order to decide whether or not to move. To do this, we can have each agent keep track of a Boolean variable, `happy?`, that is true if the agent is happy and false otherwise. In object-oriented languages, this variable should be set as a property of the agent class. In NetLogo, we can declare an agent-level variable held by all agents with the following code. Note that, in the code below, I've also introduced another agent-level variable, `prop-similar-neighbors`, which the agents will use to keep track of the proportion of their neighbors that share their group identity. The utility of having agents keep track of this number will soon become clear.

NetLogo code 3.1

```
turtles-own [
  happy? ;; are a sufficient proportion of my neighbors like me?
  prop-similar-neighbors ;;proportion of neighbors like me
]
```

We're now ready to initialize the model. By convention, we title the initialization procedure in NetLogo `setup`. This procedure needs to accomplish two things. First, we need to create the agents, place them on the grid, and assign them a group identity (represented by a color). Second, we need to have each agent assess whether or not it is happy. To do this, I have modularized the `setup` procedure so that it calls two other procedures to do each of these steps: `make-turtles` and `update-turtles`. A third procedure, `update-globals`, will update the model's outcome measures, and is described further below.

NetLogo code 3.2

```
to setup
  clear-all
  make-turtles
  update-turtles
  update-globals
  reset-ticks
end
```

The first procedure, `make-turtles`, is very simple. We consider each patch on the grid one at a time, and create an agent there with a probability equal to our parameter `density`. It is straightforward to write code that performs an action with a user-defined probability x. We do this by generating a random number from a uniform distribution between 0 and 1, and then perform the action if the random number is less than x. To see why this works, imagine a number line between 0 and 1 and let x be any number in that range. The proportion of numbers on the number line that are less than x is exactly equal to x! Thus, if we want to perform an action with a probability equal to x, we can perform it whenever a random number in $[0, 1]$ is less than x.

Whenever we create an agent, we assign it to one of two groups (or types) with equal probability. I have chosen to represent these groups by the colors blue and yellow, and to make all agents appear as squares on the grid. The NetLogo code for the creation of the agents is:

NetLogo code 3.3

```
to make-turtles
  ask patches [
    if random-float 1 < density [
      sprout 1 [;;create an agent here
        set shape "square"
        set color one-of [yellow blue]
      ]
    ]
  ]
end
```

We then have each agent assess whether it is happy in its neighborhood, which it does by assessing the group identities of its neighbors in comparison to its own group identity. If the proportion of its neighbors that share its own group identity is less than the similarity threshold (`similarity-threshold`), the agent is unhappy, and its value for `happy?` will be set to false. Otherwise, it will set its value for `happy?` to true. The procedure that accomplishes this will be called again by the procedures governing the model's dynamics, as agents will need to continually reassess their neighbors. An agent's neighbors are the set of all the other agents in its Moore neighborhood, accounting for toroidal boundaries (for example, an agent at the bottom row of the grid may have neighbors along the top row). Similar neighbors are those neighbors with the same group identity as the focal agent. If the proportion of neighbors who are similar is greater than or equal to the similarity threshold, the agent will be happy, otherwise it will be unhappy. Most programming languages can accomplish this by setting the value of a Boolean variable directly equal to the inequality, so that the former will take on the truth value of the latter.

Each agent will define local variables for the number of total neighbors and the number of similar neighbors, and will take the ratio of these to get the proportion of similar neighbors. What should we do if an agent doesn't have any neighbors? This is particularly likely in low-density populations. We want to avoid division by zero, which is mathematically undefined and will generate a code error. The way I see it, we have two options. An agent with no neighbors may be content to live apart (and therefore be happy) or may be lonely and seek to move near others. I have opted for the former in how I present the code, but as we will see, this decision does have consequences. In NetLogo, happiness is updated as follows:

NetLogo code 3.4

```
to update-turtles
  ask turtles [
    let similar-nearby count (turtles-on neighbors) with
     [color = [color] of myself]
    let total-nearby count (turtles-on neighbors)
    ifelse (total-nearby = 0)
    [set prop-similar-neighbors 1] ;;always happy if alone.
    [set prop-similar-neighbors (similar-nearby / total-nearby)]
    set happy? (prop-similar-neighbors >= similarity-threshold)
  ]
end
```

3.5.2 Dynamics

Once the model has been initialized—with our agents having been located in space, assigned a group identity, and having their happiness computed—we need to code the dynamics. By convention, we title the procedure that controls all of the model dynamics `go`. The `go` procedure will repeat in a loop until some stopping condition is met. That stopping condition should be when all of the agents are happy and therefore feel no need to relocate. If at least one agent is unhappy, three things will occur. First, all unhappy agents will relocate to a new, randomly chosen location where hopefully they will find more happiness. I have introduced a procedure called `move-unhappy-turtles` to accomplish this. Second, all agents will reassess their happiness, as their neighborhood composition may have changed due to their own movement and/or the movement of other agents. This will be accomplished simply by calling the procedure `update-turtles`, introduced above. And third, we will update any global parameters we are using to measure the state of the model. I will tackle this last procedure, `update-globals`, in the next subsection. The `go` procedure therefore looks as follows:

NetLogo code 3.5

```
to go
  if all? turtles [happy?] [stop]
  move-unhappy-turtles
  update-turtles
  update-globals
  tick
end
```

To make the dynamics work, we have to figure out exactly how to move an unhappy agent to a new location. As with many aspects of agent-based models, there are myriad ways this can be accomplished. An agent could, for example, try to find the closest available location where it would be happy.[3] Or it could ignore its likely happiness at the new location and perform a random walk until it found an empty cell. I recommend experimenting with several algorithms to see whether different choices lead to different outcomes (different choices may also require drastically different computational resources, which may influence your choices as a modeler). It is also worth thinking about whether your modeling choices reflect clear hypotheses or theories about behavior or cognition, or whether they are chosen arbitrarily due to a lack of clear hypotheses. In this case, I have opted to code the model using a very simple movement algorithm: each unhappy agent will simply choose a cell at random from the set of unoccupied cells and move there. This is easily done in NetLogo, but may be more difficult in another language without the use of an explicit modeling library.

NetLogo code 3.6

```
to move-unhappy-turtles
  ask turtles with [not happy?]
    [move-to one-of patches with [not any? turtles-here]
  ]
end
```

These dynamics will usually continue until all agents are happy and the system is at equilibrium. Is an equilibrium always reached? This is a question worth exploring with your newly coded model.

3.5.3 Outcomes

The code described above will allow you to construct a functioning model that can be initialized and run. If you are able to visualize the population (which is very easy in NetLogo), you will be able to characterize its behavior qualitatively by describing the patterns you see. But we want more than just qualitative descriptions. We're doing science here, and science

[3]This is actually the movement rule proposed by Schelling in his original 1971 paper. I encourage the reader to replicate it and note differences from the simpler movement rule used in the provided code.

needs numbers! As Adam Savage, host of *MythBusters*, once remarked, "The only difference between screwing around and science is writing it down."[4]

It is a nontrivial question to ask what sort of outcome measures one should collect from one's model. As with the model design, the choice of outcome measures should stem from the questions one wishes to ask about the system. First and foremost, the Schelling model is about segregation, and so we need a quantitative measure of segregation. We will quantify segregation by measuring the proportion of each agent's neighbors that are similar to the agent, and then averaging across all agents.[5] If we have two equally numerous types that are randomly distributed on the grid, average similarity should be around 0.5. Higher values therefore reflect stronger segregation, so that when average segregation is close to one, most agents live in homogeneous neighborhoods of the same type.

We may also wish to consider a more dynamic measure of the stability or instability of the population. The proportion of agents who are unhappy corresponds to the number of agents who will relocate on a given time step. Call this the **unhappiness** of the population. Unhappiness is a measure of the instability of the population's spatial arrangement, and can be viewed as similar to the potential energy of a physical system such as a spin glass. The population's unhappiness should gradually decrease to zero for conditions that lead to equilibrium.

There are multiple ways to keep track of outcome measures, and best practices in terms of computational efficiency depend in part on whether you want to be able to record their values at each time step or only when your simulations have finished running—this will depend on whether you want to record data about the model's dynamics or just its final state. For scalar, population-level values, it is easy to simply declare a global variable when the model is initialized. In the provided code, I have declared global variables called, fittingly, `average-similarity` and `unhappiness`. Unhappiness is calculated by taking the fraction of agents who are unhappy. At the end of both the `setup` and `go` procedures, I've added a call to a procedure that updates these variables, called `update-globals`, which updates the global parameters acting as outcome measures.

NetLogo code 3.7

```
to update-globals
  let similar-neighbors sum [prop-similar-neighbors] of
   turtles
  set average-similarity (similar-neighbors / count turtles)
  set unhappiness (count turtles with [not happy?]) /
   (count turtles)
end
```

At this point, you should have a fully functional version of the Schelling segregation model. Before we turn to rigorous quantitative analysis, it's time for a bit of play.

3.6 The Power of Play

A simulation model is a little artificial world that you create, which you hope will give you insights into the big real world by virtue of the former's similarity to the latter. In order

[4]In reality, science often *also* involves a good deal of screwing around.

[5]This measure of segregation reflects a fairly local scale of analysis. Other measures of segregation and diversity are discussed in Roberto (2016).

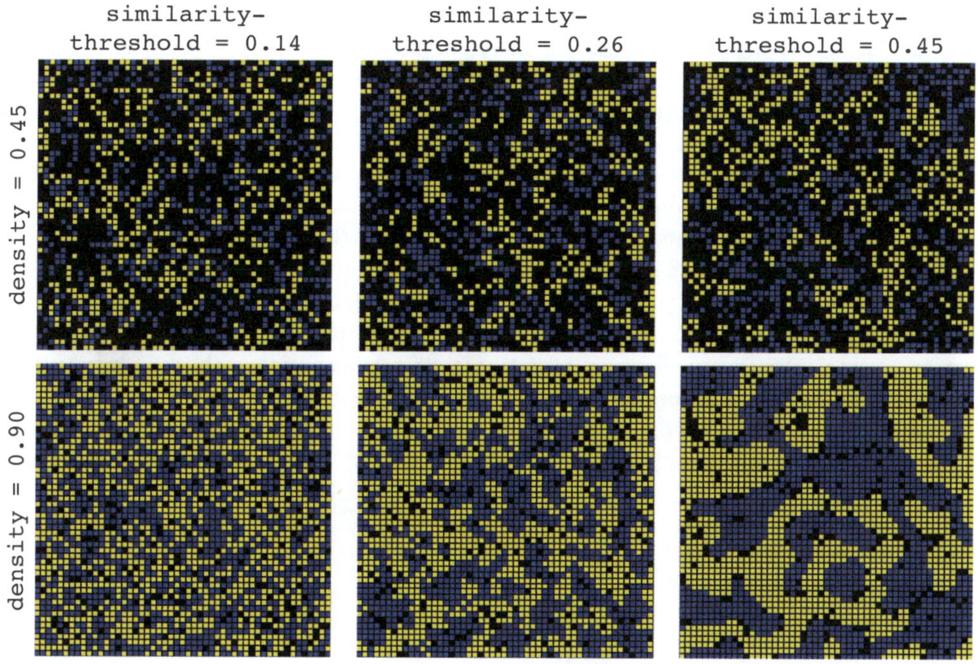

Figure 3.7 Screen captures of the segregation model at equilibrium for different values of `similarity-threshold` and `density`.

for that to happen, you have to understand the assumptions that drive your model and the consequences of those assumptions. Just as it's worthwhile to consider how your model behaves without running any computer simulations—by working out some of its dynamics by hand—it's also worth spending some time playing with your computer simulation once you've got the model coded. This is especially valuable if you have a nice visualization (as you do with the Schelling model) that allows you to see some or all of the model's information with your eyes. However, even if all you have is dynamic plots of your outcome measures, play can still be valuable. I cannot overemphasize how much you can learn by doing this. It is also, for the right kind of person, fun. Mess around with different parameter values or assumptions and see what happens. Play can give you a sense of the range of possible outcomes, and sometimes can even help you understand why certain assumptions, parameter values, or starting conditions lead to particular outcomes. Play is especially important if your main outcome measures reflect the state of your model only at the end of simulation runs. Understanding how those states arise is just as important as understanding the fact that they do arise.

For example, take a close look at the spatial patterns that emerge from different values of `density` and `similarity-threshold`, which will look something like those in Figure 3.7. Why do we see these patterns? What happens on each run? In many cases, the blue and yellow agents organized into distinct, coherent clusters. Notice how these clusters vary in shape and size as you change the model parameters using the sliders. Consider how high (or low) the similarity threshold needs to be to generate what looks like substantial segregation. The fundamental insight to draw from the Schelling model is that fairly weak preferences for similarity can generate what appears to be very strong segregation. Even when everyone is content to be in the minority, most agents end up in highly segregated neighborhoods in which they are in the majority. Such qualitative descriptions of the model behavior are often very important to gain an intuitive understanding of the dynamics. Don't underestimate their value.

To describe the model behavior more quantitatively, we have also invested in developing quantitative outcome measures for our segregation model. Exactly how weak are the preferences? Exactly how strong is the segregation? We've already seen, in the previous chapter, how to report and plot outcome measures for individual runs of the model. NetLogo makes it easy to do this with the plots and monitors you can add in the Interface tab. For example, in the provided code, I have included dynamic plots of the average similarity and unhappiness as well as monitors that report the exact values for these outcome measures. Continue to play with the model, focusing now on these quantitative data, and consider how these quantities vary in response to the model parameters. How well is the variation in the emergent spatial patterns captured by our measure of segregation, average similarity?

3.7 Model Analysis

Playing with your model and getting a qualitative sense for how it works is super important. Nevertheless, we also have to consider how to quantify the model's behavior more precisely.

3.7.1 Batch, Please

Models based on equations are often deterministic, which means that we can specify precisely how the model parameters relate to the outcomes. This is a benefit, but it is also a limitation, because mathematical tractability comes at the cost of strong assumptions that may force us to ignore important variation. Simulation models, on the other hand, capture more of the individual heterogeneity and organizational structure present in real-world systems, but at a price that includes stochasticity and uncertainty. There are many sources of stochasticity, introduced both during both initialization (such as the spatial arrangement of agent types) and during the ongoing the model dynamics (such as the order in which particular agents make decisions, or the inclusion of probabilistic decision rules). This means that individual runs can vary even if they are initialized with the exact same global parameters. This variation includes both the specific dynamic paths the system takes to reach an outcome as well as the outcomes themselves.

When we report on a model's behavior we need to account for this variation, but we also need to cut through the variation to highlight the most likely outcomes. To do this, we must run many simulations with the same parameter values in order to obtain a representative distribution of outcomes. A collection of replicated simulations is often called a **batch** or a **batch run**. Batch runs allow us to generate the representative distributions of outcomes possible under each set of parameters being investigated. We can then look at the descriptive statistics of the model outcomes across simulations to characterize the model in a more quantitative way. How many runs is enough for a batch? The short answer is: it depends. It is important to obtain limiting distributions that accurately characterize the variation in the model. So, the more variable the model outcome—the more sensitive to stochastic noise it is—the more runs you will need per batch.

A related question is: How long should you run your simulations for? For some models, there will be a natural stopping condition, such as an equilibrium, that is always reached in a reasonable amount of computational time. In other cases, the system may never settle down to a static equilibrium. In these cases, you must test the model (with the power of play) to determine some other stopping condition. This includes (but is not limited to) a fixed number of time steps, at which your model outcomes can be meaningfully described because the model seems to always have reached a relatively stable pattern. In the case of the Schelling model, the model can run until an equilibrium is reached or until enough time has elapsed to determine that no equilibrium will be reached.

3.7.2 Parameter Sweeps

The dynamics of a model are often sensitive to the parameter values being used. This means that different values lead to predictably different outcomes. This makes sense, since particular parameter values are, in fact, assumptions, and modeling is all about understanding the consequences of particular assumptions. In order to characterize the behavior of a model, we will need to systematically vary the parameter values used and record how the output subsequently responds to that variation. Obviously, we should vary parameters that are of direct interest to our theory-driven questions, using values that are meaningful or realistic for our systems of interest. For example, in the Schelling model we are interested in how strong or weak preferences for similar neighbors influence segregation. It is therefore fitting that an analysis should systematically vary the similarity threshold. We might also vary the density of agents in the space to study the effects of population density on segregation.

In general, one should vary one's parameters within reasonable ranges to test their effects on the model's behavior. Less obvious, but just as important, is to consider some of your model's more arbitrary and often implicit assumptions. For example, the Schelling model as described earlier makes many assumptions that are not always justified. Here are some:[6]

- There are only two groups in the population.
- Each group is roughly the same size.
- The quality of a neighborhood is completely determined by the trait makeup of an individual's neighbors on the nearest eight patches.
- There are no costs to moving.
- Decisions to move are all-or-none based on a threshold.
- All agents have identical thresholds for similarity.

I'm sure that, with a little effort, you can come up with other assumptions. Now, here's where it gets tricky. You will rarely be able to consider the consequences of every alternative assumption. There are too many possibilities. Instead, you should be led by your curiosity and by your skeptical scientific mind to consider the assumptions that make sense to study right now. Into this consideration you should also put the current state of knowledge of a system.

With which parameter values should you run your model? Models with more than a couple of parameters present a problem. If you have one parameter and need to test the model under five values, then you need five batches of runs. If you have three such parameters, you now need $5^3 = 125$ batches to test all possible combinations of value, which is important if the parameters interact non-additively. If you have eight parameters, you need $5^8 = 390,625$ batches of runs. If you need 100 runs per batch to establish the limiting distributions, then you need to run your model 39,062,500 times. This is the **curse of dimensionality**: the number of simulations needed increases exponentially with the number of parameters. It is less of a problem than it used to be, as our computers have grown faster and cheaper. But it is still

[6]See Bruch and Mare (2006) for a detailed exploration of some alternative assumptions for the Schelling model.

a problem, in terms of both the computational time required to run all those simulations as well as the difficulty in analyzing such a complex system.

There is no perfect solution to the curse of dimensionality.[7] However, this is another instance where play can help. Playing with your model can help you realize which parameters are probably important to explore and at which range of values, as well as which parameters you can more safely ignore, or run for only a couple of values to ensure robustness. Like anything else, an important step to getting good at this is practice.

3.7.3 Null Models

It is often useful to consider parameters that completely eliminate the mechanisms hypothesized to generate the main outcomes in your model. For example, what if agents move to new locations at random, without any regard at all for their neighbors' colors? Sometimes this just requires running the model under certain parameter values, while other times it can require writing some additional code. In our example, if agents don't use color information to determine their happiness, what initiates or halts the decision to move? Consideration of null conditions may also require the consideration of additional mechanisms like this.

Null models (also called neutral models) can be quite powerful. It is common in the experimental literature to come across "null hypotheses," which usually assume that two or more variables are unrelated, such that an alternative hypothesis is the existence of some specific relationship. However, approaches like these often ignore the importance of mechanism in generating particular relationships. The existence of a relationship does not necessarily imply a particular causal explanation. The Schelling model demonstrates this for segregation, but other examples from the modeling literature abound. For example, the biopsychologists May et al. (2006) showed that the body shape of infant rat pups could, in combination with how the pups tend to explore their space, generate observed huddling behavior despite the absence of any cognitive mechanisms for social interaction. In archaeology, Brantingham (2003) showed that the distribution of materials found in stone structures can often be explained by assuming random harvesting from normally occurring variation in natural materials, rather than by strategic behavior on the part of the people that built those structures. And in economics, Drăgulescu and Yakovenko (2000) showed that completely random transfers of wealth can generate highly skewed wealth distributions in the absence of other social forces.

3.7.4 Where Are the Stats?

As we begin to explore and analyze the results of batch runs in this book, you may notice an absence of the sort of inferential statistics common in the empirical sciences. Inferential statistical techniques, such as regressions or analyses of variance, are usually used to determine whether or not two samples differ, or more generally whether some variable (or combination of variables) predicts a directional change in another variable. In simulation models, we are also interested in whether and how our parameters influence our outcome variables. However, I will generally analyze them only qualitatively. We will revisit this decision, and the reasons for it, in more detail in chapter 10.

[7] Several *imperfect* but rigorous methods have been proposed and studied. The methods presented in this book suffice for the simple models presented here, but for more complex models the reader is directed to a recent review by Ligmann-Zielinska et al. (2020).

3.8 Analyzing the Segregation Model

Now that we've gone over some general lessons for model analysis, let's return to the Schelling model. We have observed that the population often self-organizes into segregated communities, and we have developed a measure to assess the extent of that segregation. We can now return to Schelling's original question, posed at the beginning of this chapter: To what extent do individual preferences for living among similar neighbors yield segregated communities? We can analyze our model to see how much segregation emerges under varying assumptions about the thresholds that guide individual agents' decisions about whether to relocate. From playing with the model, you may have observed that the density of the agents in the grid seems to matter, and so we can also ask how population density affects the resulting segregation.

From playful exploration you should observe that the model tends to either reach equilibrium fairly quickly (in less than 100 time steps) or not at all. It is in fact possible under some conditions for the model to reach equilibrium only after many hundreds of time steps, particularly for very high density populations and high similarity thresholds. You should verify that this is the case. Nevertheless, these cases are rare and even then, the state of the model at $t = 100$ (in terms of total segregation) is very close to the final state at equilibrium. In the interest of reducing computing time, I have opted to stop each run after 100 time steps if equilibrium has not yet been reached. A more rigorous, scientific examination of the model should probably run simulations for longer, as setting too low a time limit can lead to strange artifacts. Indeed, one could run a complementary analysis and examine how the time to reach equilibrium is influenced by the similarity threshold and the population density. This is left as an exercise.

Finally, we need to decide how many runs to repeat for each combination of parameters. When you are first exploring a model, doing only a few runs for each combination can give you a fairly good sense of the model behavior while saving valuable computational time. However, for a more accurate depiction of the model behavior, in terms of both central tendencies and the characteristic variation across runs, you'll often want more runs for each combination. We'll do 100 for each combination, which is usually plenty.[8] We will in fact examine how the number of runs per parameter combination affects our outcome estimates.

So, we will sweep over a number of values of `density` and `similarity-threshold`, running 100 simulations per parameter combination and running each simulation for 100 time steps. NetLogo has a nice tool for batch runs called BehaviorSpace, which can be found in the Tools menu (Figure 3.8). For `similarity-threshold`, we'll test values between 0.05 and 0.70 in increments of 0.05. We'll also explore values of `density` between 0.1 and 0.9, in increments of 0.1. Why are these the right values to use? They may not be, but they cover a very wide range, and in my experience they capture most of the important variation for this version of the model. As usual, this determination comes from the power of play mixed with curiosity and critical consideration. When possible, run a wider range of parameters than you think matters. The outcomes may surprise you.

---

[8]Though not always. For example, if you are interested in an outcome that only occurs on 5% of runs, you may want to run, say, 5,000 runs for each combination to get enough data on the rare cases. More generally, the more variation your model exhibits between runs, the more runs you'll need to accurately capture the distribution.

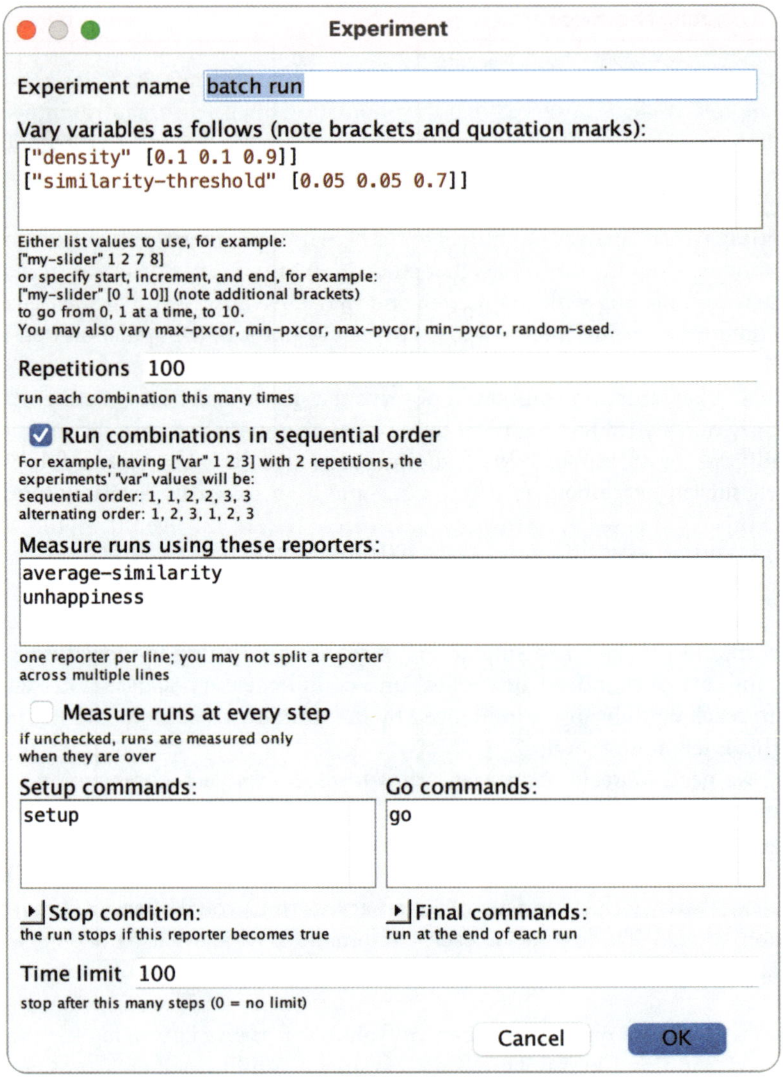

Figure 3.8 BehaviorSpace for a batch run of the Schelling model. This will yield 12,600 independent simulation runs.

In some cases, collecting data at each time step or some number of time steps may be important for understanding the dynamics of the model. For example, we might want to see how segregation changes at each time step, in order to determine the most critical time periods in the process. Here, I'll focus on the maximum segregation the model generates for each case, and so we'll look at the state of the model only on the final time step (either when the model has reached equilibrium, or after 100 time steps).

Once you begin your batch runs, it may take some time to run all the simulations you need. Our batch here uses nine values for `density` and fourteen values for `similarity-threshold`, with each combination repeated 100 times. This yields $9 \times 14 \times 100 = 12{,}600$ individual runs of the model. This model runs pretty fast—on my

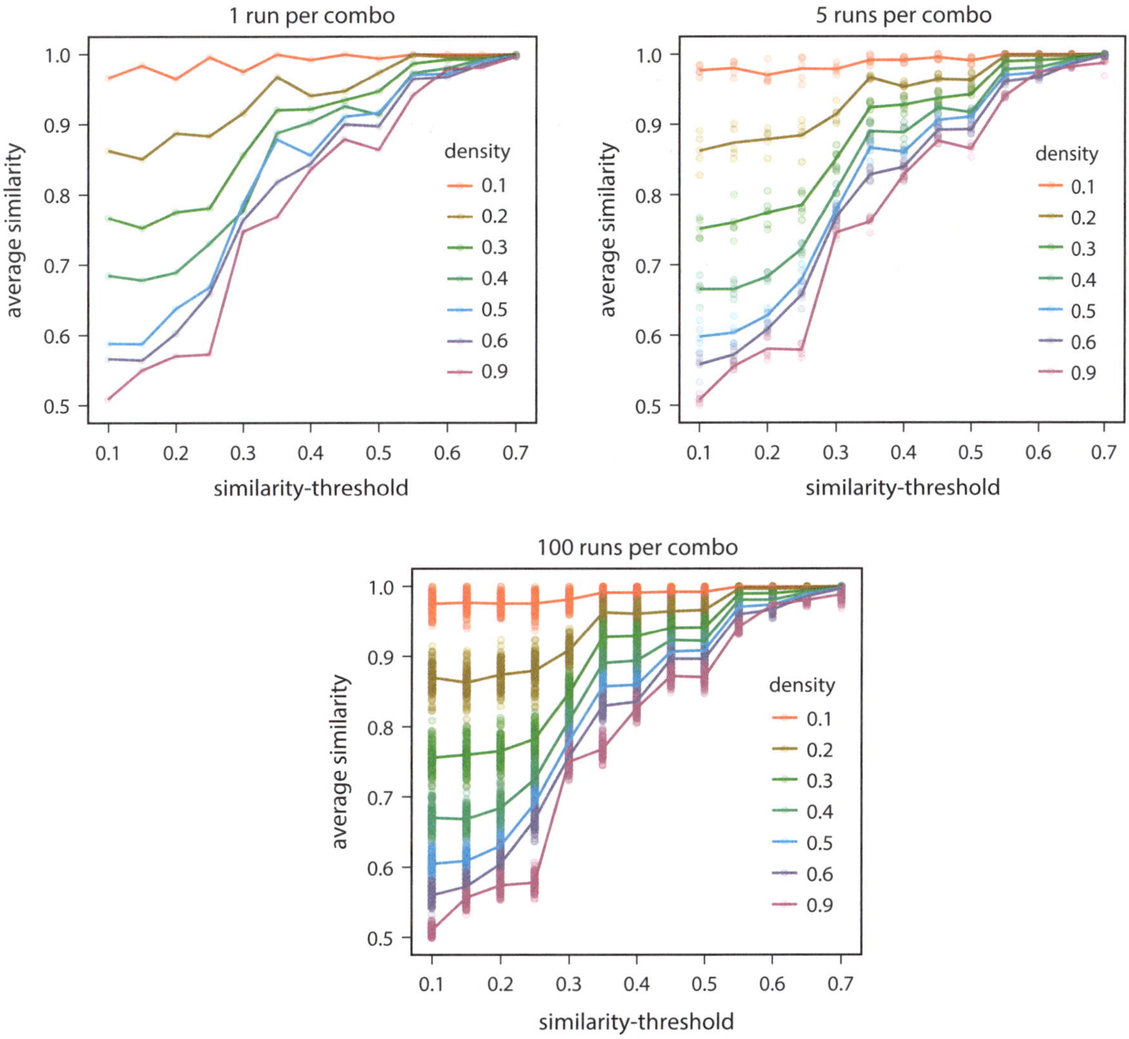

Figure 3.9 Average similarity at the end of a simulation for different values of `density` and `similarity-threshold`. Open circles are values from individual runs, solid lines are means across runs. As we simulate more runs of the model for each combination of parameters, we get a more accurate estimate of the variation and central tendency of the model behavior.

laptop the whole batch took only a few minutes to run. For some more complex models, you may need to run your computer overnight or even enlist the aid of a high-performance computing cluster (if you have access to one) to perform necessary batch runs. For most of this book, we will work with batches that can be run on a personal computer in a reasonable amount of time.

Figure 3.9 shows the output of our simulations, with a focus on segregation as measured by the average percent of neighbors of the same color each agent observes. You can see that the patterns are quite visible even when a single run is examined for each parameter combination, although the output is still noisy. The trends in the results become smoother and more accurate when more runs are considered. It is readily apparent that, as `similarity-threshold` increases, we get more segregation. Importantly, the average percent of similar neighbors is always considerably higher than the minimum acceptable percent of similar neighbors that agents will tolerate. For example, when the threshold is 0.3, so that an agent only moves if fewer than 30% of its neighbors are the same color as itself, the model yields an average similarity of over 70%. Here is the key point this model demonstrates: if agents move when their preferences are not met, even weak individual preferences (such that all agents will tolerate being in the minority) can generate strong patterns of segregation at the population level.

The other thing to observe is that the effect of population density is quite large, especially for small decision thresholds. At lower densities, the population becomes much more segregated than it does at high densities. This is partly because at lower densities, individuals are likely to have only one or two neighbors, making the threshold harder to reach. Imagine you are in an isolated cluster of three agents. If you have only two neighbors, the only way to have at least 30% of them be the same color as yourself is for at least one of them to be that color. But if the remaining agent is a different color, then 100% of *their* neighbors are a different color, and so they will move. These **tipping point** events are much more common under low densities. Look at the case where the density is 0.2 (the brown line) and agents are satisfied as long as at least 20% of their neighbors are similar to them. Even though every agent is perfectly satisfied being in the minority, the average agent lives in a community where over 80% of their neighbors are like them!

These results also reflect a particular instantiation of the model. Let us return for a moment to our rather arbitrary decision that lone agents with zero neighbors should be considered happy. Call these agents "contented loners." Because they have no neighbors, we asserted that the proportion of their neighbors who are similar was equal to one. We could also imagine a scenario in which agents do not like to be alone, and so they count as zero the proportion of their neighbors who are similar when they have no neighbors. Call these agents "lonely loners." What do we get if we re-run our simulations where the loners are lonely? The results are depicted in Figure 3.10.

A few things should jump out from the consideration of Figure 3.10, especially in comparison with Figure 3.9. First, the broad, qualitative patterns of how segregation increases with the similarity threshold and decreases with the population density are preserved, which is a good sign for the model's general robustness. Second, the need for multiple runs per parameter combination becomes more apparent here—the patterns are quite noisy when only one run is used. Third, we see a slight decrease in segregation, particularly for low densities and low similarity thresholds. Consider why this might be. Being a loner should be fairly common under very low densities. If loners are always unhappy, they are more likely to end up in mixed neighborhoods, thereby decreasing overall segregation. We have also eliminated the arbitrary total segregation of loner agents. Notice that something funny seems to be happening for the lowest density and highest similarity threshold condition, as seen in the upper right of the rightmost plot in Figure 3.10. Average segregation dips down as a result of greater variation in outcomes. Playing around with individual runs shows that this is due to longer times needed to reach equilibrium for the lonely loners, often greater than the 100 time steps we allotted. The result is that the simulation ends before the population has reached its peak level of segregation. The spatial patterns of segregation that emerge at low densities also differ between the two conditions, as depicted in Figure 3.11. This figure illustrates both the importance of examining the effects of seemingly minor assumptions, as well as the ways in which summary statistics can be limited in describing the true patterns present in a model system (or, indeed, in a real-world system).

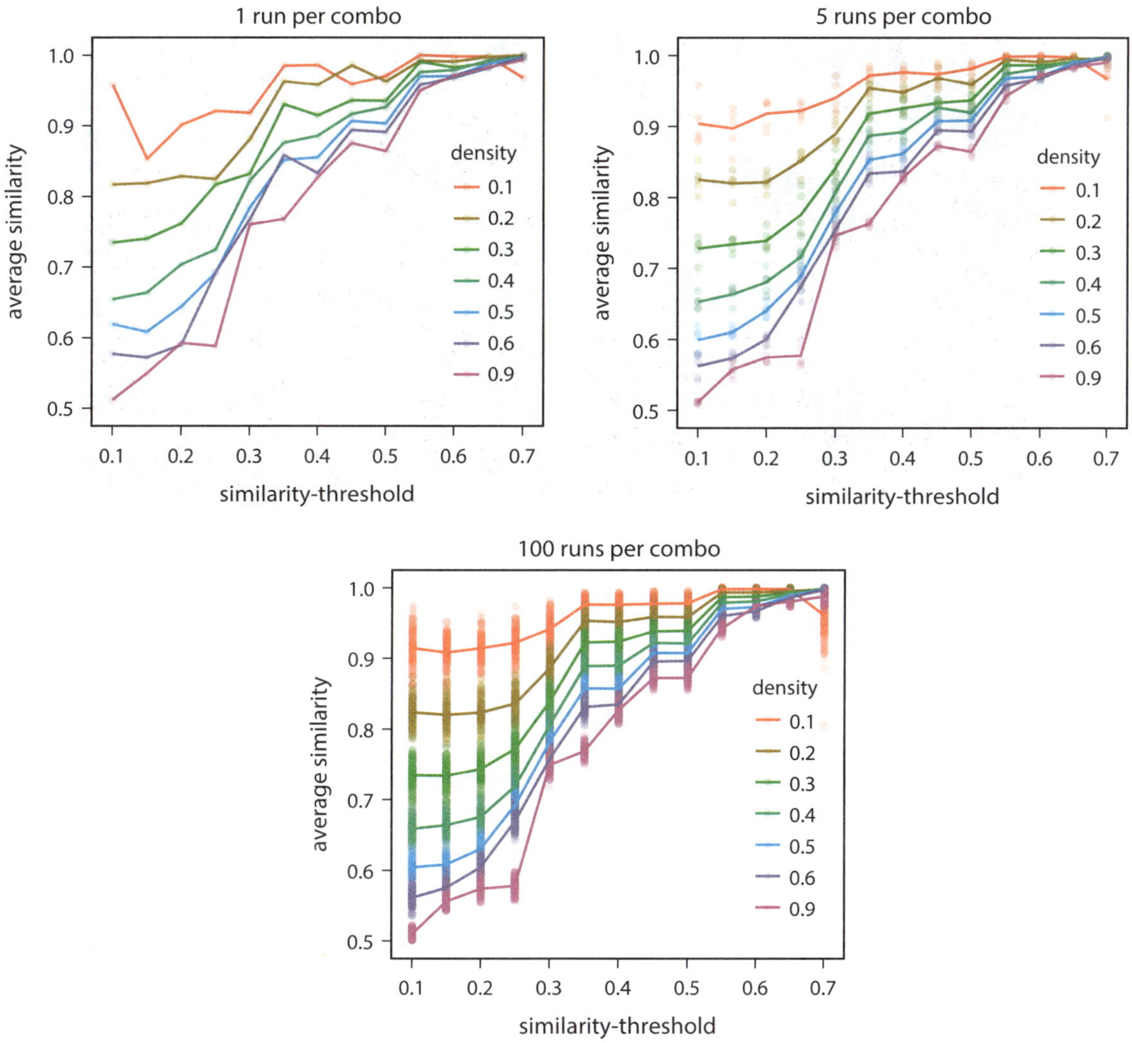

Figure 3.10 Batch runs with lonely loners. Average similarity at the end of a simulation for different values of `density` and `similarity-threshold`. Open circles are values from individual runs, solid lines are means across runs. Compare with the runs using contented loners shown in Figure 3.9.

3.9 Reflections

You have now built and analyzed the classic Schelling segregation model. If you're like me, you may feel a bit uneasy or unsatisfied with this accomplishment. The model is highly abstract, and while it yields some valuable insights into social processes like segregation, it also leaves out a lot of important stuff about social behavior and the infrastructure that supports and constrains it. To some extent, this unease is part of the growing pains involved in modeling. You may eventually become more comfortable with abstraction and oversimplification, but it can take a while. And while it is valuable to embrace the abstraction, we should also be wary of putting too much stock in the model as a complete explanation. There

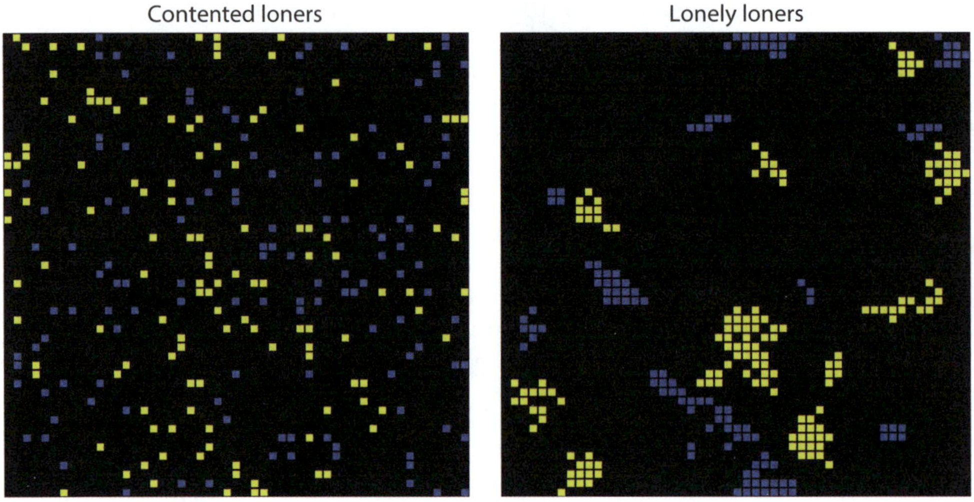

Figure 3.11 Comparing contented and lonely loners. These are screen captures of the Schelling model at equilibrium. In both cases, `density` = 0.1 and `similarity-threshold` = 0.7. Both report average similarities of 1. The times to reach equilibrium were 8 time steps for the contented loners and 250 time steps for the lonely loners.

is a lot of nuance in the real world, and that nuance often matters. The ability to feel and act on this sort of dissatisfaction is important to becoming a good modeler. Try your hand at extending and modifying the Schelling model. Add things that matter to you. Make some locations more desirable than others. Experiment with more than two groups. Add heterogeneity in preferences. Maybe some individuals actually prefer to be in the minority! There's a lot to explore.

It is not uncommon to see claims that the Schelling model explains segregation. Does it? My former postdoctoral supervisor Josh Epstein likes to say, "If you didn't grow it, you didn't explain it" (Epstein, 1999). Now that we've grown segregation, have we explained it? The inverse of Josh's statement—"If you grew it, you explained it"—is not always true. Growing it is perhaps necessary, but it is not sufficient for an explanation, at least not without considerably more constraints on model assumptions (this and related points are explored in more depth in chapter 10). Segregation is driven not only by individuals acting on their preferences, but by institutional and economic constraints, and by the affordances provided by social networks. The model does not incorporate any of these features, and so is probably a poor explanation of any real segregated community. What the model *does* show is that segregation can emerge more easily than might be suspected. The model shows that even in a perfectly egalitarian society in which economics, education, and opportunity were equally distributed, strong segregation might still arise from even weak individual preferences to be near similar others. None of this implies that we currently live in such a society, but it can help us to consider what sort of society we want to live in and how we would know when we'd achieved it.

We should, of course, question the assumptions of the Schelling model. I've mentioned some of the assumptions regarding what the agents can and can't do, perceive, or prefer. But we can also consider how the system is decomposed in the first place. The model situates agents on a broad, featureless grid. We could instead represent space in a very different way, breaking it into neighborhoods with economic and geographical features. Individuals could be represented as belonging to social networks and having mobility constraints like families and jobs. And their decisions could be made by considering and weighing multiple factors as well as by impulses or incomplete information, as real decisions surely are. The way we represent things focuses but also limits the questions we can ask. We should therefore always be reevaluating our representations in light of our continually evolving questions.

All that said, understanding how to make and analyze models like this produces a real value by helping us to better understand how complex systems work. One of the main challenges in social science is to explain how constraints, preferences, and affordances interact to produce population-level outcomes. Exploring models like the Schelling model exercises our mental muscles for understanding these processes.

3.10 Going Deeper

The Schelling model ignores the social and economic costs of mobility. Many people are constrained and therefore consigned to live in less-preferred areas. Work on urban organization indicates that economic and infrastructure constraints yield important scaling effects, whereby the opportunities in larger cities are orders of magnitude greater than in less densely populated areas (West, 2017). Infrastructure, and the related resources and socioeconomic niches to which it provides access, very likely influence mobility and hence segregation in complex ways.

While the Schelling model organizes people into discrete groups, identity is not singular but multidimensional and context dependent (Stryker and Burke, 2000; Roccas and Brewer, 2002; Smaldino, 2019). Social identity allows individuals to assort for more efficient coordination on norms and goals, and being around similar others may produce better psychological well-being, as indicated in the Schelling model. However, in diverse, cosmopolitan societies identity may also be expressed more covertly (Smaldino et al., 2018c; Smaldino and Turner, 2022), which may in turn suppress urban segregation, as you can only avoid dissimilar neighbors if you can readily identify them as dissimilar along meaningful dimensions. And while ethnicity may be largely fixed, other aspects of identity are more mutable. Iannaccone and Makowsky (2007), for example, studied an inversion of the Schelling model in which agents responded to local pressures to conform in order to investigate regional variation in religious identity. This work highlights that it is often productive to consider how existing models can be extended and adapted to tackle new questions.

Demographers have observed much greater nuance in patterns of segregation than can be generated by the simple Schelling model. For example, studies of Jewish-Arab ethnic residential distribution in Israeli cities have observed instances where members of both ethnic groups live in segregated enclaves adjacent to areas where members of both groups are well integrated. Hatna and Benenson (2012) studied several extensions of the Schelling model and determined that no straightforward variation of the model could generate patterns like these. Their work indicates that while the Schelling model was useful in explaining many empirically observed patterns, there were some patterns it could not explain. This makes sense to me. Really understanding segregation almost certainly requires taking into account the influences of organized action and socioeconomic filters that Schelling himself discussed, as well as some of the other forces mentioned above.

3.11 Exploration

1. **Exactly so.** In the code provided for the Schelling model, the parameter `density` controls the probability that any given patch will initially contain an agent. Let $p =$ `density` and let N be the total number of patches (the maximum possible population size). Rewrite the model's initialization procedures so that the population always contains exactly pN agents (rounded to the nearest integer). As is often the case, there are several possible ways to do this. One way is to use the NetLogo primitive `n-of` to select target patches upon which to create turtles. You should then create a switch for a Boolean variable called `exact?` that allows you to toggle between probabilistic and deterministic population size.

2. **Comparing metrics.** In addition to the average similarity of neighbors, Schelling also considered another metric of segregation: the proportion of agents that have strictly zero neighbors of the opposite color.

   (a) Write code to report this value using a new reporter called `prop-uniform`. One way to do this is to introduce a turtles-own variable called `uniform?` that is set to true only if all of a turtle's neighbors share its color. For low densities, you will have to decide how to count agents without any neighbors.
   (b) Add a new plot to display `prop-uniform` dynamically as the model runs, and a monitor to collect the precise value.
   (c) Run batches of 10 runs per parameter combination with values of `similarity-threshold` that vary between 5 and 75 in increments of 5, with density fixed at both 0.45 and 0.90. Stop runs either at equilibrium or after 100 ticks. Plot the two resulting segregation measures as a function of threshold for both values of density. Are the patterns of segregation as a function of similarity threshold the same for both metrics? If not, how do they differ?

3. **Make it stop.** For most cases, the Schelling model reaches a stable equilibrium fairly quickly. However, for high density, high threshold cases, it takes much longer to reach equilibrium, and in some cases never does. First, explain why, under high density, there should be a jump in time to equilibrium when the similarity threshold, S, is greater than 50%. Then, perform batch runs to quantify this. Fix density at 45% and 90%, and run batches of 5 runs per parameter combination with values of `similarity-threshold` between 40 and 85. Make sure to record the number of ticks that have elapsed when the simulation ends (NetLogo should already be recording this). Run simulations out to a maximum of 1000 time steps. How do density and the similarity threshold influence the time to equilibrium? What explains these results?

4. **Neighborhood size.** In the initial analysis, each agent only considered the group membership of the other agents in its Moore neighborhood. However, there is no reason agents could not be interested in a wider set of neighbors. Add a parameter called `neighborhood-size`. This will be an integer x such that an agent's neighborhood will be the set of all agents in the square of width $2x + 1$ centered on the focal agent (see Figure 3.3). Modify the code to adjust agents' neighborhoods accordingly. Then perform batch runs to assess the effect of neighborhood size on the resulting segregation.

5. **Minority report.** Rewrite the model so that a fraction `minority-seeking` of agents actively prefer to be in the minority. These agents behave like normal Schelling

agents except that they will move if the fraction of agents in the *same* group as they are *above* a threshold. Analyze the model in light of this addition. Does a population of minority-seeking agents lead to reduced segregation? Why or why not?

6. **Multiple groups.** Append the model so that at initialization, agents are assigned belonging to one of `num-groups` groups, where this number can be larger than two. How does the model behave in the presence of multiple groups? How does the number of groups affect the resulting levels of segregation? Does segregation increase or decrease with the number of groups?

7. **There goes the neighborhood.** Consider altering the Schelling model to reflect additional assumptions that might matter for segregation. For example, some areas might be more desirable or costly to live in, and individuals may differ in their preferences and abilities regarding mobility. Decide on one or more changes of this sort, then describe how they would change the model assumptions and how you would implement them in a Schelling-like model. What do you expect to find? (Bonus for actually implementing these changes.)

8. **Bust another move.** Study the properties of random walks using batch runs. How far will one agent tend to move from its initial location using a correlated random walk as a function of the maximum turning angle (as in the Particle World model from chapter 2)? Create a model in which a single turtle is initialized in the center of a grid that is at least 201×201 in area. Each time step, the agent will turn first to the left and then to the right, and then move forward 0.1 units. Let each left and right turning angle be determined by a random number between zero and `turning-angle` (in NetLogo, this is done using the command `random-float turning-angle`). For each simulation run, have the agent move for exactly 1000 time steps.

   (a) Create at least three trajectory plots showing the movement path of the agent through space for a few values of `turning-angle`. In NetLogo, this can be accomplished through the use of the `pen-down` procedure.
   (b) Run 100 simulations per value of `turning-angle` for a wide range of values in [0, 360]. For each run, calculate the Euclidean distance between the agent's ending position and the origin. This can be accomplished using the `distancexy` procedure. Plot these distances as a function of `turning-angle`.
   (c) What is the relationship between turning angle and distance traveled? Notice that the average distance traveled continues to decrease for turning angle values between 180 and 360 degrees, even though all possible turning angles are available for this entire range. Explain why.

4 Contagion

> I have your disease in me now.
>
> —Dorothy Vallens, *Blue Velvet* (1986)

Usually, when you give something to someone, you don't have it anymore. But there are some things that you can keep and still give to others. Diseases. Knowledge. Beliefs. Behaviors. We call such things **contagions**. How contagious things spread one from person to another is an important area of research in many fields. Epidemiologists and ecologists want to understand how a disease will spread through a population, as well as what might impede its spread. Social scientists want to understand the spread of ideas or behaviors. Marketers want to know what influences the adoption of products and innovations.

In this chapter, we will consider the dynamics of how contagions spread. As with the previous chapters, we'll mess around with agent-based models. However, this chapter also represents a watershed in the course of our study of social behavior, because it is in this chapter that we begin to incorporate the use of purely mathematical models. For those who cringe at the thought of doing math, I cannot help you—math is wonderful, and you should feel bad for your discrimination against one of the great intellectual achievements of human civilization. For those who worry that your math skills are rusty or underdeveloped, I promise to go slowly and to restrict our analysis to mathematical techniques you should remember from high school—algebra and functions. For those of you with advanced math training who scoff at simple approaches, I urge you to try your hand at it and see how powerful such approaches can be. If you crave more of a mathematical challenge, there are many resources available that can help you to go deeper.[1]

The coronavirus pandemic of 2020–2023[2] made nearly everyone in the world aware of how important it is to understand contagion processes. And in this chapter, we *will* cover dynamics specific to infectious diseases. To start, however, I want to focus on the diffusion of information or innovations. I have two reasons for this. First, it's not entirely obvious that the

[1] For example, see Keeling and Rohani (2008) or Bjørnstad (2018). For a more lighthearted approach, see also Smith? (2014).

[2] One can only hope that some sort of end is in sight.

adoption of a new innovation actually *is* anything like a contagion, and modeling may help to convince you that it indeed is. Second, the innovation diffusion models are formally quite simple yet provide good scaffolds for the more complicated contagion models that are introduced later in the chapter. We'll start with models of spontaneous adoption and of adoption by social influence. After that, we'll tackle more disease-relevant issues related to recovery and immunity, as well as the importance of public health measures such as vaccination and social distancing to reduce the severity of an epidemic.

4.1 The Diffusion of Innovations

How do new innovations spread, or diffuse, through a population? In 1962, the first edition of Everett Rogers's now-classic book, *Diffusion of Innovations* was published.[3] Rogers studied how a variety of innovations spread, with examples ranging from the adoption of hybrid seed corn by Midwest farmers to the adoption of ham radio among tech enthusiasts to the adoption of new ideas among French intellectuals. Rogers found that in all cases, when the cumulative adoption rate was plotted over time, it typically yielded a sigmoidal or S-shaped curve. That is, adoption increased slowly at first, then increased rapidly, then slowed down again.

What explains this apparently universal pattern of adoption? Rogers proposed an explanation based on the idea that individuals differ in their proclivity for adoption. He posited that a population could be reliably partitioned into several types: initial innovators, early adopters, early majority, late majority, and laggards. To make this taxonomy work, he also had to posit fairly specific proportions of the population that fall into each category: approximately 2.5% innovators, 13.5% early adopters, 34% early majority, 34% late majority, and 16% laggards (Figure 4.1).

Rogers's explanation involves two major assumptions. First, all individuals are assumed to have access to the same information about the availability of a new innovation. Second, individuals are assumed to vary in a very particular way: in terms of how long that information takes to provoke adoption of the innovation. The first assumption seems plausible; after all, if information spreads rapidly through a population—much more rapidly than the time it takes to actually be influenced by that information and adopt the innovation—then the exposure to information might be effectively simultaneous. The second assumption strikes me as more problematic. It assumes an awful lot about why people adopt and why those differences might occur, and requires very specific distributions of personality types to produce the sigmoid adoption curve. Is it plausible that these same distributions occur across contexts and cultures? It's possible, but unlikely. For now, let's reject the second assumption and assume instead that individuals respond to information pretty much identically. We can then ask if the first assumption of identical exposure to information is sufficient to generate the sigmoid adoption curve.

4.2 Spontaneous Adoption

We're going to build a model in which a new innovation is introduced into a population. Everyone perceives the innovation in the same way, and everyone has the same proclivity to adopt. In this model, there is no social influence, so there is no need to model social

[3] The fifth edition was published in 2003, shortly before Rogers's death.

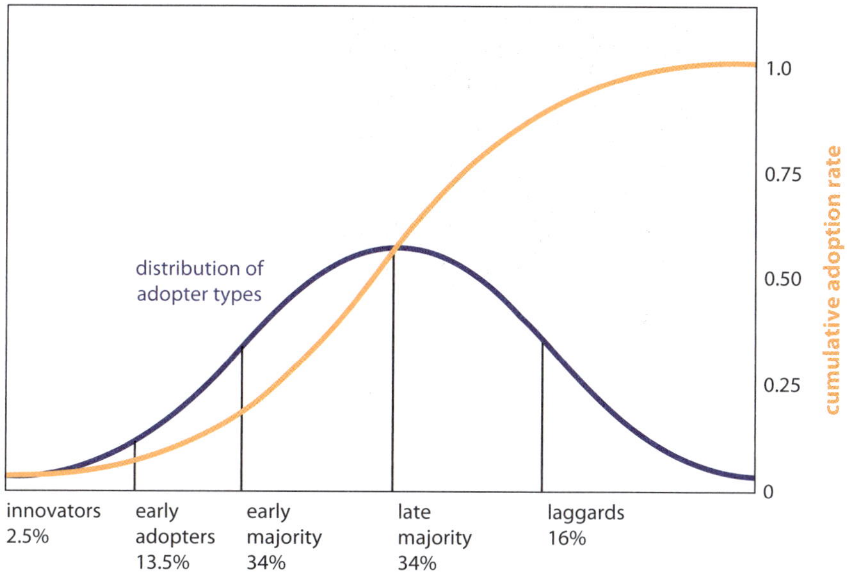

Figure 4.1 Rogers's diffusion model. The yellow curve shows the overall adoption rate over time. The blue curve shows the distribution of adopters according to the timing of their adoption.

structure. However, we might want to add social influence in the future so that we can allow agents to come into contact with one another and also vary the rate of that contact. Luckily, we have a simple model that can do the trick: it's our basic Particle World model from chapter 2. In that model, agents are embodied on a 2-D space with toroidal boundaries and move around using a correlated random walk. The speed and turning angle can be adjusted to affect the rate at which agents contact new social partners. For simplicity, we'll ignore collisions and flocking. In this way, we have a world in which agents move around in physical or social space and come into contact with others at varying rates.

In the Particle World model, agents differed from one another only in terms of their current position and directional heading. Now we'll add one more agent-level property: infection status. Agents in our new model are in one of two possible states: **susceptible** or **infected**. This terminology is taken from the idea of disease contagion, but it can be applied to the adoption of products or behaviors if you think of "infected" as simply indicating that the agent has adopted the product or behavior. We can initialize the model with a small number of infected agents; if we are thinking about innovation diffusion, these are our innovators.

The spontaneous adoption model works as follows. We consider a population of N agents, with some number $n_0 < N$ of agents as initial adopters (or innovators). At each time step, every agent who has not yet adopted will adopt with probability α. That's it. That's the entire model description, at least as far as assumptions that influence the dynamics of adoption. Our code also allows agents to move around, but because social influence is (for now) irrelevant to adoption, movement doesn't matter. Notice that there's no way for an agent to become *uninfected*, though we'll consider that later on as well. If agents cannot become uninfected, the population is always going to end up with everyone infected. However, exactly *how* that happens isn't necessarily obvious, so we'll focus on the temporal dynamics of infection. Will we see the telltale sigmoid adoption curve under spontaneous adoption?

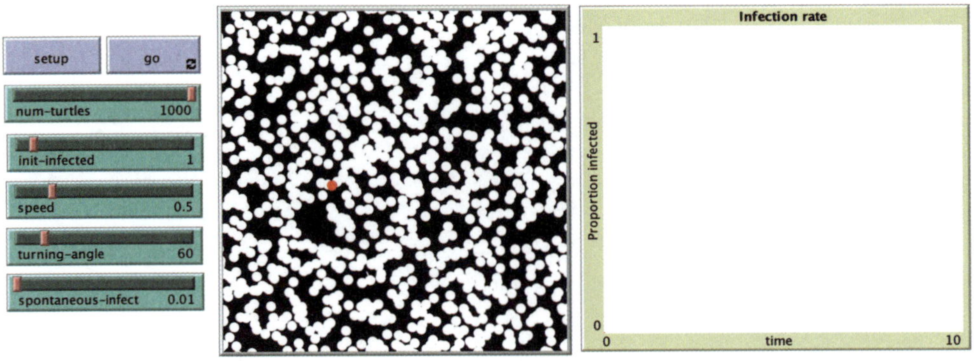

Figure 4.2 The Interface window for the spontaneous adoption model. Uninfected agents are white, and there is a lone red infected agent.

4.2.1 Coding the Model

The NetLogo code for this model is **contagion_spontaneous.nlogo**. Because much of the code will be carried over from the Particle World model, I'll focus on what is new here. Figure 4.2 depicts the model's Interface window.

As before, we will include global parameters for the population size and for the speed and turning angle of the random walk. In addition, we'll also add parameters for the number of initially infected agents (`init-infected`) and the rate of spontaneous infection (`spontaneous-infect`). Each agent will keep track of its own infection status, which we can code as an agent-level Boolean variable, as follows:

NetLogo code 4.1

```
turtles-own [
  infected?
]
```

At initialization, the agents are created and placed in random locations on the grid. We'll initialize each agent as uninfected, and then infect an initial few. We could infect initial agents probabilistically, but I have chosen to control the precise number of initially infected agents. What *is* probabilistic is exactly *which* agents become infected. This is controlled in the code by the procedure `setup-infected`, which is called from `setup`, and which chooses n_0 agents at random and infects them:

NetLogo code 4.2

```
to setup-infected
  ;;can't infect more agents than exist
  if init-infected > num-turtles
  [set init-infected 1]
  ask n-of init-infected turtles [
    set infected? true
  ]
end
```

For the purposes of visualization, I have made infected agents red and uninfected agents white. This is controlled by a procedure called `recolor`. The colors do not influence the model dynamics, but they do help us keep track of which agents are and aren't infected. Later, we'll see how we can use color as redundant information about infection status to avoid undesired modeling artifacts regarding the timing of infection and recovery.

The model dynamics work as follows. The simulation stops if everyone is infected, since no further infections are possible. This is useful for two reasons: first, it reduces computational time for running batches of simulations, and second, it allows us to record the time it takes for the contagion to diffuse throughout the population. If at least some agents are susceptible, the procedure `infect-susceptibles` is called. This causes each susceptible agent to become infected with probability `spontaneous-infect` (technically, we don't need to differentiate between infected and uninfected agents, because agents who are already infected cannot change their state, but this won't be the case when agents can recover or disadopt). Again, we model a probabilistic event by comparing a random draw from a uniform distribution to some threshold.

NetLogo code 4.3

```
to infect-susceptibles
  ask turtles with [not infected?] [
    if random-float 1 <  spontaneous-infect
    [set infected? true]
  ]
end
```

The provided code then recolors agents based on their infection status and has them move using a random walk, but neither of these are essential to the model dynamics. With our model now coded, it will be useful to be able to plot the proportion of the population that is infected as a function of time. In NetLogo, this is done by inserting the following command into a plot in the Interface window. We can then take a look at the temporal dynamics of adoption under spontaneous adoption.

NetLogo code 4.4

```
plot (count turtles with [infected?]) / num-turtles
```

4.2.2 Analyzing the Model

When every agent has the same information and the same proclivity for adoption, it readily becomes apparent that the time course of adoption is *not* described by a sigmoid curve. Instead, we observe an r-shaped curve, characteristic of asymptotic growth (Figure 4.3). The rate of adoption is maximal at first and then slows as there are fewer and fewer individuals who haven't yet adopted. This makes sense. When all susceptible agents become infected at the same rate, the number of new infections will always be proportional to the number of uninfected agents, which is maximal at initialization. As this number declines, so too does the rate of new infection.

Because local interactions don't actually matter in this model, its dynamics can be computed much more simply than the way we have done it. Instead of modeling the individual agents, we can use straightforward numerical simulation of a **recursive function**, or more simply, a **recursion**. Similar to a recursive procedure in programming, a recursive function takes its own prior value as input to compute the next value. Consider a population of size

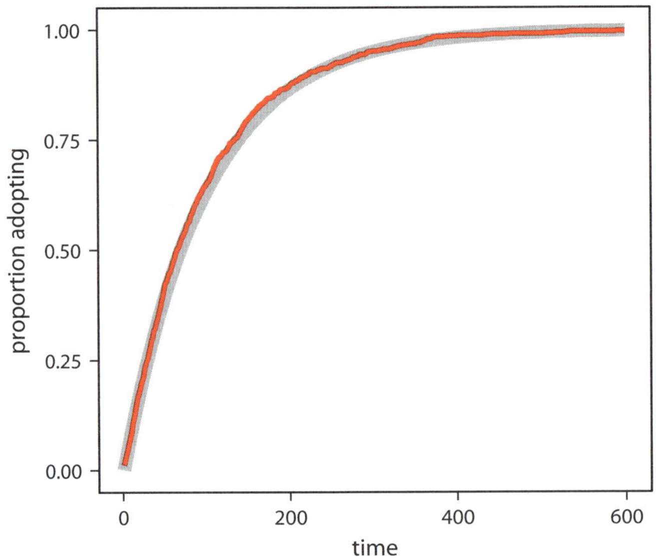

Figure 4.3 Adoption dynamics under spontaneous infection, with a population of $N = 1000$ and a spontaneous adoption rate of $\alpha = .01$. The red line shows the results of a single run of the agent-based model. The grey line shows the results of numerically simulating the mathematical model of spontaneous adoption.

N, where the number of infected agents at time t is given by I_t, and where the rate of spontaneous adoption is α. The number of infected agents at time $t+1$ is given by the following equation:

$$I_{t+1} = I_t + \alpha(N - I_t) \tag{4.1}$$

It is sometimes useful to represent these dynamics using a **difference equation** instead of a recursion. Difference equations place the focus on how the function changes (differs) between consecutive time steps. We can rewrite the spontaneous adoption model as a difference equation by simply subtracting the value of I_t from I_{t+1}:

$$\Delta I = I_{t+1} - I_t = \alpha(N - I_t). \tag{4.2}$$

The capital Greek letter Δ (Delta) is very commonly used to represent change, and so this equation dictates how the value of I changes from one time to the next. Note that this model, like all the models in this book, uses **discrete-time dynamics**. This means that we define some discrete increment of time, and then specify how the system changes between two moments in time separated by this increment. We can make the time between events as small as we like, but the increments are nevertheless discrete. Models of dynamical systems can also be studied with continuous-time dynamics using differential equations rather than difference equations. Such an approach has several advantages, notably that one can sometimes obtain closed-form solutions for differential equations. Discrete-time modeling is nevertheless quite powerful, and also maps well onto our agent-based models, which also use discrete time. The interested reader can learn more about modeling with differential equations in Box 4.1: Coupled Differential Equations.

Numerically computing the temporal dynamics of the model from our difference equation can be easily done in any programming language and requires only that we provide values for N, α, and I_0. While NetLogo isn't particularly advantaged for this type of analysis, it can still be done with relative ease by using lists. These allow us to store an ordered list of values, which we can use to represent the number of infected agents at each point in time. The following code does the trick:

NetLogo code 4.5

```
to analytical-solve
  let num-infected []
  set num-infected fput 0 num-infected ;;set first entry to 0
  let i 1
  while [i < 1000] [
    let x (item (i - 1) num-infected)
    set num-infected lput (x +  spontaneous-infect *
        (num-turtles - x)) num-infected
    set i (i + 1)
  ]
  show num-infected
end
```

We first create an empty list called `num-infected` and initialize its first value as zero. For 999 more time steps, we add a proportion α of the currently susceptible agents to the previous number of infected agents. Figure 4.3 shows that the curve resulting from the numerical simulation very closely matches the curve resulting from the agent-based simulation. The ABM introduces more stochasticity, and so the resulting curve is a little less smooth (and thus probably more realistic) than the one from the numerical simulation, but it is also gratifying to see that the numerical simulation accurately predicts the results of a more stochastic process. In general, when mathematical approaches are possible for the models we examine, I will usually try to include some discussion of those approaches. It is also worth noting that, while detailed descriptions are often necessary for communicating agent-based models, mathematical models are usually much easier to describe because the models are fully captured by the variables and their governing equations.

The r-shaped curve resulting from the spontaneous adoption model is distinctly different from the sigmoid adoption curve that is observed empirically for the diffusion of innovations. So we can confidently say that the sorts of innovations studied by Rogers don't diffuse this way, with everyone having the same proclivity for adoption and the same information. Must we therefore posit strongly prescribed individual differences in adoption proclivity, as Rogers did? As we will see, there is another way.

4.3 Social Influence: The SI Model

For our next model, let's stick with the assumption that everyone has the same proclivity to adopt. Wait a minute, you say, we already showed that doesn't work. Not so fast. A key assumption in the spontaneous adoption model was that all individuals had access to the same information *at the same time*. What if *that* assumption no longer held?

A key assumption of the spontaneous adoption model is that simply knowing about an innovation's existence or utility is what drives adoption. Instead, let us assume that people are influenced primarily by direct interpersonal interaction or exchange—such as directly witnessing the innovation in use by another person. Even though we will continue to ignore

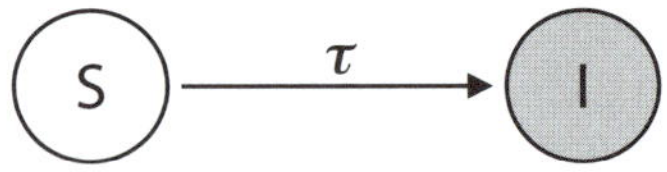

Figure 4.4 The SI model, in which τ is the transmissibility of the contagion.

differences in individuals' proclivity to adopt, this new assumption permits heterogeneity among individuals in their social interactions. If I meet lots of people who have already adopted and you do not, then I will be more likely than you to adopt the innovation. This sort of heterogeneity is not intrinsic to the individual agents in the sense of psychological characteristics, but rather emerges through variation in their social encounters. Models of this sort, in which susceptible agents become permanently infected through social influence, are sometimes called SI models (the letters stand for susceptible-infected, not social influence; see Figure 4.4).

The key idea here is that innovations might spread sort of like diseases. Assume that every exposure to an infected individual carries some risk that a susceptible individual will be convinced and adopt the innovation. To build this model, we need to consider just *how* exposure translates into adoption. In our agent-based model, agents move through space using random walks, and we can use proximity in this space to represent opportunities for exposure. Unlike in the spontaneous adoption model, the rate of interpersonal interactions matters, because agents will be influenced by encounters with their spatial neighbors. Location also matters, because some areas of space may have more infected individuals than others. The strength of social influence can be dictated by the **transmissibility** of the innovation, similar to how contagious a disease is. Presumably, not all interactions lead to transmission. Transmission depends on a number of factors. If we are talking about an innovation, these factors might include the intrinsic attractiveness and affordability of the innovation. If we are talking about a disease, factors influencing transmissibility include the health of susceptible individuals, the physiology of the pathogen, and the duration of contact between susceptible and infected individuals.

4.3.1 Transmissibility

Let τ be the transmissibility of the contagion, the probability of being influenced to adopt by someone who has already adopted. Consider a susceptible agent surrounded by several infected agents at the same time. What is the probability that this susceptible agent will become infected? If the agent has only one infected neighbor, then we're set: the probability of adopting is τ. However, things get more complicated if there are multiple neighbors who have adopted. The joint probability[4] of being infected by each of two infected neighbors is τ^2. Because τ is a probability, it's bounded in [0, 1], which implies that $\tau^2 \leq \tau$. This makes sense, because being infected by each of two individuals is less likely than being infected by only one of them. But it doesn't matter, for present purposes, which neighbor successfully infects an individual or whether the infection spreads from one or more neighbors. The joint probability τ^2 is therefore not what we need in order to model how an individual with multiple infected neighbors becomes infected.

What we really want is the probability that *at least one* infected neighbor transmits the contagion. To do this, we first need to calculate the probability that the contagion *doesn't*

[4]Joint probability refers to the probability that each of two events both occur. For more on probability theory, see chapter 8.

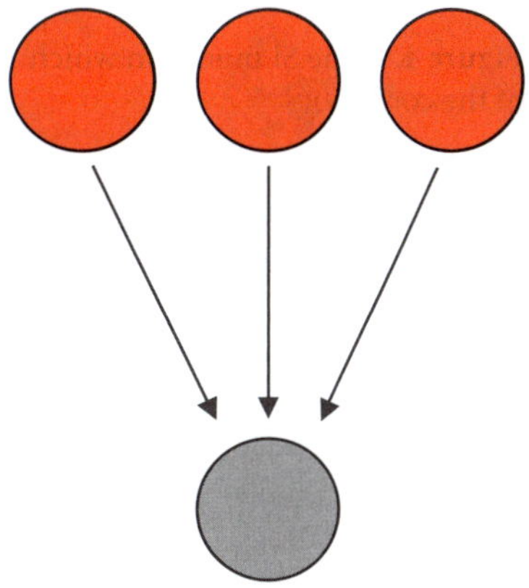

Figure 4.5 A susceptible agent (grey) encounters three infected agents (red).

spread. With multiple infected neighbors, this is equivalent to the joint probability that all of them fail to transmit the contagion. The probability that one infected neighbor fails to transmit the contagion is $1-\tau$. If three neighbors have adopted (Figure 4.5), the probability of the contagion simultaneously failing to spread to a susceptible agent from any of its neighbors is $(1-\tau)(1-\tau)(1-\tau)=(1-\tau)^3$. More generally, with n infected neighbors, the probability of the contagion failing to spread is $(1-\tau)^n$. So, the probability of this *not* happening—that is, of at least one neighbor transmitting the infection—is simply the inverse:

$$\Pr(\text{infection}) = 1-(1-\tau)^n \tag{4.3}$$

With this in mind, we can update our code, building on the code for the spontaneous adoption model. In our new model, agents will move around as before, using a random walk. When agents are sufficiently close, within distance r from one another, infected agents can transmit the contagion to susceptible agents. It's worth noting the importance of physical proximity in this model—other assumptions would be required to accurately model behavioral contagions that can spread through long-distance communication or diseases that spread via vectors such as biting flies.

4.3.2 Coding the Model

The NetLogo code for the SI model is **contagion_SI.nlogo**. To our previous model parameters we need only add the `transmissibility` of the contagion. The initialization of this model is exactly the same as the spontaneous adoption model: agents are placed in random locations on the 2-D space, and a select few of them are infected. The dynamics of the model work somewhat differently, however, with space and movement becoming critical.

At each time step, the simulation once again ends if all agents have become infected. If at least one susceptible agent remains, each susceptible agent counts the number of infected agents nearby. Of course we have to define more precisely just what is meant by "nearby." I have defined nearby agents as those whose locations are less than one spatial unit away from the focal agent, which captures the physical closeness needed for some contagions (though not others). Once we know how many infected agents are nearby, the agent becomes infected with a probability dictated by Equation 4.3. The new `infect-susceptibles` procedure is written as follows:

NetLogo code 4.6

```
to infect-susceptibles
  ask turtles with [not infected?] [
    let infected-neighbors (count other turtles with
     [color = red] in-radius 1)
    if random-float 1 <  1 - (((1 - transmissibility) ^
     [infected-neighbors) * (1 - spontaneous-infect))
    [set infected? true]
  ]
end
```

Each agent is then recolored according to its infection status and takes a step using the `move` procedure we've seen before, which involves turning a random angle dictated by `turning-angle` and then taking a step of size `speed`. The model dynamics continue until all agents are infected.

4.3.3 Analyzing the Agent-Based SI Model

Let's now examine what happens when adoption spreads via direct social influence rather than by a fixed response to global information. In the beginning of a simulation run, when few agents have adopted, most interactions are between pairs of susceptible agents and there are few new adoptions. The rate of growth will therefore be slow. As more agents adopt, the adoption rate will increase until a majority of agents have adopted. After this, adoption will slow down again since most interactions will be between agents who have already adopted. In other words, the model generates just the sort of S-shaped, or sigmoid, adoption curve observed by Rogers and others (Figure 4.6).

This simple model points to a potential explanation for why so many different innovations all have sigmoid adoption curves, and suggests that social influence likely does drive

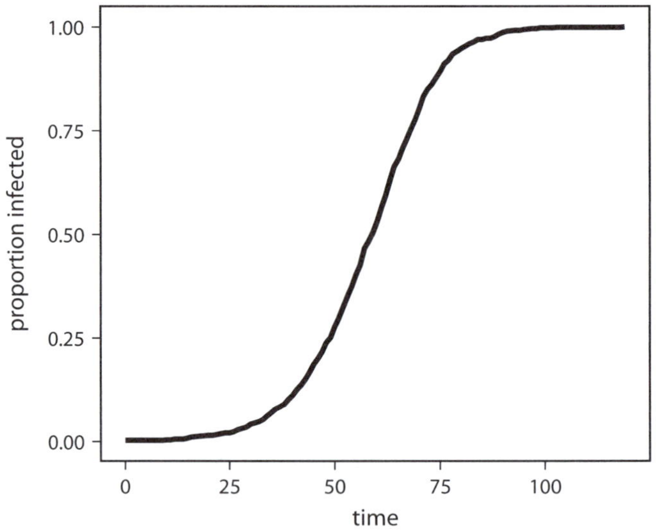

Figure 4.6 Adoption dynamics under social influence. For this run, I used 1000 agents on a 33 × 33 continuous square grid with toroidal boundaries, `speed` = 0.5, `turning-angle` = 60, and `transmissibility` = 0.05.

adoption. Indeed, SI models are regularly used to model the diffusion of innovations and behaviors. What else can we learn from this model? As I've said, it's often valuable to simply play with a model to get a feel for the sort of dynamics that occur when we change the parameters. I encourage you to do so with this model. Even experiments where the results are obvious are valuable as a check to make sure your model is running as it should. For example, if you vary the transmissibility of the contagion, you should observe that the contagion spreads much faster when `transmissibility` is 0.1 than when it is 0.01. Obviously, innovations that are more intrinsically attractive or otherwise contagious will diffuse more rapidly through a population.

Other outcomes are perhaps less obvious. The fewer agents there are—the less dense the population—the fewer interactions will occur per unit time, and the longer it will take for the innovation to diffuse. This might be counterintuitive if it seems like innovations should take longer to spread in larger populations with more people. But if adoption events are independent and rely on the density of social networks, then denser populations will adopt more rapidly. This is certainly the case with the spread of infectious diseases. Relatedly, the speed of diffusion depends on how much mixing goes on—the rate at which adopters will interact with new people who haven't yet adopted. We can use the `speed` and `turning-angle` parameters to control the contact rate and ultimately the speed of diffusion. Play around and see what kinds of relationships you can uncover.

It may seem like we have built a fairly complicated model to generate fairly simple insights. Of course, one person's simple is another person's complicated. However, the fact that the models generate smooth curves and make only simple assumptions about agents' states and dynamics is a good indication that our system may be a good candidate for analytical mathematical modeling.

4.4 The Analytical SI Model

Agents in our agent-based SI model interact at random, moving around using a random walk. This sort of random mixing is often modeled mathematically as a **well-mixed** population, in which interactions are completely random and there is no consideration of geography or social relationships. If there are multiple types of individuals in the population (such as infected and uninfected individuals), then the probability that a randomly encountered individual is of type A is simply the frequency of type A individuals in the population. You might picture a bustling city in which people are constantly moving all over the place and encountering one another willy-nilly. In later chapters we will consider group- and network-structured interactions. For now, we will derive a mathematical SI model in a well-mixed population.

Assume a large population of N individuals, for which some number, I, are infected with the contagion. The remaining $S = (N - I)$ individuals are uninfected but susceptible to infection. If two susceptible individuals meet, nothing happens, since neither is infected. If two infected individuals meet, nothing happens, since both are already infected. The interesting scenario is when a susceptible individual meets an infected individual. We therefore need to ask how often a randomly selected interaction will be one of these interesting cases that involves the possibility of a new infection. We can consider a focal individual and ask about their probability of infecting a new individual. By considering infections only from this perspective, and not the perspective of the focal individual becoming newly infected, we avoid double counting when we sum over the entire population. Because of our assumption that individuals meet at random, the probability of an encounter that could lead to a new infection in this way is the joint probability that the focal individual is infected and the individual they encounter is susceptible:

$$\Pr(S, I) = \frac{I}{N}\left(\frac{N-I}{N}\right) \tag{4.4}$$

Note that this is the probability that the interaction *could* lead to a new infection; it says nothing about whether it will.

As before, let τ be the transmissibility of the contagion. The probability that a random interaction leads to a new transmission is the joint probability that the interaction is between an infected individual and a susceptible one and also involves a transmission event:

$$\Pr(\text{new infection}) = \tau\frac{I}{N}\left(\frac{N-I}{N}\right) \tag{4.5}$$

This equation characterizes a new infection resulting from a single interaction. However, we are interested in changes to the population at large—the sum total of all interactions. We therefore need to describe how the number of infected individuals changes over time.

There are N individuals who could be the focal agent in an interaction. For each of those N individuals, we take the joint probability that the focal agent is infected, the individual they encounter is susceptible, *and* such an encounter leads to a new infection. The number of new infections at each time step is therefore given by

$$\#\text{ new infections} = \Delta I = N\tau\frac{I}{N}\left(\frac{N-I}{N}\right) \tag{4.6}$$

To model the dynamics of infections over time, we can turn this equation into a recursion for the number of infected individuals at time $t+1$:

$$I_{t+1} = I_t + N\tau\frac{I_t}{N}\left(\frac{N-I_t}{N}\right) \tag{4.7}$$

This equation is still a bit busy, so let's do some algebra to simplify it. There are a lot of N's in the equation above. Using the magic of mathematics, we can cross some of them out, leaving us with:

$$I_{t+1} = I_t + \tau I_t\left(1 - \frac{I_t}{N}\right) \tag{4.8}$$

Inspecting Equation 4.8 tells us a few things. First, the infection will spread faster if τ increases. This makes sense, and it also fits with what we observe in our simulations. If you have a more contagious infection, the number of infected individuals will increase more quickly. How does the overall time trajectory of the contagion play out? When I_t is very small, few interactions will lead to new transmissions, because most interactions will involve two susceptible individuals, neither of whom can spread the contagion. When I_t is large and close to N, the term $(1 - I_t/N)$ will be close to zero, and the rate of new transmissions will similarly be small because most individuals will already be infected. This indicates that the rate of change should be largest when an intermediate fraction of the population is infected, and indeed this is exactly what we find if we run numerical simulations of Equation 4.8. Figure 4.7 shows that the infection curve takes a nice sigmoid shape, where the overall time to universal infection is determined entirely by τ.

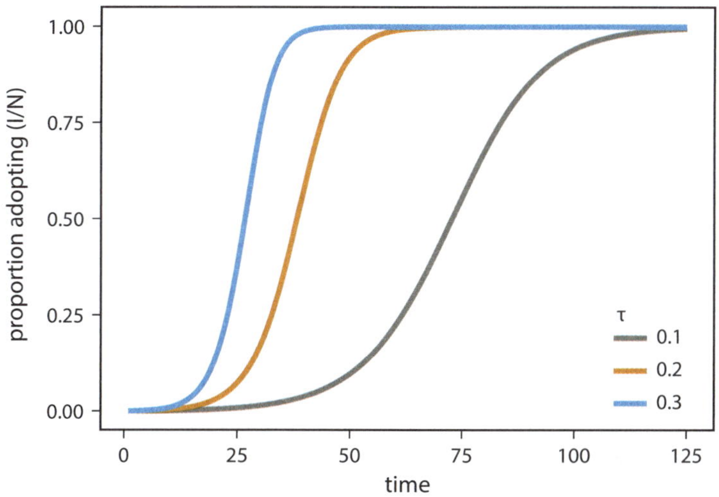

Figure 4.7 Dynamics of the SI model for varying τ.

Notice how the factors related to agent movement that we discussed in terms of the agent-based model—density, speed, and turning angle—appear to be absent from our analytical model. Indeed, all the factors contributing to variation in transmission are captured by the single variable, τ. In some models, this term is explicitly decomposed into two components: τ is used to represent the direct transmissibility that occurs during a contact, and another term, c, is used to represent the per capita contact rate, so that the **effective transmissibility** of the contagion, β, can be characterized as $\beta = c\tau$. This is the joint probability that an individual has a new encounter *and* that the encounter leads to a new infection (conditional on it being between susceptible and infected individuals). Framing it this way highlights that one could decrease the rate of new infections either by decreasing the probability of transmission during an interaction or by decreasing the rate at which interactions occur. This formulation also has the advantage of neatly capturing key sources of variation in transmission as well as mapping more directly onto our agent-based model. That said, the agent-based model still affords us a more nuanced view of how interpersonal contact can vary. Movement speed might represent the duration of individual interactions, while turning angles might indicate the degree to which social networks are cliquish and assortative. In general, analytical and agent-based models are often complementary, and insights are often most abundant when the two approaches are combined.

4.5 Getting Better: The SIS Model

A key assumption of the SI model is that once people are infected—or once they have adopted whatever product or behavior we're talking about—that's it. They remain infected for as long as they live (or at least for as long as the model is relevant). This is a pretty strong assumption. In terms of infection, sometimes people recover and are no longer infected. In terms of products or behaviors, sometimes people decide that they don't want or need them anymore and so stop using them. They might give up the product or behavior forever, or they might simply take a hiatus. Let's first focus on the latter case, so that people can recover (or disadopt), and thereafter return to a renewed state of susceptibility. For the

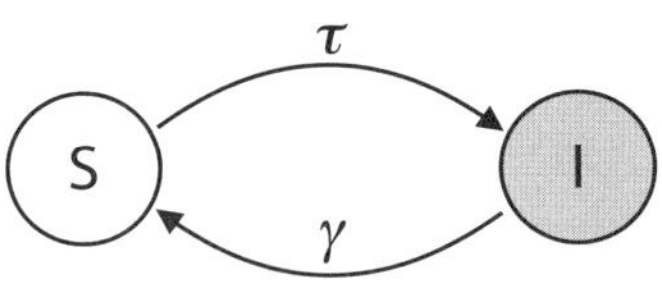

Figure 4.8 The SIS model.

remainder of this chapter, I will default to talking about contagions in terms of infectious diseases. However, many of the lessons will continue to apply to the spread of innovations or behaviors.

In technical terms, we now move from the SI model to the SIS (susceptible-infected-susceptible) model (Figure 4.8). How does recovery work? The simplest assumption is that infected individuals recover with a fixed probability for every unit of time they are infected. We therefore need to add just one additional parameter to our model: γ, the *recovery rate.* Every time step, a proportion γ of infected individuals recover and become susceptible again, so that the number of susceptible individuals is increased by γI and the number of infected individuals is reduced by that same amount. This means that the expected duration[5] of an infection is $1/\gamma$. We can update our recursion for the number of infected individuals at time $t+1$ accordingly:

$$I_{t+1} = I_t + \tau I_t \left(1 - \frac{I_t}{N}\right) - \gamma I_t \tag{4.9}$$

At each time step, susceptible individuals are becoming newly infected, but infected individuals are also recovering and becoming newly susceptible. So what happens? If the contagion effectively spreads, the population will settle to an equilibrium where the rate of new infections equals the rate of recovery. Because of this, the entire population never becomes fully saturated with the contagion, as it did in the SI model, because infected individuals are recovering at the same time as new individuals are getting infected. This model achieves what is known as a **dynamic equilibrium**, in which a fixed proportion of the population remains infected even though the individuals in the population continue to change their infection status.

We can use our mathematical model to calculate the precise proportion of the population that will remain infected at equilibrium. This occurs when there is no longer any change in the total number of infected individuals; that is, when $I_{t+1} = I_t$. We begin by rewriting Equation 4.9 and letting $I = I_{t+1} = I_t$:

$$I = I + \tau I \left(1 - \frac{I}{N}\right) - \gamma I \tag{4.10}$$

I've omitted the time subscripts here because we are considering the case where there is no change between times t and $t+1$, and so time, in a sense, becomes irrelevant. We can cancel out some of the I terms, which leaves us with:

$$\tau I \left(1 - \frac{I}{N}\right) - \gamma I = 0 \tag{4.11}$$

[5]Using a fixed probability of recovery implies that infection durations will be exponentially distributed, which in turn implies that the most common infection duration will be a single time step. This is an unrealistic but mathematically useful assumption. See Yan (2008) for a discussion of models incorporating other distributions.

Now let's rearrange the equation by moving γI over to the right side of the equals sign, and have some fun with algebra:

$$\tau I\left(1-\frac{I}{N}\right)=\gamma I \tag{4.12}$$

$$\tau\left(1-\frac{I}{N}\right)=\gamma$$

$$1-\frac{I}{N}=\frac{\gamma}{\tau}$$

One more rearrangement gets us to

$$\boxed{\frac{I}{N}=1-\frac{\gamma}{\tau}} \tag{4.13}$$

This is the equilibrium infection rate, the proportion of the population that always remains infected. We can use this equation to check our intuitions about the model. The equilibrium infection rate goes down when transmissibility is lower or when recovery rate is higher, which makes sense.

It's worth taking a bit of time to get some additional intuition about why this equilibrium occurs. One approach is to plot the difference equation as a function of the infection rate. Dividing all sides by N to get rates rather than quantities of infection, we get the difference equation

$$\Delta I=\tau I(1-I)-\gamma I \tag{4.14}$$

If we plot ΔI as a function of I, we end up with the plot in Figure 4.9 (left side). This sort of graph is very useful. Note that when I is below 0.75 (the equilibrium infection rate for the values of τ and γ used in this example), the rate of change is positive. This means that the number of infections will increase. However, when I is above 0.75, the rate of change becomes negative, meaning that infections will decrease. In this way, a stable equilibrium is

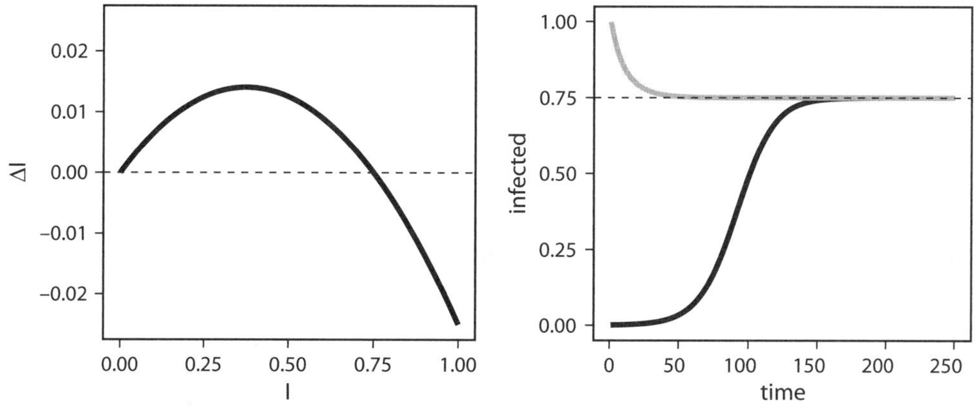

Figure 4.9 Equilibrium dynamics of the SIS model. Left: the difference equation for ΔI plotted against I, where I is the rate of infection. Right: infection dynamics starting at either $I(0)=0.001$ or $I(0)=0.999$. In all cases, $\tau=0.1$ and $\gamma=0.025$.

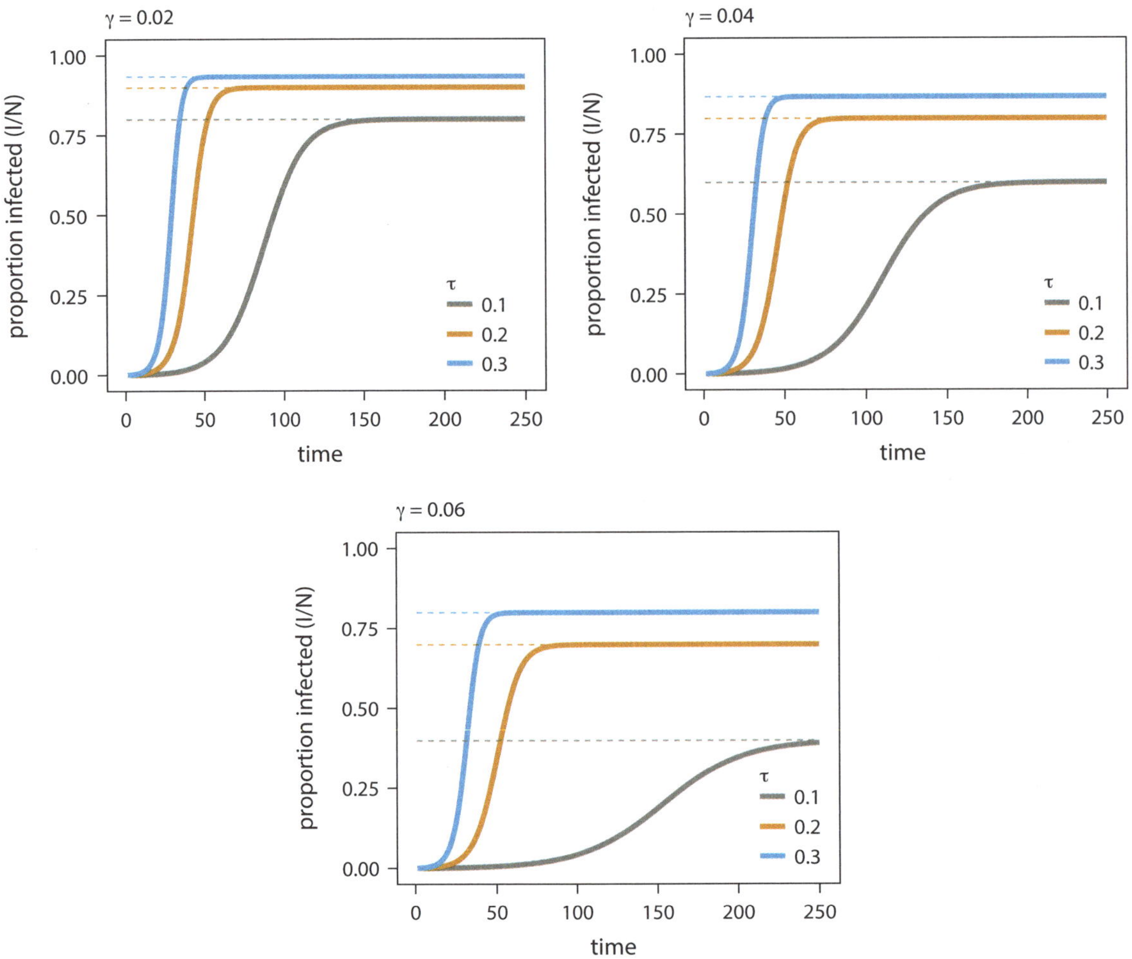

Figure 4.10 SIS model dynamics for several values of effective transmission rate ($c\tau$) and recovery rate (γ). The dashed lines indicate the equilibrium frequencies of infection.

reached at $I = 0.75$, where the rate of change is zero. Note also that there is another value of I for which the rate of change is zero: $I = 0$. This makes sense, because there can be no new infections when no one is infected. However, this is an **unstable equilibrium**, because a slight perturbation (that is, the introduction of new infections) will yield a positive rate of change and push the system to its stable equilibrium. This situation is further illustrated by the right side of Figure 4.9, which plots numerical simulations of the infection rate starting when the infection rate is either close to zero or close to one. In both cases, the infection rate returns to its stable equilibrium value.

While the ratio γ/τ determines the equilibrium proportion of the population infected, each parameter contributes to the dynamics with which the system reaches that equilibrium. Figure 4.10 shows the dynamics for the same parameters used above in our analysis of the SI model (Figure 4.7), for different values of γ. Compare this figure with the SI model in Figure 4.7 (the equivalent of the SIS model in which $\gamma = 0$). Notice that, in addition to the equilibrium infection rate dropping, the speed at which the infection reaches that equilibrium is slower for higher recovery rates, and the overall course of the contagion through the population is slower. This makes sense if you consider that individuals are recovering as well as becoming infected at each time step.

The analytical model we have just derived indicates that if we know the effective transmissibility rate and the recovery rate for the contagion, we can determine the equilibrium frequency of infection (or adoption) with confidence, and that infection frequency should remain stable. Let's investigate this claim further and see what happens when we model individuals explicitly as agents moving in space who become infected due to physical proximity with an infected agent. This will introduce a very small amount of the heterogeneity that real-world systems have in spades.

4.5.1 The Agent-Based SIS Model

The NetLogo code for this model is **contagion_SIS.nlogo**, and it builds upon the code for the SI agent-based model. We must add one new global parameter, the recovery rate (`recovery-rate`). The initialization of this model is the same as the SI model, and the dynamics differ primarily in the inclusion of recovery, though I have also added the provision that the simulation will stop if zero agents are infected, indicating a failed outbreak.

The principal dynamics in the SI model involved three stages at each time step. First, susceptible agents could become infected from contact with infected agents. Second, agents were recolored based on infection status. And third, agents moved using a correlated random walk. To this we will add a new stage before the recoloring: recovery. During this stage, infected agents may recover and become susceptible again. Importantly, we don't want a new infection to be immediately negated by having a newly infected agent recover before the agent has the opportunity to infect its neighbors (or, more importantly, before it has actually been infected for any amount of time). Here is where careful consideration of one's code is crucial. If we use only the value of an agent's `infected?` parameter, an agent could flip from susceptible to infected and back to susceptible on a single time step. One way to avoid this is to use two different markers of agent infection. Luckily, we already have this redundancy built into our code, since an agent's infection status is represented both by the Boolean variable `infected?` and by the agent's color. If newly infected agents remain white until the end of the time step, we can restrict recovery to agents who are both infected and red. The code for the procedure `recover-infecteds` is as follows:

NetLogo code 4.7

```
to recover-infecteds
  ask turtles with [infected? and color = red]
  [
    if random-float 1 < recovery-rate [
      set infected? false
    ]
  ]
end
```

When simulations of this model are run, we should notice that the dynamics of infection do resemble those of the analytical model, in the sense that the proportion of infected individuals in the population climbs to its equilibrium value and then hovers around this

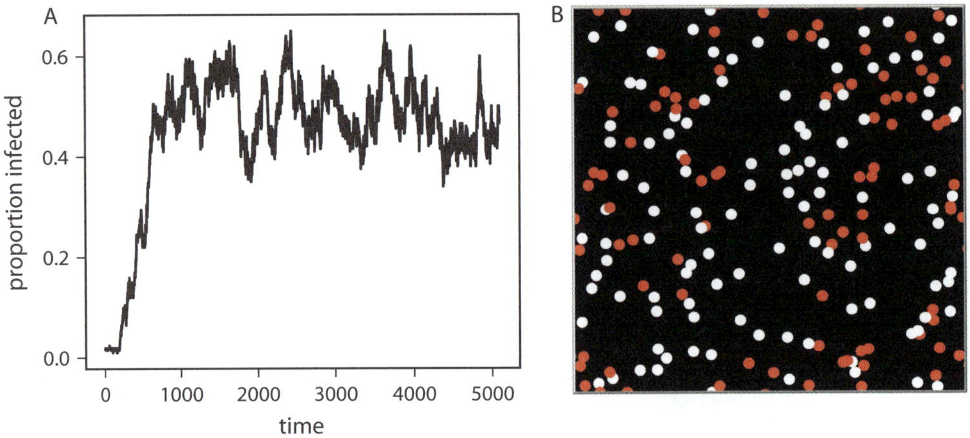

Figure 4.11 Agent-based SIS model dynamics. (A) The proportion of infected agents plotted over time. (B) Spatial positions of infected (red) and uninfected (white) agents at $t = 3400$. For this simulation, `num-turtles` = 200, `speed` = 0.5, `turning-angle` = 360, `transmissibility` = 0.1, and `recovery-rate` = 0.02.

value indefinitely. An important *difference* between the agent-based and analytical models is that the infection rate in the former does not completely stabilize as neatly as predicted by the latter. Rather, the infection rate oscillates around the equilibrium value due to stochastic fluctuations in the spatial distribution of infected agents. Decreasing the mobility of agents by decreasing the population density and increasing their maximum turning angle exacerbates this effect and can create dynamic spatial clusters of infection and non-infection (Figure 4.11).

This disparity again highlights the complementary nature of the analytical and agent-based approaches. The analytical model shows us how to calculate the equilibrium infection rate. The ABM shows us that this calculation is merely an estimate of the average, and that fluctuations around that average are likely and will probably cluster in nonrandom ways on social networks. In general, it is almost always useful to be able to look at the same problem from multiple angles. Agent-based models of contagion are especially useful when one wants to consider highly structured interactions and/or heterogeneity in individuals or environments. Analytical models are useful because one can calculate certain parameters exactly and explore the consequences of other parameters instantly without needing to run computationally demanding simulations. In the next subsections, we will consider two useful values that can be derived analytically: the basic reproduction number, R_0, and the critical vaccination threshold needed for herd immunity.

4.5.2 Will It Spread? Calculating R_0

The preceding analyses implicitly assumed that, following an outbreak, a contagion will spread until it reaches its equilibrium infection level. However, another outcome is possible: the initial outbreak may fizzle out before it spreads very far. This can happen if the first individuals to become infected recover before they have a chance to transmit the disease further. What factors might determine whether this happens? After some thought, your intuition should hopefully be that whether or not an initial outbreak spreads will depend on the contact rate, the transmissibility of the contagion, and the recovery rate. If contact rate is low (so that people have few chances to transmit the contagion), transmissibility is low (so that contact events rarely lead to new transmissions), and the recovery rate is high (so that infected individuals recover quickly before they can spread the infection), an epidemic might be avoided. These intuitions are valuable, but they can be improved by quantifying them.

Recall that the dynamics of the SIS model are captured by Equation 4.9:

$$I_{t+1} = I_t + \tau I_t \left(1 - \frac{I_t}{N}\right) - \gamma I_t$$

The term in parentheses, $(1 - \frac{I_t}{N})$, will be very close to one at the start of an outbreak in a large population, because $\frac{I_t}{N}$ will be close to zero (that is, almost everyone is uninfected). Thus, we can rewrite the recursion equation as:

$$I_{t+1} \approx I_t + I_t(\tau - \gamma) \tag{4.15}$$

We can rewrite this as a difference equation, which allows us to drop the subscripts:

$$\Delta I \approx I(\tau - \gamma) \tag{4.16}$$

This equation should make it clear that the number of infected individuals will increase only if $\tau - \gamma > 0$. Rearranging this, the infection will spread if

$$R_0 = \frac{\tau}{\gamma} > 1, \tag{4.17}$$

where R_0 (usually pronounced "R-naught") is known as the **basic reproduction number**. This is the expected number of new infections an infected person will generate in a fully susceptible population before they recover. Calculating this confirms our intuition that an infection is more likely to spread if the effective transmission rate is high and the recovery rate is low. Specifically, it will spread if the rate of new infections is greater than the rate of recovery. Note that R_0 is emphatically *not* purely a property of the contagion. Rather, it is affected by *all* the factors that influence transmission and *all* the factors that influence recovery. It is an emergent property of the interaction between a contagion and the environment, including the social environment. Thus, social and behavioral measures can change R_0. If social behaviors are relatively predictable, at least in the aggregate, then R_0 can be—and often is—calculated for specific contagions, which can help researchers to prioritize certain outbreaks over others. Calculating R_0 is additionally useful because it gives us access to another tool. It allows us to study a phenomenon called **herd immunity**.

4.5.3 Vaccines and Herd Immunity

We have just learned that we can decrease the chance that an infection will spread throughout a population by decreasing its effective transmissibility. One way to reduce transmission is through vaccination. Of course, not all infections have available vaccinations, and not all vaccinations are perfect. Vaccines are a politically loaded topic in some circles, and there are important ethical and medical concerns surrounding vaccines. Vaccines have saved millions of lives, and as such their benefits usually outweigh their societal costs. However, we are not here to debate the benefits of vaccines but instead to do what modelers do best, which is to consider the consequences of particular assumptions.

We will assume that (1) vaccines are administered at random to a proportion V of the population before the initial outbreak of infection, and (2) vaccines are perfectly effective, so that a vaccinated individual can neither become infected nor transmit the infection to others. These assumptions can later be changed as needed, but making them simplifies the math and allows us to establish a baseline. We can now write a new difference equation for the change in the number of infected individuals that incorporates vaccinations:

$$\Delta I = \underbrace{\tau\left(1-\frac{I}{N}\right)}_{A}\underbrace{(1-V)}_{B} I - \gamma I \tag{4.18}$$

In this equation, A is the probability of encountering a susceptible individual and transmitting the infection, and B is the probability that the susceptible individual is not vaccinated and therefore *becomes* infected. At the onset of an outbreak (when I is small), this equation can be approximated as

$$\Delta I \approx I\,(\tau(1-V)-\gamma) \tag{4.19}$$

Following the same logic as in the previous section, the infection will spread when rare if and only if

$$\frac{\tau}{\gamma}(1-V) > 1 \tag{4.20}$$

The left side of this equation can be rewritten as $r_0 = (1-V)R_0$, where r_0 is known as the *effective* basic reproductive number.

What good is this? Well, if we have information about the basic reproduction number for an infection and we have an effective vaccine, we can calculate the proportion of the population we need to vaccinate to stop the spread of disease, V^*. Importantly, we don't need to vaccinate everyone, just enough to drop the rate of transmission below the rate of recovery. This phenomenon is known as *herd immunity*. The unvaccinated proportion of the population is kept safe by vaccinating enough individuals to stop the disease from spreading. Note that herd immunity does not mean that unvaccinated individuals will never become infected, just that only a small number of infections will occur before the infection dies out.

We can calculate the threshold vaccination rate for herd immunity by letting V^* denote the smallest value for which $r_0 \leq 1$. We can derive this as follows:

$$\begin{aligned}(1-V)R_0 &\leq 1 \\ R_0 - R_0 V &\leq 1 \\ R_0 - 1 &\leq R_0 V \\ 1-\frac{1}{R_0} &\leq V\end{aligned} \tag{4.21}$$

The threshold vaccination rate for herd immunity is therefore:

$$V^* = 1 - \frac{1}{R_0} \tag{4.22}$$

It is clear that when the basic reproduction rate, R_0, is larger—that is, when the disease will tend to spread more rapidly—larger proportion of the population will need to be

Figure 4.12 The SIR model, in which τ is the transmissibility of the contagion and γ is the recovery rate.

vaccinated to avoid an outbreak. If we combine this information with what is known about the transmission rates of certain diseases in high density areas (e.g., where R_0 will tend to be highest), we can calculate the minimum vaccination rate needed for herd immunity for different diseases. If vaccine rates are high enough, the disease will fizzle out via recovery before it can infect a large number of individuals. If the proportion of vaccinated individuals drops below V^*, however, an outbreak can occur in which most or all of the unvaccinated individuals will contract the disease. This is why vaccination cannot be viewed merely as a choice for individuals to make. Vaccination only provides herd immunity—which is especially important for at-risk people who either cannot be vaccinated or for whom vaccines are ineffective—if sufficiently high numbers of people vaccinate.

Calculations of quantities like R_0 and V^* should be treated with caution, however, because they stem from simple models that make a lot of strong assumptions. For example, the models we have examined assume that individuals don't change their behaviors when they are infected, that infected individuals are instantly contagious, that vaccines are perfectly effective, and, most importantly, that populations are perfectly well mixed. Incorporation of these and other factors into a model's assumptions may change the conclusions one can draw from that model. We will return to this discussion of model complexity at the end of the chapter. For now, we will continue extending our simple model for one last hurrah.

4.6 Staying Better: The SIR Model

The SIS model assumes that a recovered individual can immediately become infected once again. This isn't the case for many contagions. Antibodies may provide immunity. More pessimistically, infection may lead to death or removal from the population by other means. In the case of innovations or behaviors, an individual may abandon them for good.

In the SIR model, individuals can be in one of three states: susceptible, infected, or removed.[6] Susceptible agents can become infected at a rate proportional to the transmissibility τ, and infected agents become removed at the recovery rate γ, just as in the SIS model. The main difference is that, in the SIR model, removed individuals remain removed and cannot again become infected (Figure 4.12).

The SI, SIS, and SIR models belong to a family of models known as **compartment models**, because they can be viewed as keeping track of flows between different "compartments," which represent the different states in which individuals can be found. With only two compartments, as in the SI and SIS models, systems can be modeled using a single difference equation, because individuals who are not in one compartment are necessarily in the other one. With three compartments, we need at least two equations, with the third compartment implied by the proportion of the population not in the other two compartments. For clarity,

[6]The "R" sometimes stands for "recovered," with the understanding that recovered agents are immune. However, I think this too easily leads to confusion between the SIS and SIR models, so I have chosen to use "removed," which is also consistent with most common usage in mathematical epidemiology. In its most general sense, removal from the population of susceptible or infected individuals can occur via recovery with immunity as well as death.

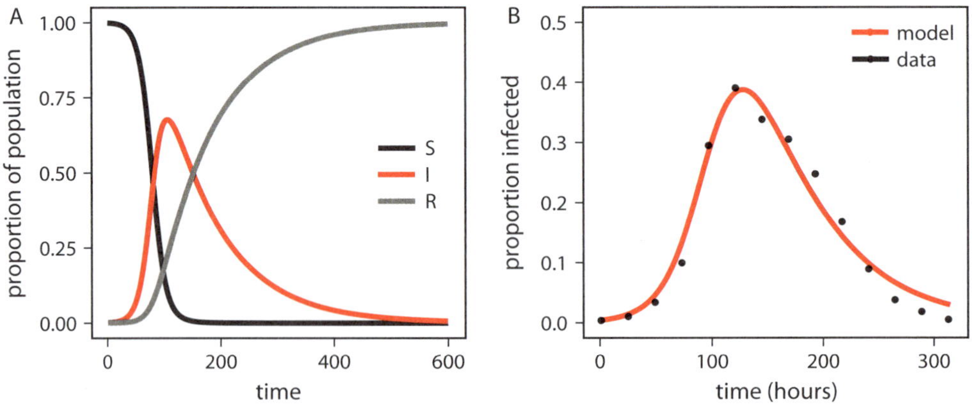

Figure 4.13 (A) Dynamics of the discrete-time SIR model from numerical simulation, using $\tau = 0.1, \gamma = 0.01$. (B) Fit of an SIR model to data from an influenza outbreak in a British boarding school in 1978.

I will use three equations. I have chosen to show difference equations rather than recursions, in order to place the focus on how the size of each compartment changes over time:

$$\Delta S = -\tau S \frac{I}{N} \tag{4.23}$$
$$\Delta I = \tau S \frac{I}{N} - \gamma I$$
$$\Delta R = \gamma I$$

The number of susceptible individuals decreases by the expected number who encounter an infected individual, times the probability of transmission. The number of infected individuals increases by this same number and decreases by the expected number of already-infected individuals who recover or are removed. And the number of removed individuals increases by the same number.

The only equilibria for this model are for everyone to either remain susceptible or become removed—infection is an unstable state of being. As such, the number of infected individuals will first rise and then fall. I've numerically simulated Equations 4.24 with some arbitrary values for τ and γ; Figure 4.13A shows the path of the resulting epidemic.

The SIR model is foundational for mathematical epidemiology, because it is the simplest model that can capture the full time course of an epidemic. As mentioned, it and the other contagion models we've considered make the simplifying assumption of a well-mixed, closed population. This assumption will often fail to hold in a strong sense, though extensions can be made to the model to accommodate many common objections. However, the model does provide a good baseline. There's a well-known case in epidemiology of an influenza outbreak in a British boys' boarding school in 1978 (Murray, 1989). A boy came back sick from a holiday, and infections were tracked during which time boys neither entered nor left the school. As seen in Figure 4.13B, an SIR model can be made to very closely match the true path of the outbreak. There are, however, differences between the data and the model. One source of difference stems from the fact that the mathematical model allows for fractional individuals to be infected, which thereby increases the infection rate and prolongs the time course of the epidemic. This type of assumption is reasonable in a

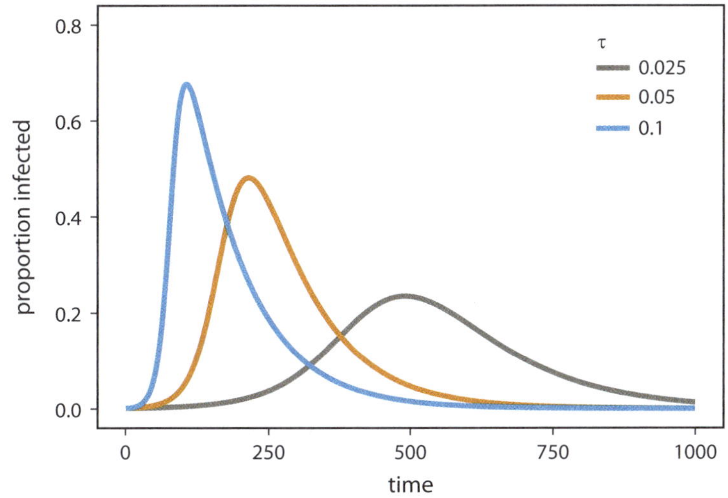

Figure 4.14 Temporal dynamics of infection in the analytical SIR model for different values of transmissibility, τ. In all cases $\gamma = 0.01$.

very large population but leads to issues in a relatively small population like a school (which had a total enrollment of 763). Agent-based models avoid this problem, but at the cost of lowered mathematical tractability.

4.6.1 Flattening the Curve

Starting in 2020, the COVID-19 pandemic brought considerable attention to the importance of "flattening the curve," referring to a reduction in transmissibility in order to decrease the maximum number of individuals infected at any given time. This allows more time for public health researchers to develop treatments, and also decreases the chance that health care facilities will become overwhelmed with too many patients. The SIR model helps us to understand exactly what those advocates were talking about. Consider Figure 4.14, which plots infection trajectories for three different values of τ, calculated via numerical simulation of the SIR recursions.

It should be clear that measures that reduce effective transmissibility are a good thing, all things being equal. It *should* be clear. However, it is not always easy to imagine how various factors really do influence transmissibility. Individuals are not a monolith. Rather, each of us operates in our own private world, aware primarily of our own existence and that of our closest social relations. It can be difficult to see how our own behavior contributes to the whole. In the case of adopting measures to reduce transmissibility, the assumption of homogeneity has another glaring mismatch with reality, which is that public health measures are almost never adopted uniformly or instantaneously. It is difficult to capture these heterogeneities in an analytical model. For a complementary approach, we will once again turn to our agent-based model.

4.6.2 The Agent-Based SIR Model

The NetLogo code for this model is **contagion_SIR.nlogo**. We need only to make very minimal changes to our code for the SIS model. In fact, it will be useful to be able to easily toggle back and forth between the two models, since their only difference is whether or not recovered individuals become immune to further infection. This is generally a good strategy when working with a family of related models—it is advisable to create code where different model assumptions can be toggled by varying the value of certain parameters. In this case, we'll introduce a global Boolean variable called `remove-recovered?`. When this is true, we'll be working with the SIR model, in which recovered agents are "removed" and so become immune to further infection. When it is false, we'll revert to the SIS model in which recovered agents return to a susceptible state. The SIR model does not require the addition of any new global parameters.

Previously, we added an agent-level Boolean variable that tracked whether or not the agent was infected. To this, we'll add another agent-level Boolean variable that tracks whether an uninfected agent can become infected—I've called it `immune?`. When this is true, the agent is immune and cannot become infected. At initialization, `immune?` will be false for all agents. In NetLogo, the declaration of the agent-level variables looks like this:

NetLogo code 4.8

```
turtles-own [
  infected?
  immune?
]
```

Other than the addition of the `immune?` variable, the initialization of the SIR model is identical to the SIS model. The stages of the model dynamics are also the same: first, susceptible agents can be infected; second, infected agents can recover; third, agents are recolored based on their infection status; and fourth, the agents move. The devil, however, is in the details, and we will need to update the first, second, and third stages to account for removed agents. Let's start with recoloring. Since color is used not only for visualization but also as information to guide agent behavior, we need a third color to signify removed status. I have chosen to make removed agents grey. Here is our simple procedure to recolor agents, now only slightly less simple:

NetLogo code 4.9

```
to recolor
  ask turtles [
    ifelse infected?
    [set color red]
    [ifelse immune?
      [set color grey]
      [set color white]
    ]
  ]
end
```

The procedure to infect susceptibles is exactly the same as before, but we now have to exclude removed agents from the set of potential new infections. Here is the code to accomplish this:

NetLogo code 4.10

```
to infect-susceptibles
  ask turtles with [not infected? and not immune?] [
    let infected-neighbors (count other turtles with
     [color = red] in-radius 1)
    if (random-float 1 <  1 - (((1 - transmissibility) ^
      infected-neighbors) * (1 - spontaneous-infect)))
    [set infected? true]
  ]
end
```

Similarly, we have to update the recovery procedure so that agents who recover become immune, but only if we are using an SIR model. Recall that we introduced a Boolean variable, `remove-recovered?`, in order to toggle between the SIR and SIS models. When this Boolean variable is true, we need to set recovered agents to become not only uninfected, but also immune.

NetLogo code 4.11

```
to recover-infecteds
  ask turtles with [infected? and color = red]
  [
    if random-float 1 < recovery-rate [
      set infected? false
      if remove-recovered?
      [set immune? true]
    ]
  ]
end
```

That's it. You should confirm the model works as expected, and play around with it. Notice that there are sometimes a few agents who never become infected. The number of such agents should increase as the effective transmissibility of the infection decreases and as the recovery rate increases. With this in mind, I want to return to our earlier discussion about flattening the curve.

Recall that in the analytical model, τ is really the *effective* transmissibility, which combines the probability that an interaction occurs with the probability that it leads to a transmission. In the ABM, these factors are explicitly separated. The probability that contact leads to a transmission event is captured by our `transmissibility` parameter. The probability of that contact occurring in the first place emerges endogenously from agents' movement strategies (controlled by the speed and turning angle parameters) and the population density (controlled by the population size and the size of the space). This helps bring into focus the fact that many things contribute to the spread of a contagion by virtue of their contributions to its effective transmissibility. As an example, let's focus on the parameter `speed`. I ran several simulations of the SIR ABM, holding all parameters fixed except `transmissibility` and `speed`. Figure 4.15A shows that, as expected, reducing direct transmissibility can flatten the curve. The probability of interactions stays the same, but the probability of an interaction leading to a new infection is diminished. You can think of this as akin to wearing face masks that block the spread of airborne droplets. Figure 4.15B shows that, even without reducing direct transmissibility, reducing agents' speed—which in turn

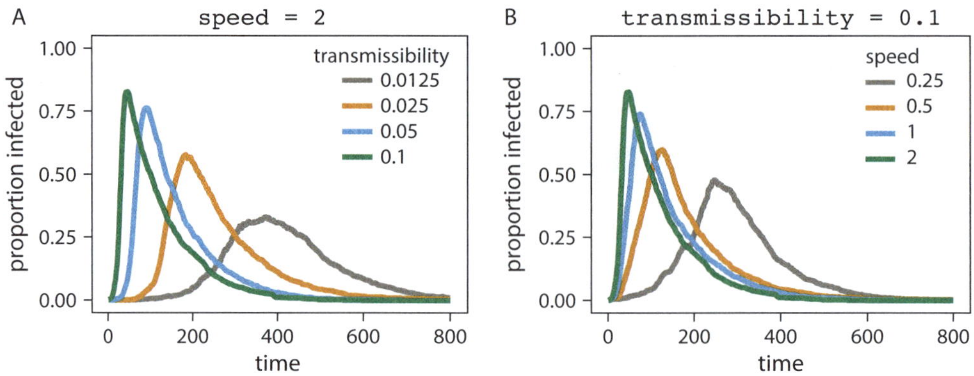

Figure 4.15 Temporal dynamics of infection in the agent-based SIR model. (A) Reducing transmissibility flattens the curve, holding speed constant. (B) Reducing speed flattens the curve, holding transmissibility constant. In all cases, `num-turtles` = 1000, `init-infected` = 1, `turning-angle` = 360, `recovery-rate` = 0.01.

reduces the number of new individuals an agent interacts with in a given time period—also flattens the curve. Think of this as akin to social distancing and avoiding crowds. It is left as an exercise to the reader to show that when both transmissibility and speed are reduced, the effect can be additive, but also that when adoption of these measures is sufficiently low, the effects can be muted.

BOX 4.1: Coupled Differential Equations

We have modeled the change in the relative sizes of the S, I, and R compartments under the assumption of discrete time units. This assumption is intuitive and allows us to analyze complex social dynamics using simple algebraic manipulation. However, dynamical systems are often modeled over *continuous* time, using **differential equations**. If you have ever studied physics or had a course in differential calculus, you will be familiar with the idea that a differential equation can specify a function for the instantaneous rate of change in some variable for any arbitrary time t. What makes differential equations powerful is that one can define the rate of change in a variable as a function of its own current value. When the differential equations defining how multiple quantities change over time include one another as variables, the equations are said to be *coupled*. For example, our SIR model can be rewritten as a set of three coupled differential equations like this:

$$\frac{dS}{dt} = -\tau S\frac{I}{N} \tag{4.24}$$

$$\frac{dI}{dt} = \tau S\frac{I}{N} - \gamma I$$

$$\frac{dR}{dt} = \gamma I$$

Note that the right side of each equation is exactly the same as in the discrete-time difference equations, but now each defines the *instantaneous* rate of change at any time t.

What do we get out of this? Well, other than being pretty, differential equations lend themselves to a number of useful analysis techniques, and can sometimes even yield closed-form analytical solutions by which we can obtain the state of the system for any arbitrary time t without numerical simulation. Consider our simple model of spontaneous adoption. The differential equation defining this system is:

$$\frac{dI}{dt} = \alpha \left(N - I(t)\right) \tag{4.25}$$

Knowledge of differential equations allows us to obtain a closed-form solution for the infection rate:

$$I(t) = N\left(1 - e^{-\tau t}\right) \tag{4.26}$$

This means that if we know that a system is governed by Equation 4.26 and we know the initial conditions, we can precisely predict the infection rate at any other time. Coupled differential equations can also generate cyclical or even chaotic dynamics. Their use is worth mastering.

Despite their value, I have avoided the use of differential equations throughout this book. I do this for several reasons. First, one of the goals of this book is to make the skills and ideas presented here available to those with minimal mathematical training. Second, social behaviors are often best represented as discrete events rather than as things unfolding in continuous time. And third, discrete time is the default in most agent-based models, so connecting analytical models to agent-based simulations is more intuitive using discrete-time models. For an excellent introduction to the mathematical and numerical analysis of dynamical systems, with a heavy emphasis on differential equations, see Strogatz (2015). For innovative social science approaches, see Turchin (2003).

4.7 Reflections

This chapter has provided a brief introduction to a very large domain of modeling. The field of mathematical epidemiology is built on sophisticated extensions to the sort of compartment models we examined here, as well as on the use of agent-based models that incorporate things like geography, network topology, demography, and life history. The use of contagion models to study behavior change and the diffusion of innovations is also widely used in sociology, marketing, and other areas of computational social science.

The first draft of this chapter was begun in 2018, when contagion seemed like an important topic with which students of human social behavior should engage. I then revised this chapter throughout 2020, 2021, and 2022, while the world was reeling from the COVID-19 pandemic. Understanding contagion now seems absolutely paramount. The treatment it has been given here is very introductory. There are many books written on the mathematics of disease contagion alone,[7] and there is a growing literature on its relation to "epidemics" of information and misinformation.[8] Nevertheless, I hope that I have provided you with a firm

[7] For example, see Bjørnstad (2018).

[8] For example, see O'Connor and Weatherall (2019).

foundation from which to learn more elsewhere. Indeed, this is a primary goal for many of this book's chapters.

Contagion is a complex phenomenon, and the models presented here are, as usual, radical simplifications. We can and should probe more deeply at multiple levels. Here are a few noteworthy limitations that can and should be addressed by more advanced models. First, individuals vary in their ability to transmit and adopt contagions, as well as their ability to recover or become reinfected. Second, individuals do not meet up at random, but live and move in structured populations. Indeed, people move both within and between communities in nonrandom ways. Third, individuals are affected by cultural norms, socioeconomic circumstances, and identity. This affects who they interact with and who they learn from. All of these factors (and others) affect how contagions spread.

In the models I have presented I have treated the spread of behaviors and innovations as perfectly analogous to the spread of disease, but they are not really the same. Ideas are not discrete packets that are transmitted whole cloth between minds. Information is multidimensional, multimodal, and influenced by a host of cognitive and social forces. Engaging with some of that complexity is the topic of the next chapter, where we will encounter the spread of continuous opinions.

4.8 Going Deeper

Contagion is a rich topic, and our simple models barely scratch the surface. Here are some brief descriptions of further directions to be explored when considering contagion.

The models we considered in this chapter assumed that interactions are either well mixed or, at minimum, random, with contagions spreading perfectly and instantaneously. In reality, interactions are often constrained by physical space, institutions, and social relationships. How interactions are structured can play an important role in the spread of a contagion. One of the most important developments in the study of social interactions has been the rise of network theory beginning in the late 1990s. This has given rise to both the subfield of network epidemiology (Danon et al., 2011) as well as the study of behavioral contagions on networks (Young, 2006).

Many contagion models, even those considering structured populations, assume that the ability or opportunity to transmit or contract a contagion is unaffected by one's infection status or by variation in informational states. In reality, these all interact. Disease state may influence behavior, as when sick individuals stay home and refrain from mixing with others. Intelligent organisms respond to new information, further influencing disease transmission. Contagion of a disease might interact with contagion of the *knowledge* of that disease. We may avoid overtly sick individuals. Awareness that an epidemic is spreading may further alter behaviors in ways that can either impede or hasten the spread of a disease (Funk et al., 2010). Social influences on behaviors like vaccine refusal (Mehta and Rosenberg, 2020) or mask wearing (Smaldino and Jones, 2021) might also spread like contagions. **Coupled contagion** models, which incorporate both behaviors and diseases, can be useful to study the simultaneous spread of multiple phenomena when each influences the spread of the others (Bedson et al., 2021).

Another aspect to consider, particularly in terms of informational or behavioral contagions, is that not all individuals are influenced in the same way by the same people. You may like a product, especially when your friends also adopt it (producing synergies like reduced costs and coordination benefits), but you may also dislike adopting the product if it becomes associated with a group that you do not identify with. In marketing experiments, subjects disadopted or rated poorly products that were associated with outgroup individuals, even if those outgroups were not particularly disliked (Berger and Heath, 2007, 2008). If the probability of adoption is positively influenced by ingroup adoption but negatively influenced by outgroup adoption, a number of outcomes are possible. These include delayed adoption by one group, suppressed adoption (in which one group initially begins to adopt and then fully disadopts), or polarization, in which products are associated with different groups in different regions (Smaldino et al., 2017; Smaldino and Jones, 2021).

The models we considered in this chapter assume that contagions spread through direct exposure. Some pathogens are transmitted through other means, such as by sexual contact or by vectors like mosquitos. For beliefs and behaviors, adoption may require multiple exposures to the contagion from multiple "carriers." There is a large and growing literature on the variety of social learning strategies used across contexts for the adoption of adaptive behaviors (Laland, 2004; Kendal et al., 2018), and these are sometimes incorporated into contagion models under the banner of **complex contagion** (Centola and Macy, 2007; Eckles et al., 2019). The dynamics of such models can be quite different from those in the "simple" contagion models seen here. We will revisit contagion dynamics under different social learning strategies when we tackle networks in chapter 9. Other researchers have criticized the notion that the spread of cultural artifacts like beliefs and behaviors can be usefully modeled as contagions at all, and argue that incorporating more sophisticated cognitive mechanisms is needed to model social transmission; this is an exciting area of research at the boundaries of the cognitive and social sciences (e.g., Goldberg and Stein, 2018; Rabb et al., 2022; Falandays and Smaldino, 2022).

Finally, the models we considered in this chapter assumed that, for a behavioral contagion, exposure to one adopter was equivalent to exposure to any other. But there are good reasons why this won't always be the case. In an uncertain environment, conformity to the majority behavior can be beneficial (Boyd and Richerson, 1985). If the value of behaviors is opaque, copying successful or prestigious individuals can be a good strategy (Henrich and Gil-White, 2001). An extensive literature has considered the dynamics and evolutionary implications of various transmission biases, also called **social learning strategies** (Laland, 2004; Kendal et al., 2018), which may be important to include in models of behavioral contagion. Models have also indicated how empirical patterns in the rate of adoption of a particular behavior might be used to test for the psychological biases being used for adoption decisions, such as a bias for conformity (Smaldino et al., 2018a).

4.9 Exploration

1. **It's not my thing.** Consider the contagious adoption of consumer products. What factors influence the adoption of products or behaviors? How well do the SI or SIS models capture these contributing factors? If you have concerns, how might you extend or alter one of these models to address your concerns?

2. **Spontaneous adoption.** In most of the models considered in this chapter, we assumed that the only way to pick up an infection was to catch it from direct contact. This is true for most diseases, but for products or behaviors, an individual might also adopt spontaneously even in the absence of social influence. This indicates we might want to combine the spontaneous adoption and SI models.

   (a) Assume that once individuals are infected, they stay infected. However, susceptible individuals can become infected either spontaneously or by contacting an infected individual. Write a recursion equation that describes the change in adoption rate over time, such that susceptible agents spontaneously adopt with a probability α and become infected upon contact with an infected individual with probability τ.

(b) Using either numerical simulation or an agent-based model, compute the time it takes the infection to reach saturation in a population as a function of effective transmissibility (τ, in the range $[0.001, 1]$) for several values of α: $\{10^{-4}, 10^{-3}, 10^{-2}, 10^{-1}\}$. Plot the relationship between τ and time to saturation on a log-log plot. What pattern do you observe?

3. **Don't move.** Modify the SIS ABM so that infected individuals change their behavior and move differently from susceptible individuals. They either stop, or slow down and increase their turning angle (so that they minimize their movement). Explore variations on this assumption. What happens to the contact rate when infected agents restrict their movement? How is the trajectory of the contagion affected by these changes? Explain why this violates the assumptions of the analytical SIS model.

4. **3-2-1 Contact.** Starting with the SIS ABM, write new code and calculate the rate of contact in the SIS agent-based model. That is, determine the probability that a given agent interacts with another agent, and thereby has an opportunity to acquire the contagion, per unit time. The actual number of contacts will vary over time, so you will probably want to get an average.

5. **Zombie invasion.** Modify the spatial SI ABM to study how to handle the *ZOMBIE APOCALYPSE*. In this model, infected agents become zombies. A healthy agent bitten by a zombie becomes a zombie. Zombies move more slowly than uninfected agents, and eventually die if they go too long without feasting on human flesh. You will need to add at least two new parameters: one controlling the movement speed of infected agents, the other controlling the death rate of infected agents. Holding the movement behavior of healthy agents constant, investigate the extent of the zombie apocalypse (how many agents become zombies) as a function of the zombie movement speed and death rate.

   (a) Write up a formal description of the model, including your new parameters.
   (b) Write code for your new model, including any necessary outcome measures.
   (c) Perform batch runs to answer the research questions, and plot the results. How do you interpret your results?

6. **Flatten the curve.** Use the SIR ABM to demonstrate how reducing disease transmissibility and factors influencing mobility both serve to "flatten the curve" of infection over time.

   (a) Write code to produce a reporter called `prop-infected` that reports the proportion of agents that are infected when called.
   (b) Use BehaviorSpace to run simulations varying `transmissibility`, `speed`, and `turning-angle`. Plot the resulting dynamics of infection for at least two values of each variable. What do you observe?

7. **Thank you for your compliance.** Use the SIR ABM to study how compliance with social distancing measures influences disease transmission. Start with a population of $N = 500$ and one initially infected individual. Let `turning-angle` $= 360$ and `recovery-rate` $= 0.01$. Define two classes of behavior. *Safe* agents wear face

masks and practice social distancing, which we will model so that they move with `speed` $= 0.1$ and, if infected, transmit the disease with probability $\tau = 0.01$. *Unsafe* agents do none of these things. They move with `speed` $= 1$ and, if infected, transmit the disease with probability $\tau = 0.1$.

(a) Write code for a model that implements these changes. Include a slider for a variable called `prop-safe` that determines the proportion of agents who are safe (the rest are unsafe).
(b) Write code to track the maximum infection rate over the course of an epidemic. Plot the trajectory of the population's infection rate over time for several (at least 3) values of `prop-safe`, and report the maximum infection rate for each run. How much does widespread compliance seem to matter? Note that here, the infection rate refers to the proportion of agents who are infected, not to how that proportion changes over time.
(c) Use BehaviorSpace to run simulations to record the maximum infection rate, varying `prop-safe` between 0 and 1 in increments no greater than 0.1. Do this for at least 3 population sizes: $N = \{50, 200, 500\}$. Run at least 5 simulations per condition. Plot your results. What do you conclude about the importance of compliance and its relationship to population density?

8. **Vaccinating agents.** Modify the SIS ABM so that a fixed proportion of the population is vaccinated. Start with a single infection and consider whether the infection spreads or dies out (so that no agents are infected). Run a small batch of runs to consider the proportion of runs for each parameter condition in which the contagion failed to spread, varying the transmissibility and recovery rate for at least two values each. Do your simulation results match the analytical predictions?

9. **Herd immunity.** Using the analytical SIR model, consider an outbreak of a new disease. Given that some of the infected individuals may recover before they can spread it to others, we want to calculate the conditions for when the outbreak will or will not spread.

(a) Assume that when an outbreak is new, $1 - I/N \approx 1$. Show that the infection will spread only if $\frac{\tau}{\gamma} > 1$. This quantity on the left side of the inequality is sometimes called the "basic reproduction number," R_0.
(b) Assume we have a vaccine that works perfectly, so that anyone who is vaccinated is protected against the outbreak. However, vaccinating everyone is infeasible for a variety of reasons. Luckily, we don't need to vaccinate everyone to prevent a disease from spreading, even if $R_0 > 1$. If enough people are vaccinated, the disease will be unable to spread effectively before it dies out, having only infected a few people. This is known as *herd immunity*. How many do we need to vaccinate? It depends on R_0. Let V be the proportion of individuals vaccinated, so the change in the number of infected individuals at any time is given by

$$\Delta I = \tau\left(1 - \frac{I}{N}\right)(1 - V)I - \gamma I.$$

Show that the minimum proportion of the population that must be vaccinated for herd immunity is

$$V^* = 1 - \frac{1}{R_0}.$$

10. **Vaccine efficacy.** Consider the analytic SIS model with vaccines. Assume a proportion V of the population has been vaccinated, and that both interactions and vaccinations are uniformly distributed at random in the population. In the chapter, we showed that under the assumption of perfectly effective vaccines, the vaccine threshold for herd immunity is $V^* = 1 - \frac{1}{R_0}$. Now, let's assume that the vaccine has an efficacy e, meaning that the vaccine prevents infection with probability e. The probability that an interaction between an infected individual and a vaccinated susceptible individual leads to a new infection is therefore $\tau(1-e)$. In this case, the difference equation governing the rate of new infections is:

$$\Delta I = \tau I \left(1 - \frac{I}{N}\right)((1-V) + V(1-e)) - \gamma I.$$

    (a) Prove that, when the outbreak is new, the threshold vaccination rate for herd immunity is $V^* = \frac{1}{e}\left(1 - \frac{1}{R_0}\right)$.
    (b) Plot the equation for the threshold vaccination rate as a function of the efficacy, e. Use $\tau = 0.1$ and $\gamma = 0.05$. Plot both e and V^* in the range $[0.5, 1]$. What does the resulting curve imply for the relationship between herd immunity and vaccine efficacy?

11. **Complex contagion.** Adoption of some behaviors or technologies may require exposure to multiple "infected" individuals. Some sociologists have referred to this phenomenon as a "complex contagion." This differs from "simple" contagions in which contact with a single infected individual is sufficient for spread.

    (a) Create a new NetLogo model on a 121 × 121 grid, using patches only. Patches can be either infected or susceptible. Initialize the model so that all patches are uninfected, except for a 3 × 3 square of patches in the center of the grid. Assume each patch is in contact with its nearest eight neighbors (Moore neighborhood). Establish a slider for a parameter called `threshold` that varies from 1 to 20. At each time step, each uninfected patch that has at least `threshold` infected neighbors becomes infected (when `threshold` > 1, the contagion is considered "complex"). Is there a maximum threshold for the spread of infection? If so, why?
    (b) Plot the proportion of patches that are infected as a function of time for several values of `threshold`. How long does it take for the infection to completely diffuse through the population in each case, if indeed it *does* diffuse? (Use BehaviorSpace to run multiple simulations as needed.)
    (c) Create a switch for a Boolean variable called `asynchronous?`. If `asynchronous?` = true, patches can become infected and immediately infect other agents on the same time step. If false, all patches first determine if they will become infected on a given time step, then all of those who will change status (from uninfected to infected) do so at the same time. Does this switch change the spatial dynamics of the contagion?
    (d) The maximum threshold for the spread of infection may depend in part on the network connectivity—that is, how many other agents each is connected to. Let's introduce a more strongly connected network. Create a new reporter called `big-neighbors`, which when called by a patch returns the set of patches corresponding to the focal patch's nearest 24 neighbors (its Moore neighborhood

with $r = 2$; you may want to use the NetLogo primitive `patch-set`). Create a switch for a Boolean variable called `bigger-neighborhood?`. If this is true, agents will use these bigger neighborhoods instead of just their closest 8 neighbors. Initialize the simulation so that the initially infected agents form a 5×5 square. How does the network connectivity (8 vs. 24 neighbors) influence the maximum value of `threshold` regarding whether or not the contagion will diffuse?

5 Opinion Dynamics

> Yeah, well, you know, that's just like, uh, your opinion, man.
> —The Dude, *The Big Lebowski* (1998)

While our big brains allow us to figure out lots of things on our own, we are inherently social creatures. Much of what we know and believe is learned from others. The words you are reading now, you learned from others. The behaviors that govern your daily life, from crossing the street to brushing your teeth, you learned from others. The opinions and attitudes you hold, from politics to pop culture, from religion to fashion, were shaped by the opinions of those around you.

In this chapter we'll focus on the social dynamics of beliefs, attitudes, and opinions, including how they change in response to interacting with other people. The contagion models we studied in the previous chapter are also models of how social information spreads. A limitation of those models is the assumption that contagion is binary. You're either infected or you're not. You either adopt a belief (or a behavior or a product) or you don't. The binary aspect of contagion works well as a model for some situations but not for others. In particular, if we're interested in how opinions, beliefs, and attitudes change, we probably have to get a bit more nuanced than a simple yes/no question.

Opinions, beliefs, and attitudes are all somewhat different from one another, but for the sake of tractability here, I'll treat them equivalently and stick with the term "opinion" for the remainder of the chapter. We can view an opinion as a position on some issue, and formalize that position as a number that varies between two extremes. *Is cake delicious? Is a hotdog a sandwich? Is Batman really a hero? Should countries open their borders to unlimited immigration? Is the theory of natural selection true?* One can be ruthlessly for or against a particular opinion, but one can also be ambivalent or lean weakly for or against. Through social interaction, an individual holding some opinion can influence others to adjust their opinions and in turn be influenced to change his or her own.

Sometimes a diversity of opinion may persist, with many different views represented. If everyone comes to hold the same opinion on some issue, the population has reached a **consensus**. When individuals' opinions gravitate toward either of two extremes—and we can refer to those individuals as **extremists**—a population can become **polarized**. Our focus in this chapter will be on the dynamics of opinions at the population level, examining the conditions that lead to consensus, extremism, and polarization.

The view of opinions described above is a somewhat limited one. A few possible objections come to mind. First, opinions and beliefs are not fixed values but are instead complex cognitive artifacts that emerge dynamically from the interplay of memory, mood, and context. Second, opinions are not shaped solely by social influence but also by evidence. Whether Batman is really a hero should depend not only on the beliefs held by others in my social network, but also on my personal understanding of the nature of heroism and on which particular incarnation of the character we are discussing (and, I suppose, on whether a fictional character can truly be a hero). Third, not all opinions are equivalent in their consequences. Some have very immediate or socially important consequences (Is access to affordable healthcare a human right? Should I cut the red wire or the blue wire on this ticking bomb?!) while some are relatively inconsequential (Is ankylosaurus the coolest dinosaur? Is it a travesty that the sort of gastrically distressing doughy cheese casserole served in Chicago is inexplicably referred to as "pizza?"[1]).

The study of opinion dynamics is a relatively immature research area. Unlike models tackled in other chapters of this book—like those dealing with segregation, contagion, or cooperation—models of opinion dynamics have retained a somewhat more abstract relationship with empirical data. This may be partly due to the fact that while segregation, contagion, and cooperation involve directly observable patterns or behaviors, opinions are invisible, cognitive phenomena. To complicate things further, publicly stated opinions don't always align with privately held opinions (Kuran, 1995). However, just like those other topics, opinions are also essentially social. This means that progress in understanding opinion dynamics requires contributions from both social scientists and cognitive scientists to create models that best represent how opinions are represented, expressed, and influenced. It is worth keeping these and other caveats in mind as we proceed with our study of opinion dynamics models. The simple models we tackle in this chapter will help us to gain intuition about the social dynamics of opinions and beliefs, and will provide a foundation upon which one can eventually incorporate more complex instantiations of opinions, scaffolding the development of richer, more realistic models.

5.1 Building a Model of Opinion Dynamics

Many models exist that examine the dynamics of opinions among interacting individuals.[2] Many of these models, perhaps most of them, focus on two prominent social forces: **positive assortment** and **influence**. Positive assortment refers to the tendency for similar individuals to interact preferentially. Influence simply means that interactions between individuals can lead to nonrandom changes in their opinions. In this chapter we'll consider a series of models that make varying assumptions about assortment and influence. We'll examine how those assumptions affect the subsequent distributions of opinions in a population, paying particular attention to consensus, extremism, and polarization.

In our contagion models from chapter 4, contagion was represented by a simple binary trait (agents were either infected or not), social influence occurred by physical proximity (agents were infected by others nearby), and the population structure involved interactions

[1] In my opinion, yes and yes.

[2] Reviewed in Castellano et al. (2009), Flache et al. (2017), and Galesic et al. (2021).

Figure 5.1 A representation of agent opinions. Here, two agents differ in their assessment of whether ankylosaurus is the best dinosaur.

occurring at random. A model of opinion dynamics similarly requires assumptions of three categories:[3]

1. A representation of opinions.
2. A mechanism for social influence.
3. A population structure.

Let's examine each of these in more detail.

A representation of opinions

Real opinions are emergent products of a complex mind, or worse, complex interactions between complex minds. A model of opinions needs to be simpler to be tractable. We can imagine an opinion as a position on some issue. For example, let's consider the issue of whether ankylosaurus is the best dinosaur. An individual could be strongly in favor of this position (much like your author, dear reader). An individual not fully convinced could lean weakly in favor, or, having been misled, lean weakly opposed. An uninformed individual could be completely neutral, while a villainous scoundrel, probably a stegosaurus fan, might be strongly opposed.[4] We can represent these positions numerically. Imagine a number line where zero is the neutral point, so that positive numbers represent positions in favor of ankylosaurus, and negative numbers represent positions opposed, and maximally extreme positions are represented as positive and negative one. Figure 5.1 shows how two agents with differing opinions on this important issue can be represented by positions on the number line. Of course, individuals generally hold lots of opinions simultaneously.[5] In such cases, an individual's opinions can be represented by a vector, with each element representing an opinion on a different issue. This is only one way to represent opinions, and I encourage you to think about others.

[3]These considerations are discussed here in terms of opinion dynamics, but most of them apply to models of social interaction and influence more generally.

[4]While it's not relevant to this book in any way, I am constantly awed by the knowledge that the time separating the existence of ankylosaurus, which lived during the Cretaceous period, and stegosaurus, which lived during the Jurassic period, is approximately eighty-five million years—substantially more than the time separating humans from the last dinosaurs. We are but specks on the map of time.

[5]At minimum, there are a *lot* of dinosaurs to consider.

A mechanism for social influence

Agents don't just hold opinions, their opinions are informed and influenced by others. If two agents interact, *how* do they influence one another? That is, how do their opinions change as a result of their interaction? In our contagion model, an "infected" agent could transmit the infection to another agent through physical proximity. Discrete transmission doesn't really make sense for continuous-valued opinions, though. Before you read on, think about your position on some issue, and how it might change through interaction with someone whose opinion is either somewhat similar or quite different from your own. Psychologists and other social scientists have produced a large research literature on the factors involved in social influence,[6] and we can formalize some of those ideas here. The way we have modeled opinions somewhat constrains the sort of influence that is possible, because an opinion's value can only travel up or down the number line. With that constraint in place, agents whose opinions differ can influence each other in several ways:

- *Positive influence.* When agents interact, their opinions become more similar to one another. This could represent mutual persuasion or desires for conformity or conflict avoidance. In the simplest version, the influence is symmetrical (so that each agent's opinion moves toward the other's to the same degree), but it need not be.
- *Bounded confidence.* Also called biased assimilation. Individuals are positively influenced by those with whom they are already in substantial agreement, but are more skeptical of opinions shared by individuals with whom they differ. We can represent this with a threshold for influence. If agents' opinions are sufficiently different (as dictated by the threshold), they do not change in response to the interaction.
- *Negative influence.* Rather than ignoring one another, individuals who differ may actually grow more dissimilar after interacting. In other words, interactions with individuals whose opinions differ substantially from their own can push individuals toward more extreme versions of their initial views. The empirical support for this effect is mixed.

In this chapter we will explore models of opinion dynamics with all three of these forms of influence. Note that we use the terms "positive" and "negative" to refer to the directionality of influence and not to imply any moral valuation of that influence. Remember also that these are simplified, operational definitions of influence mechanisms. Our purpose in this chapter is not to evaluate the validity of these mechanisms but to examine the consequences of assuming their existence. That said, if the consequences of our assumptions are unrealistic, our results may be telling us something about the validity of those assumptions.

A population structure

What determines who interacts with whom? In the previous chapter, we saw that the rate of population mixing could influence the spread of a contagion. When mobility was low and agents tended to interact with the same individuals over and over again, contagions spread more slowly and sometimes simply fizzled out. When mobility was high and agents regularly interacted with new partners, contagions spread rapidly. In the opinion dynamics models we will consider in this chapter, we'll focus on two extremes. At one extreme, we'll consider

[6]In particular, there is a useful emerging literature on *social learning strategies* that enumerates how individuals learn from one another and when various strategies might be adaptive (Laland, 2004; Kendal et al., 2018).

well-mixed populations in which individuals pair up for interactions at random. At the other extreme, we'll consider a rigid, unchanging network structure, in which individuals only interact with their closest neighbors. This contrast will give us some intuition about how population structure influences opinion dynamics.

5.2 Opinion Dynamics under Positive Influence

Now that we have the broad strokes of our model's components, let's start building. We'll begin by constructing a simple model in which opinions change exclusively through positive influence. That is, every interaction between two individuals with differing opinions brings them closer together. You might think you know what will happen, and you are probably correct. Nevertheless, the positive influence model forms the basis for more complicated models in which your intuition may fail you, so it is worth starting out as rigorously as we wish to end up. It is also useful to have several competing models when one is trying to explain specific patterns in empirical data—this sort of explanation is discussed further in chapters 10 and 11.

Consider a population of N agents so that each agent i is characterized by a single opinion, x_i, which is a real number in $[-1, 1]$. For simplicity, we'll initialize our agents so their opinions are uniformly distributed across the range of possibilities. In other words, each opinion is initially equally likely. If we wish to incorporate population structure, we also have to make a choice about the nature of that structure. I have chosen to use one of the simplest possible structures: a fully occupied $L \times L$ square grid with toroidal boundaries. Fully occupied means that each cell of the grid contains exactly one agent, so that $N = L^2$. Each agent is fully characterized by its opinion and its spatial position.

The dynamics of this model are also very simple. Each tick of the simulation clock, two randomly chosen agents will interact, and their opinions will become more similar to one another. Exactly *how* shall this positive influence work? We will borrow a mathematical formalism from research on reinforcement learning, which has shown that the amount of learning about the value of a stimulus is often proportional to the difference between the initially predicted value and the subsequently observed value (Silvetti and Verguts, 2012; Sutton and Barto, 2018). Consider the two agents in question, characterized by opinions x_1 and x_2, and a learning rate γ, denoting the magnitude of social influence. We can bound γ in $(0, 0.5]$ in order to avoid the strange situation in which agents appear to swap opinions. The agents will update their opinions as follows:

$$\begin{aligned} x_1 &\leftarrow x_1 + \gamma(x_2 - x_1) \\ x_2 &\leftarrow x_2 + \gamma(x_1 - x_2) \end{aligned} \tag{5.1}$$

The left arrows denote an update rule and are simply another way to write recursions. Each agent updates its opinion by a fraction γ of the difference between its current opinion and that of its interaction partner. The model dynamics will continue until the agents are sufficiently close to an equilibrium. Because opinions are represented as real numbers, the system is never purely at equilibrium, but the average opinion change can get arbitrarily close to zero, and we can choose the model's stopping condition based on how close we think is close enough.

Before we code this model, let's think a bit about what will happen to the distribution of opinions after we start with an initial population uniformly distributed in $[-1, 1]$. In fact, the population will always reach a consensus near the initial mean of zero. Consider this verbal proof. The expected value of a randomly selected agent's opinion is the population mean, which is zero at initialization. Thus, the average agent will move toward this mean by an amount determined by its current opinion and the learning rate, γ. Although some agents will move away from the mean, many others will move toward it. The further away from the mean an agent is, the more likely they are to interact with another agent whose opinion is in the direction of the population mean. Moreover, agents with opinions further from the mean will take larger steps toward the mean than agents closer to the mean, because agents update their opinions in proportion to the difference between their current opinion and the one they've newly encountered. In this way, the population's opinion variance shrinks. Because there are no asymmetries in terms of how opinions change when they are above versus below the population mean, that mean will remain static. As the process repeats, the variance will continue to shrink while the mean will remain the same. Eventually, all opinions will get arbitrarily close to the initial mean, though the exact value of this convergence will vary slightly between individual simulations due to stochasticity.[7] When positive influence is the only force at work, the population will always achieve universal consensus.

We will now move on to coding the model. Although it's pretty clear what will happen, having the code allows us to extend the model and add additional complexity. It also provides us with an opportunity to talk about how to visualize the model dynamics.

5.2.1 Visualizing the Model

There are usually many ways to visualize the dynamics of an agent-based model. Although the opinion dynamics model we've described does not need to be spatially explicit, it is often valuable to build into one's code the *opportunity* to extend the model to a spatial version, and (spoiler alert) that is what we will do. Having agents represented visually as individuals also provides the opportunity to use color to indicate their opinions. I have chosen to color agents black for an opinion of -1 and white for 1, with shades of gray in between.

Color can provide us with some nice intuition about opinion change, but these values are also imprecise and don't necessarily communicate much about the dynamics of opinion change. An elegant solution is to plot the opinion of each agent on the y-axis of a graph, with time represented on the x-axis. When extended to variations of the opinion dynamics model, this sort of scatterplot will allow us to see the emergence of phenomena such as consensus, clique formation, and polarization. An alternative way to convey information about the distribution of opinions is with a histogram. One histogram won't provide information about dynamics, but it provides a nice redundancy concerning agent opinion positions. In the NetLogo code, I have provided all three displays (see Figure 5.2).

5.2.2 Coding the Model

The NetLogo code for this model is **opiniondynamics_posinfluence.nlogo**. The model is quite simple. I have used a 21×21 lattice for $N = 441$ agents. The only other global parameter we need to worry about is the learning rate, γ, designated in the code as `learning-rate`. Each agent keeps track of its own `opinion`, which we add in the Code window as an agent-level parameter.

[7] For a mathematical proof of consensus for a class of similar models, see DeGroot (1974).

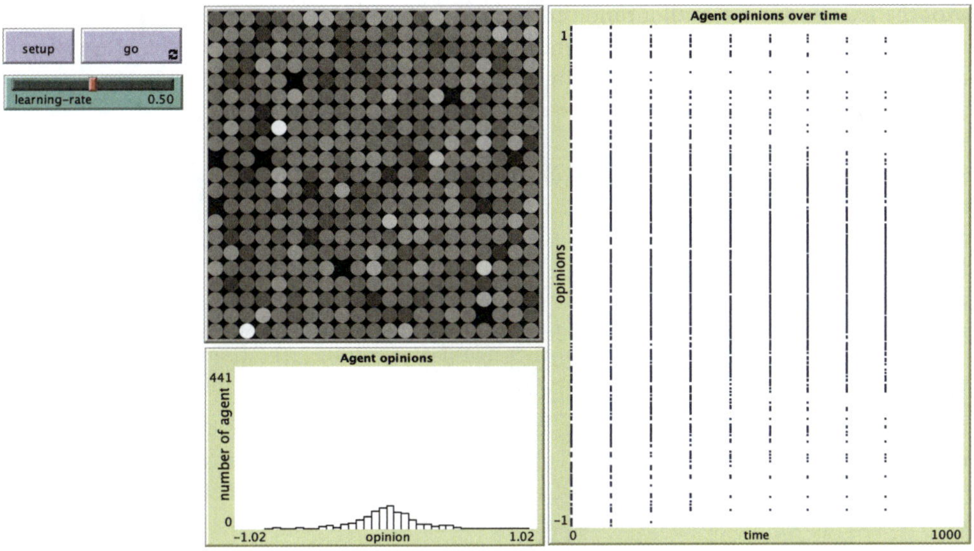

Figure 5.2 NetLogo Interface tab for the model of opinion dynamics under positive influence. Three representations of agent opinions are shown: a spatial lattice with shaded agents (top left), a histogram of agent opinions (bottom left), and a scatterplot of agent opinions over time with values shown every 100 ticks (right).

NetLogo code 5.1

```
turtles-own [opinion]
```

Initializing the model is straightforward. On each patch of the grid we create an agent, which is then assigned an opinion from a uniform distribution in $[-1, 1]$. Like most programming languages, NetLogo has a random number generator that can produce a random floating point number between zero and some upper bound. To produce a number in the desired range, we first generate a random number between zero and two, and then subtract one from the result. The following code is run for each newly created agent:

NetLogo code 5.2

```
set opinion (random-float 2) - 1
```

I've opted to have the agents appear as circles. Once opinions are assigned, the color of each agent is updated to reflect its opinion. This is similar to how we recolored agents in the contagion models in the previous chapter. However, in this case we need to assign a color based on a continuously varying opinion. Most programming languages have various ways to represent colors numerically. In NetLogo, black is represented by zero and white is represented by 9.9, with the numbers in between representing increasingly light shades of grey. We therefore need a function to map opinions onto colors. We can represent the required function mathematically for opinion x as:

$$\text{color} = \frac{x+1}{2} \times 9.9, \tag{5.2}$$

with the corresponding NetLogo procedure being:

NetLogo code 5.3

```
to update-colors
  ask turtles [
    set color (opinion + 1) * 9.9 / 2
  ]
end
```

On to the dynamics. Each time step, one agent is selected at random, and its opinion is recorded. That agent then chooses another agent at random to be its interaction partner, whose opinion is also recorded. The two agents then update their opinions using Equation 5.1, and update their colors accordingly. This dynamic is fully captured by the go procedure below:

NetLogo code 5.4

```
to go
  ask one-of turtles [
    let x1 opinion ;;my opinion
    let other-turtle one-of other turtles
    let x2 [opinion] of other-turtle ;;other agent's opinion
    let x1-new (x1 + learning-rate * (x2 - x1))
    let x2-new (x2 + learning-rate * (x1 - x2))
    set opinion x1-new
    ask other-turtle [ set opinion x2-new]
  ]
  update-colors
  tick
end
```

Finally, we need to set up our additional visualizations: our scatterplot and histogram. Creating these involves a little more effort than the simple line graphs we used in the previous chapters, but I like to think if a thing's worth doing, it's worth doing right. Again, NetLogo makes it quite easy to add dynamic plots, and you should examine the provided code to see how it's done. The finished product is shown in Figure 5.2.

The scatterplot is the most informative display, particularly for the well-mixed model. The x-axis is time and the y-axis is the agent opinions, so each point represents a particular agent's opinion at a particular point in time. In order to reduce the memory demands and the size of the resulting data file, I've chosen to plot new points only every 100 time steps. This is easy to do using the modulo operator, which in NetLogo is mod. If you are unfamiliar with this operator, now is a good time to familiarize yourself with it and with **modular arithmetic** more generally, because it is very handy for coding all sorts of things. Modular arithmetic is sometimes called "clock arithmetic," because adding numbers "wraps around" from the highest number back to the lowest, as with an analog clock (though modular arithmetic usually assumes a starting index of zero rather than one). The modulo operator essentially returns the remainder from integer division. For example, $426 \bmod 100 = 26$, because after removing all the 100s from 426, only 26 remains, while $13 \bmod 3 = 1$, because after eliminating the largest integer multiple of 3 less than or equal to 13 (that is, 12), we are left with 1. The relevant plotting command is:

NetLogo code 5.5

```
if (ticks mod 100 = 0) [
   ask turtles [
      plotxy ticks opinion
   ]
]
end
```

Some information *is* lost in the scatterplot. In particular, the points are small and completely opaque—NetLogo does not allow you to manipulate the visual properties of the points (this is one of the many reasons I have replotted much of the model output in this book using more flexible software[8]). As such, it's not always easy to see the extent to which opinions cluster around particular values. We can recover this information with a histogram. You can examine the model code to see how this is accomplished in NetLogo. As you run simulations of this model, you can see how the histogram transforms dynamically from a uniform distribution to a single central peak.

With working code and visualization in hand, try running the model. Confirm that the model always reaches a consensus close to zero. Consider what happens when the learning rate is increased or decreased. This model implies that if all interactions involve only positive influence, a population will always achieve a global consensus in the long run. Clearly, that is *not* the case, as disagreements in opinion often persist. Moreover, different communities often converge on different opinions and beliefs. Is there a simple mechanism that can produce distinct communities that share distinct opinions?

5.3 Bounded Confidence

Here we'll extend the positive influence model by adding **bounded confidence**. The Bounded Confidence (BC) model was introduced in a 2000 paper by Deffuant et al., but its structure is similar to ones used in other models of social influence (Carley, 1991; Axelrod, 1997b; Mark, 1998; Hegselmann and Krause, 2002). In this model, agents who are sufficiently dissimilar simply ignore one another. The implication is that when we interact with someone with a fairly similar opinion, our opinion and theirs become more similar to one another, just as in the previous model. However, if someone holds an opinion that is wildly different from our own, then we ignore them—neither our opinion nor theirs changes. Perhaps we don't see their opinion as relevant or trustworthy.

There is some empirical justification for bounded confidence (also called biased assimilation), suggesting that people may find arguments that come from similar individuals more compelling or persuasive (Lord et al., 1979; Dandekar et al., 2013). Of course, two people may be broadly similar without necessarily being similar on some particular opinion. We will address this concern later in this chapter. For now, let's incorporate bounded confidence into our model.

Formally, we extend our positive influence model by adding a new global parameter: the confidence threshold, d, which you can think of as the minimum *difference* in opinions to warrant social influence. Consider again two agents with opinions x_1 and x_2. The agents update their opinions according to Equation 5.1 if and only if the absolute difference

[8]For the curious, I have used ggplot in R.

between x_1 and x_2 is strictly less than d. Otherwise, the two agents ignore one another. The BC model's initialization and dynamics are otherwise identical to the positive influence model. As we code up the model, think about how this new assumption is likely to change the dynamics and the resulting patterns of opinions in the population. Agents will only be influenced by other agents whose positions in opinion space are close to theirs. Will the population still reach consensus?

5.3.1 Coding the Model

The NetLogo code for this model is **opiniondynamics_BC.nlogo**. To our positive influence model, we add a new global parameter for the confidence threshold, d, coded as `confidence-threshold`. The only other change needed is to add a conditional statement to our `go` procedure, instructing the agents to update their opinions only if the difference between their opinions and those of their interaction partners is less than this threshold. If this condition is not met, the agents do nothing. That's it.

Actually, that's not it. Let's add one more thing. In the previous model, we only examined the case where agents interact at random. However, in many cases individuals do *not* interact at random. Their interactions are constrained by their social networks and physical locations. We can adopt an extreme version of this constraint by taking advantage of the lattice structure we have already coded for our population. Instead of choosing interaction partners at random from the entire population, we will add a spatial condition to our model in which interaction partners are restricted to an agent's closest four neighbors (i.e., from their von Neumann neighborhood). To do this, we can add a global Boolean variable, `spatial-interactions?`, which will allow us to toggle between the spatial and nonspatial versions of the model. We can then alter our `go` procedure once more so that when `spatial-interactions?` is true, the focal agent selects an interaction partner from among the agents in its von Neumann neighborhood. Try to imagine what will happen when social influence is restricted to local neighbors. The `go` procedure now looks like this:

NetLogo code 5.6

```
to go
  ask one-of turtles [
    let x1 opinion
    let other-turtle one-of other turtles
    if spatial-interactions?
    [set other-turtle one-of other turtles-on neighbors4]
    let x2 [opinion] of other-turtle

    if (abs (x1 - x2) < confidence-threshold) [
      let x1-new (x1 + learning-rate * (x2 - x1))
      let x2-new (x2 + learning-rate * (x1 - x2))
      set opinion x1-new
      ask other-turtle [ set opinion x2-new]
    ]
  ]
  update-colors
  tick
end
```

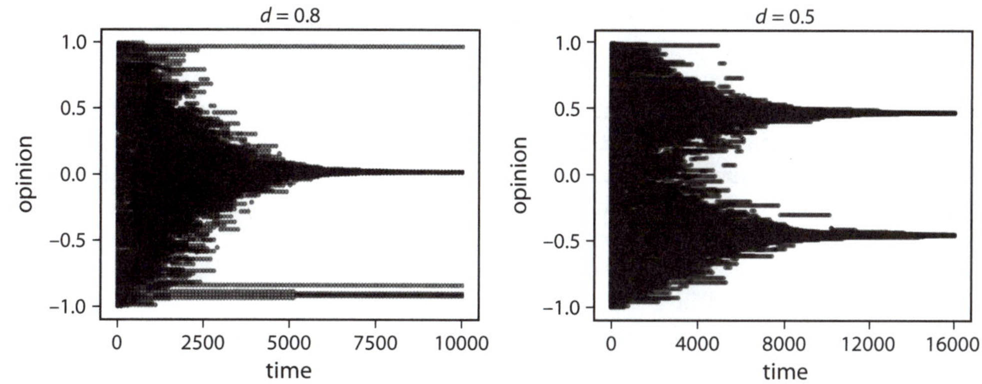

Figure 5.3 Temporal dynamics of opinion change under bounded confidence for $d = 0.8$ and $d = 0.5$. In both cases, $N = 441$ and $\gamma = 0.3$.

5.3.2 Analyzing the Model

Recall the power of play! Play around with the model and see what happens, starting with the nonspatial version. First, you may have noticed that the learning rate has little influence other than on the time it takes the model to converge. For this reason, I'll focus on the effect of the confidence threshold, d. When the confidence threshold is large ($d \approx 1$), we end up with consensus around a moderate opinion, just as in the positive influence model. This is because everyone is sufficiently close in opinion to enough agents that the population is still pulled together, even though agents close to opposing extremes initially ignore one another. As the confidence threshold decreases, however, consensus is no longer guaranteed. A little smaller (say, $d = 0.8$), and we start to see the persistence of a few "extremists." These agents can become isolated, so that no one influences them or is influenced by them. As individuals become increasingly selective and only let themselves be influenced by increasingly similar others, we start to see the emergence of distinct "cliques"—clusters of agents who all hold the same opinion, though not necessarily the population's average opinion (Figure 5.3).

When agents only care about similar opinions, the population organizes into several distinct cliques of like-minded members. From playing around with the model, you should observe that the number of cliques tends to increase as the confidence threshold decreases. What exactly is the relationship between the confidence threshold, d, and the number of emergent cliques? Does a particular value of d always lead to the same number of cliques? To investigate these questions, we need a way to measure the number of cliques. We could simply look at the results of each simulation one at a time and manually count the peaks. However, it would be much better to have a way to automate this process so we can run large batches over a range of the parameter space.[9] Perhaps we could simply count the number of distinct opinions in the population at the end of a run? This won't work; recall that small differences will remain even among members of a clique due to the continuous nature of opinion space and on the way that opinions are updated. Instead, we'll develop an algorithm to count the resulting clusters of opinions, to be implemented once a simulation run has converged to a number of distinct peaks.

[9] Also, come on, we're trying to do science here.

We'll start out by ignoring extremists, which we can define as those with opinions below $-1 + d/2$ or above $1 - d/2$. These agents will tend to have few other agents within their confidence interval—they will be loners, and not really part of cliques.[10] Next, we'll sort the opinions of our remaining agents, ordering them according to their value, and find the smallest opinion above $-1 + d/2$. Call this lowest opinion value x_{min}, the lowest value in our first clique. The number of cliques we've found so far is $c = 1$. Since any opinion less than d away from x_{min} would have been influenced by it, we can then look for the existence of an opinion that is larger than $(x_{min} + d)$ but smaller than $1 - d/2$. If one is found, we increment c by 1 and set x_{min} to the value of the new opinion we found. We repeat this until the algorithm fails to find anymore opinions. In NetLogo, the algorithm to count cliques is implemented as follows, as a reporter called `num-cliques`. We can call this reporter in batch runs to record the number of cliques that emerge in each run.

NetLogo code 5.7

```
to-report num-cliques
  let cliques 1
  let d confidence-threshold ;;shorten name for convenience.
  let min-value min [opinion] of (turtles with
    [opinion > (-1 + .5 * d)])

  while [any? turtles with [opinion > (min-value + d) and
    opinion < (1 - (.5 * d))]] [
      set min-value min [opinion] of (turtles with [opinion >
        (min-value + (.5 * d))])
      set cliques (cliques + 1)
      ]
  report cliques
end
```

I ran simulations of the model long enough for cliques to emerge (80,000 time steps for $\gamma = 0.5$) for a range of confidence threshold values, running 30 replications for each value of d. The results are plotted in Figure 5.4A. The relationship we intuited from playing with the model becomes quite clear. For larger values of the confidence threshold (around $d > 0.65$), the population still converges to universal consensus. As individuals restrict the set of individuals by whom they'll be influenced to those who are already similar—with smaller and smaller d—the population organizes into more and more distinct cliques. Thus, by simply paying attention to (or, equivalently, interacting with) only those who are already similar, populations can maintain discrete communities demarcated by gaps in existing opinion values.

Another thing to notice is that although the number of cliques is strongly correlated with the confidence threshold, it is not fully determined by it. For a given confidence threshold, there is a range to the number of cliques that emerge. What determines this number? And why do agents end up in one clique or another? It comes down to two sources of stochasticity in the model: the precise distribution of opinions in the initial population and the order in which individuals interact. Nevertheless, the model dynamics still involve quite a bit of **path dependency**—their final positions being connected to where they started. Early

[10]You are welcome to re-run this analysis with the inclusion of extremists.

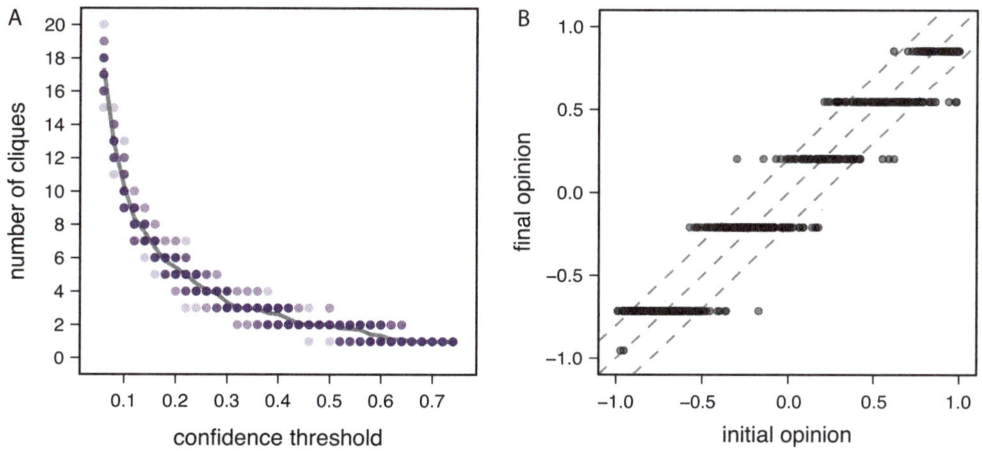

Figure 5.4 Results for the non-spatial BC model. (A) The number of cliques that emerge under bounded confidence as a function of the confidence threshold, *d*. Open circles are individual runs, the grey line connects the means across 30 runs for each condition. (B) Agents' initial opinions plotted against their final opinions from a single run of the bounded confidence model, using $d = 0.2$. The outermost dashed lines have their y-intercept 0.2 units above or below the origin. In all cases, $N = 441$, $\gamma = 0.5$.

interactions set the stage for the position of cliques and which agents will be influenced by which other agents. We can see this if we focus on a single run, for which I've used $d = 0.2$ as an example (Figure 5.4B). Here, five cliques emerged (plus a very small sixth clique of two extremists near -1). Agents' final positions were strongly correlated with their initial positions, but were hardly determined by them. Because of stochastic events, some agents barely changed their opinions from their initial values, while others changed theirs quite a bit. Moreover, two agents who started right around the same initial position could wind up in totally separate cliques.

5.3.3 The Spatial BC Model

So far, we've assumed that interactions occur at random in the population. However, the entire gambit of social network analysis—and indeed, one of the driving motivations in the social sciences—is that interactions are *not* random but rather occur based on social ties. Consider, for example, the lessons from chapter 3 regarding how local neighborhoods can become segregated, and then think about how spatial and network sorting can be based on opinions and beliefs as much as (or more than) ethnicity.[11] The square lattice is admittedly not the most realistic network structure, but it is among the easiest to model computationally, and it can provide us with some intuition for how structure influences social dynamics. So now let's consider simulations for the spatial version of the BC model.

Playing around with the model, we see that for larger confidence thresholds ($d > 0.65$), the results are very similar to the non-spatial version of the model in that we usually get consensus. A noteworthy difference from the non-spatial model is the increased likelihood of

[11]For example, Bishop (2009) documents how the United States has become increasingly segregated by political affiliation.

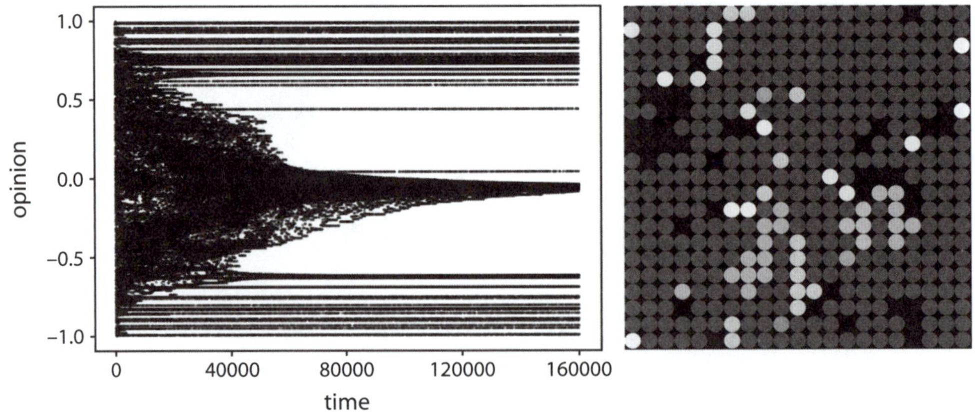

Figure 5.5 Left: Temporal dynamics of opinion change for the spatial bounded confidence model, for $d = 0.5$, $N = 441$ and $\gamma = 0.5$. Right: Corresponding spatial distribution of agents.

seeing "extremists" who maintain opinions close to ± 1. It also takes considerably longer for the population to reach consensus, because opinions must percolate through local channels. For smaller d, the dynamics can look quite different from the non-spatial model. Instead of several cliques of roughly the same size, we often see one big cluster dominate the network, while a number of isolated agents or small clusters with differing opinions can be maintained if the agents in these clusters have no neighbors with similar opinions (Figure 5.5).

In this model, much greater diversity of opinions can persist in the population than in the non-spatial version, as many people will have only a few local individuals who are sufficiently similar to them for influence to occur. As such, opinions cannot easily spread through the network. So, if we assume strong network structure, we get two seemingly contradictory consequences in comparison with the non-spatial model: more broad consensus in the population among the majority, but more small factions with extreme or idiosyncratic opinions.

Both the non-spatial and spatial versions of the BC model show how group differences can persist even if all influence is positive, as long as having sufficiently different opinions results in a lack of influence. Next, we'll consider the case of negative influence, in which sufficiently large differences are exacerbated.

5.4 Negative Influence

There is some support for the claim that when people are exposed to the opinions of other people whom they view as different, they are negatively influenced such that their opinions and attitudes become *more* different. There are at least two intuitive ways to extend the BC model to include negative influence. First, we could restrict negative influence to those pairs of agents who are very different indeed, such as those who differ by at least $1 - d$. There is a nice symmetry to this proposal, but it has a conceptual problem. It means that when individuals are less tolerant (and so are only influenced by those to whom they are very similar), there will simultaneously be a smaller range of people by whom they are negatively influenced. An alternative is to assume that all interactions lead to either positive or negative influence, with positive influence by those whose opinions differ by

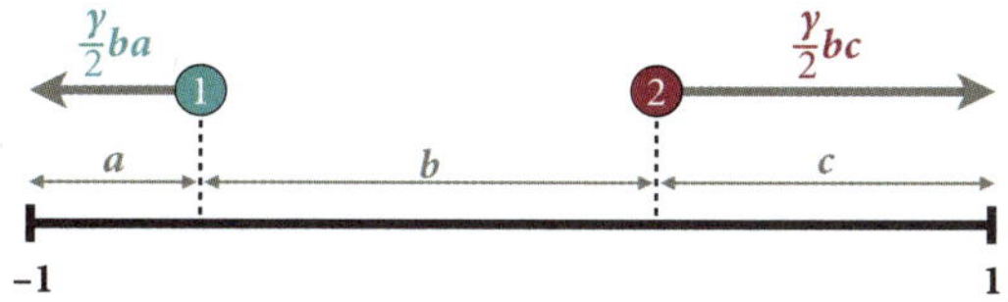

Figure 5.6 Negative influence. Agent 1 has an opinion of $-1+a$ and Agent 2 has an opinion of $1-c$, with the difference between them equal to b. The colored expressions indicate the amount each agent updates its opinion.

a magnitude smaller than d, and negative influence otherwise. For simplicity, we'll start with this assumption. Indeed, I will focus here on the symmetric threshold of $d=1$, so that an agent is influenced both positively and negatively by roughly half the range of possible opinions.

If the difference between two agents' opinions is within the confidence threshold, the model works as before, with positive social influence as dictated by Equation 5.1. However, if their opinions are sufficiently different—that is, if $|x_2 - x_1| \geq d$—they will exert negative influence, becoming more dissimilar. To do this, we need to include provisions so that (a) a larger difference translates to greater influence, but also that (b) opinions cannot exceed the boundaries of $[-1, 1]$. We accomplish this as follows:

$$x_1 \leftarrow \begin{cases} x_1 + \frac{\gamma}{2}(x_1 - x_2)(1 - x_1), & \text{if } x_1 > x_2 \\ x_1 + \frac{\gamma}{2}(x_1 - x_2)(1 + x_1), & \text{otherwise} \end{cases}$$

$$x_2 \leftarrow \begin{cases} x_2 + \frac{\gamma}{2}(x_2 - x_1)(1 + x_2), & \text{if } x_1 > x_2 \\ x_2 + \frac{\gamma}{2}(x_2 - x_1)(1 - x_2), & \text{otherwise} \end{cases} \tag{5.3}$$

Consider Equation 5.3 carefully. The last parenthetical term in each expression is the distance to the extreme opinion value (± 1) that is in the opposite direction of their interaction partner's opinion. The agent's opinion moves a fraction of this distance. What fraction? Well, as before, it is scaled by γ as well as by the distance between the agents' current opinions. The maximum distance between opinions in this model is 2, and so we divide by 2 in order to bound the proportion of the distance to the nearest extreme in $[0, 1]$. In other words, due to negative influence, an agent's opinion will move away from the other agent's opinion by a fraction of the distance to the most extreme opinion in that direction, scaled by the learning rate and by the current difference between the two opinions (Figure 5.6).

5.4.1 Coding the Model

The NetLogo code for the model with negative influence is **opiniondynamics_neginfluence.nlogo**. It is easily adapted from the BC model described previously. In order to be able to compare the negative influence model with the BC model, we add a global Boolean variable, `repulsion?`, which indicates that a difference in opinion greater than d will lead to negative influence rather than inaction. When `repulsion?` is false, we recover the BC model. We then just need to update the `go` procedure to allow for negative influence, as per Equation 5.3.

NetLogo code 5.8

```
to go
  ask one-of turtles [
    let x1 opinion ;
    let other-turtle one-of other turtles
    if spatial-interactions?
    [set other-turtle one-of other turtles-on neighbors4]
    let x2 [opinion] of other-turtle

    ;;positive influence
    if (abs (x1 - x2) < confidence-threshold) [
      let x1-new (x1 + learning-rate * (x2 - x1))
      let x2-new (x2 + learning-rate * (x1 - x2))
      set opinion x1-new
      ask other-turtle [ set opinion x2-new]
    ]
    ;;negative influence
    if (abs (x1 - x2) > confidence-threshold) and repulsion? [
      ifelse (x1 > x2)
      [
        let x1-new (x1 + learning-rate * (x1 - x2) * (1 - x1) * .5)
        let x2-new (x2 + learning-rate * (x2 - x1) * (1 + x2) * .5)
        set opinion x1-new
        ask other-turtle [ set opinion x2-new]
      ]
      [
        let x1-new (x1 + learning-rate * (x1 - x2) * (1 + x1) * .5)
        let x2-new (x2 + learning-rate * (x2 - x1) * (1 - x2) * .5)
        set opinion x1-new
        ask other-turtle [ set opinion x2-new]
      ]
    ]
  ]
  update-colors
  tick
end
```

5.4.2 Analyzing the Model

Playing around with our simulations, we can observe that if agents are inclusive and accept positive influence from most other agents ($d > 1$), we end up with consensus even if some interactions do initially lead to negative influence. When the confidence threshold is just a little bit smaller (say, $d = 0.9$), consensus still usually occurs, with a few extremists persisting from time to time. However, due to negative influence, these few extremists can have a large effect over time. Occasional interactions with extremists push the rest of the population slowly but inevitably toward the opposite extreme. Eventually, even a small number of extremists can push everyone toward an extreme opinion if social influence is sufficiently widespread such that everyone can influence everyone else (Figure 5.7, left). This result is probably not realistic, but it is important to understand that it is the inevitable consequence of our model assumptions.

If the confidence threshold gets a little bit lower (e.g., $d = 0.8$), the population no longer reaches consensus, even a consensus of extremism, but is instead pushed rapidly toward polarization, with roughly equal numbers at both extremes (Figure 5.7, middle). This

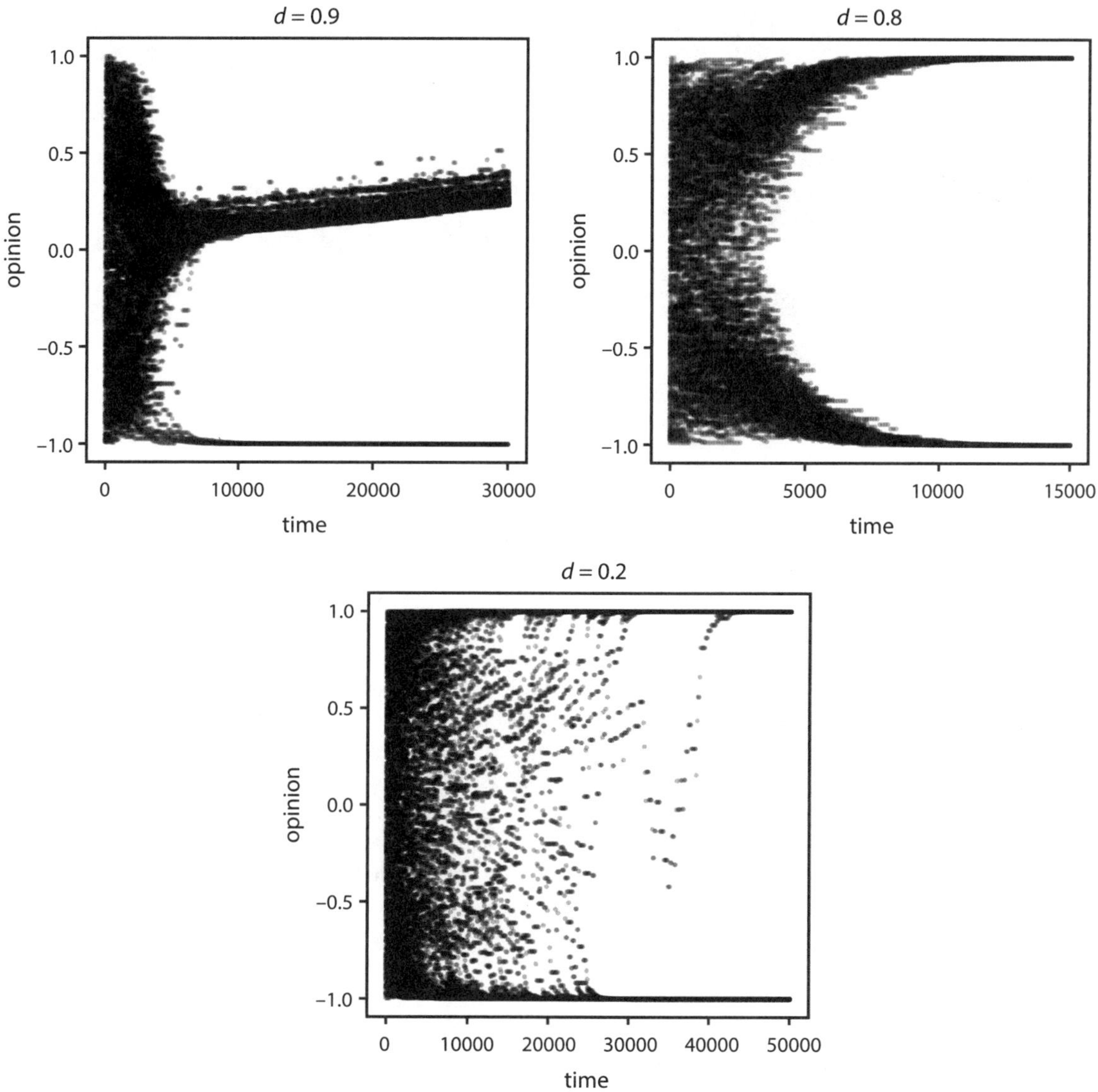

Figure 5.7 Example temporal dynamics of opinion change for the non-spatial model with negative influence, for different values of the confidence threshold, d. In all cases, $N = 441$ and $\gamma = 0.3$.

polarization occurs because every agent is repelled from more than half of the others and driven toward the rest. When the confidence threshold is very low (e.g., $d = 0.2$), each agent is repelled from most other agents. Here, something interesting occurs. The overall result is once again polarization to the extremes, but the path to polarization is the prolonged perseverance of many intermediate opinions, because those with moderate opinions are constantly pushed in both directions. This can even result in short-term oscillation for some agents. Nevertheless, the only stable state of the model is extreme polarization (Figure 5.7, right).

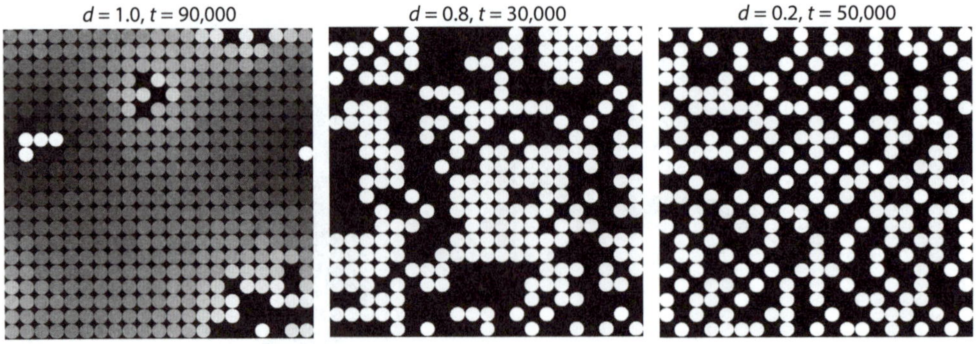

Figure 5.8 Examples of emergent demographic organization of agent opinions in the spatial model with negative influence, for different values of the confidence threshold, d. Agent color signifies their opinion: black is -1, white is 1. In all cases, $N = 441$ and $\gamma = 0.3$.

5.4.3 The Spatial Model with Negative Influence

Consider now the negative influence model with spatially constrained interactions. If we play with the model, we often observe the emergence of distinct demographic zones in which spatial clusters of individuals all share similar opinions. When d is close to one, this process occurs very slowly, intermediate opinions persist for quite a long time, and the emergent zones are quite large (Figure 5.8, left). For smaller values of d, the zones get smaller and polarization occurs more quickly (Figure 5.8, middle). When d is very small, polarization occurs once more, but individuals tend to differentiate themselves from their neighbors, as most agents are initially dissimilar. The result is a highly polarized population where, although most agents share similar opinions with many others, they also tend to be quite different from their spatial neighbors (Figure 5.8, right). Everyone is isolated, at least socially, in a community where few share their opinions. In this version of the model, everyone is essentially an anticonformist who has distinguished themselves from the others in their social network, and in so doing has ended up just like many others.[12]

Once again, this outcome is probably not realistic, but we must ask ourselves: Why not? One reason may be due to the simplistic ways in which opinions are modeled. As I noted earlier in the chapter, it is possible to be very similar to someone while still differing on a few particular opinions, just as it is possible for two people who mostly disagree to share certain opinions. Friends occasionally disagree, while enemies sometimes agree. Can we account for multiple simultaneous opinions in our model? What happens when we do?

5.5 Multiple Opinions, Polarization, and Extremism

The models of opinion dynamics we've looked at so far can be illuminating, particularly in terms of how simple assumptions about the nature of social influence among pairs of individuals can produce a variety of patterns at the population level. However, these models are

[12]This sort of "hipster effect," in which anti-conformists or distinctiveness seekers end up looking alike, has several known generative mechanisms; see Smaldino and Epstein (2015), Touboul (2019), and Golman et al. (2021).

Figure 5.9 Individuals simultaneously hold opinions on many topics.

also limited in a number of ways. A glaring one is the assumption that everyone is defined entirely by their position on exactly one issue, and that the way an individual will influence someone on that issue is determined solely by their position on that very issue. That's probably not the way most social interactions work. However, as I will continue to do throughout this book, I want to defend spending time with these unrealistic models. In order to understand the dynamics of complex systems, we need to understand how simpler systems work. Just as physics students start by analyzing systems without friction or air resistance, we have started with social systems governed by a single opinion. Although our end point in this chapter will still be a simplified system, we will see that things can get quite complicated just by adding a few more opinions.

People generally have opinions on more than one issue. They may have opinions not only about ankylosaurus, but also about the merits of keytars, soft serve ice cream, and democratic socialism (Figure 5.9). People also make judgments about their overall similarity to someone else based on an amalgam of *many* opinions and beliefs (among other factors), and generally do not completely reevaluate their relationships each time a new topic is discussed. Here, we'll extend our opinion model to allow agents to hold multiple opinions. There are many possible ways to do this; this is just one of them.[13]

Consider our population, as before, full of agents defined by their opinions and, perhaps, by their position in a social network. The difference now is that each agent has not one but K opinions. You can think of K as the number of opinions that matter for the majority of

[13]The multiple-opinions model studied here is based primarily on work presented by Flache and Macy (2011). The model in this section differs from theirs in terms of the network structure used and in allowing an agent only one interaction per time step rather than having simultaneous interactions with all neighbors. The results of these two models are qualitatively very similar.

social interactions.[14] In other words, an agent i is defined by an opinion vector of length K,

$$\vec{x}_i = (x_{i1}, x_{i2}, \ldots, x_{iK}), \tag{5.4}$$

so that x_{ik} represents the kth opinion of agent i. As before, each opinion is modeled as a real number between -1 and 1. If you are so inclined,[15] you can think of an agent's opinion as a position inside a K-dimensional hypercube, which we might dub "opinion space." An agent holding only extreme opinions (that is, only those with values of ± 1) occupies one of the corners of the hypercube.

There are still a number of simplifying assumptions in this new model that may generate some unease. All opinions are completely independent of one another, all opinions are weighted equally in assessing similarity, and all opinions are just as easily influenced. I urge you to hold on to this sense of unease, because it will come in handy when it comes time to draw conclusions about the real world based on our analysis of the model. For now, let us see what happens when we work with the model as given.

5.5.1 Coding the Model

The NetLogo code for this model is **opiniondynamics_multi.nlogo**. This code will build on much of the code for the previous models in this chapter. In one sense, the extension should be uncomplicated; we simply need to assign each agent multiple opinions. However, things actually do get quite complicated, because, as we shall see, mo' opinions means mo' problems. In the interest of clarity, I've interwoven the updates to the model description with the changes to the code required to execute those changes.

First, we need to introduce a new global parameter to reflect the number of opinions, K, which I've called `num-opinions`. We'll need to give each agent an opinion vector, which I've named `opinions`. This vector replaces our previous agent-level variable, `opinion` (singular), which was a simple scalar variable. Each element of `opinions` will be initialized with a random value in $[-1, 1]$. The code to initialize the model is as follows. It uses the NetLogo primitive `n-values`, which generates a list (NetLogo's version of a vector) of n values according to some set of instructions given in brackets.

NetLogo code 5.9

```
to setup
  clear-all
  ask patches [
    sprout 1 [
      set opinions n-values num-opinions [1 - (2 * random-float 1)]
      set shape "circle"
    ]
  ]
  update-colors
  reset-ticks
end
```

[14] Flache and Macy (2011) referred to K as the "cultural complexity" of the population, but this term strikes me as unnecessarily loaded.

[15] And you know you are.

Notice that we are still updating agents' colors based on their opinions, by calling `update-colors`. How does this work with more than one opinion? The short answer is that although it entails a severe loss of information, it can nevertheless be informative to know how agents vary on just one of their opinions. I have therefore adjusted the code so that an agent's color reflects the value of the first item in its `opinions` vector.

NetLogo code 5.10

```
to update-colors
  ask turtles [
    set color ((item 0 opinions) + 1) *  4.95
  ]
end
```

5.5.2 Social Influence with Multiple Opinions

With our model now initialized, it's time to talk about the new dynamics of social influence. Here is where things really get more complicated. When building a model of social behavior, it's often useful to articulate in words what you are trying to capture. Let us therefore assert that interactions between similar individuals should result in positive influence between the individuals' opinions, and that this positive influence should extend even to those opinions on which the individuals initially differ substantially. If your best friend has a wildly different opinion than you on something, you are likely to seriously consider their perspective. In contrast, you are less likely to be persuaded by a similar opinion held by your bitter enemy; in fact, you may even reconsider your view once you know it's shared by someone you dislike! It is clear that we need a way to quantify overall similarity that applies to vectors of opinions.

With just one opinion, the distance between two agents' opinions was simply the scalar difference in their values. With multiple opinions, we need something new. A simple approach is to take the average distance between each of their opinions.[16] We will define the *distance* in opinion space between agents i and j as the average absolute difference between each of their opinions:

$$D_{ij} = \frac{1}{K}\sum_{k=1}^{K} \left|x_{jk} - x_{ik}\right| \tag{5.5}$$

A pair of agents who agree on everything will have a distance of zero, while a pair who disagree on everything and are at opposite polar extremes of opinion space will have a distance of 2. This range $[0, 2]$ affords us a convenient way to think about positive and negative influence. Agents that are more similar than different will exert positive influence on one another's opinions, while agents that are more different than similar will exert negative influence. We can make this easier by translating the distance between two agents into an *influence weight* that will determine the direction and strength of influence:

$$w_{ij} = 1 - D_{ij}. \tag{5.6}$$

This weight will vary in $[-1, 1]$, with positive values indicating positive influence and negative values indicating negative influence.

[16] This measure is sometimes called the Manhattan distance. Other normalized measures of vector distance, such as Euclidean distance or cosine similarity, could also be used.

At each time step in the model, each agent will choose one other agent for an interaction. I've chosen to have *all* agents initiate an interaction on each time step. This is done primarily for computational efficiency—it is not substantively different from a model in which only one interaction occurs each time step. One at a time, each agent chooses a partner and calculates the influence weight between itself and its partner. Both agents then update their opinions accordingly. As before, positive influence causes opinions to become more similar, and negative influence causes opinions to become more differentiated. In this model, however, positive influence implies that *all* of the agents' opinions become more similar to one another, even the opinions that are initially quite far apart. Likewise, negative influence implies that *all* of the agents' opinions become more dissimilar, even the opinions that are initially quite close together. Being able to calculate the distance between agents will come in handy later when we analyze the model, so I have created a procedure to do just this called `trait-distance`, which takes as its arguments two agents and returns their opinion distance.

NetLogo code 5.11

```
to-report trait-distance [agent1 agent2]
  let t 0 ;;the index of the trait
  let total 0
  while [t < num-opinions] [;; loop over each trait
    set total total + abs (item t ([opinions] of agent1) -
       item t ([opinions] of agent2))
    set t t + 1
  ]
  report (total / num-opinions)
end
```

We can then calculate the influence weight between two agents by subtracting the distance from one.

After calculating the influence weight, we loop over each of the agents' opinions and influence them accordingly. Just as with our single-opinion model with negative influence, we must take care to keep the opinions in the range $[-1, 1]$. Rather than have two separate rules for positive and negative influence, we can generalize our influence rules to incorporate both simultaneously. For each opinion k of agents i and j, the opinions are updated as follows:

$$x_{ik} \leftarrow x_{ik} + \frac{1}{2} w_{ij}(x_{jk} - x_{ik})(1 - |x_{ik}|)$$

$$x_{jk} \leftarrow x_{jk} + \frac{1}{2} w_{ij}(x_{ik} - x_{jk})(1 - |x_{jk}|) \quad (5.7)$$

There are a couple of things to note with these equations. We have lost our learning rate, γ. Instead, the strength of influence is determined by the product of the influence weight, the distance between the opinions, and an opinion's distance from the nearest extreme value. Individual opinions that differ greatly will tend to exhibit a larger magnitude of change, but that magnitude is also scaled by the agents' overall similarity. In order to keep the opinions within the allowable range, we multiply the total change by the distance to the closest extreme. This is the final parenthetical term in each expression. The result is that more extreme opinions change less than more moderate opinions when faced with influence of the same weight. This assumption is probably reasonable—zealots and other committed individuals are more difficult to influence than those whose views are more ambivalent. Nevertheless, it is important to notice not only the explicit but also the implicit assumptions of a model's formalization.[17] The NetLogo code for the opinion update algorithm is below. Note that we loop over each opinion and apply the update rule in Equation 5.7 to each in turn.

NetLogo code 5.12

```
set t 0
while [t < num-opinions] [
  let x1 item t ([opinions] of self)
  let x2 item t ([opinions] of other-turtle)
  let x1-new (x1 + (0.5 * weight * (x2 - x1) * (1 - abs (x1))))
  let x2-new (x2 + (0.5 * weight * (x1 - x2) * (1 - abs (x2))))
  set opinions (replace-item t opinions x1-new)
  ask other-turtle[
    set opinions (replace-item t opinions x2-new)
  ]
  set t t + 1
]
```

5.5.3 Measuring Opinion Distributions

The existence of multiple opinions makes it a bit more difficult to quantify things like polarization and extremism. We can display *one* of each agent's opinions as their color, and we can also use a scatterplot to capture the relationships among agents for *two* of their opinions. We will find each of these useful for gaining intuition about the model dynamics (I have included such a scatterplot in the NetLogo code for the model). But such visual examination will not be sufficient if we are dealing with large vectors of opinions, where K can be much larger than one or two. The complexity of the model leads to a problem that social scientists have been dealing with for ages: the need for summary metrics.

In considering what sorts of metrics we might need, we can learn from our single-opinion models. We saw that opinions could reach consensus, or they could become polarized. Let's start there. Consensus is easy: it's when all the opinions are the same. But how do we characterize polarization among many issues simultaneously? There are many possible measures of polarization (reviewed in Bramson et al., 2017). One convenient measure of **polarization** is to use the variance among the pairwise distances between agents:

$$P = \text{var}\left(D_{ij}\right) \tag{5.8}$$

[17] My former graduate student Matt Turner has hypothesized that this "stubborn extremist" factor may explain the phenomenon of *group polarization*, in which communication among partisans causes the average opinion to shift toward the closest extreme; see Turner and Smaldino (2020).

Recall that variance is the expected value of the squared deviation from the mean—it is a measure of how dispersed a sample is. Variance has the nice property of being equal to zero when all agents agree on all opinions (that is, they have reached consensus), and of being maximized when agents' opinions are evenly spread among all possible extremes (that is, when there is maximal disagreement between agents).

The code to calculate the variance of the distances takes advantage of our procedure `trait-distance`, which reports the distance between two agents. In order to create a list of the distances between each pair of agents, we have to loop over each pair. If we simply looped over each agent and added its distance to every other agent to the list, we would end up counting each distance twice (the distance between *i* and *j* as well as the distance between *j* and *i*). To avoid doing this, we take advantage of the fact that each agent has an index denoting the order in which it was created during the model's initialization. NetLogo indexes agents automatically, but you may need to do it deliberately if you are coding in another language. We loop over each agent and have it consider only its distance to the agents with a larger index than itself, thereby avoiding double counting. Once we have our list, we simply calculate its variance. Methods for calculating basic statistics like variances are prepackaged with most program languages, but if yours does not have one they are relatively easy to produce on your own.

NetLogo code 5.13

```
to-report polarization
  let distances []
  let n 0
  while [n < count turtles]
  [
    let m (n + 1)
    while [m < count turtles]
    [
      let d trait-distance (turtle n) (turtle m)
      set distances lput d distances
      set m m + 1
    ]
    set n n + 1
  ]
  report variance distances
end
```

This metric will be useful, but it does not capture all the information we might want to know about the distribution of opinions in a population. In particular, a population in which every position is perfectly moderate (with an opinion value of zero) would exhibit the same low polarization as one in which every position was extreme, as long as they were all extreme in the same direction. It seems useful to differentiate between these cases. To do this, we can consider the proportion of agents with extreme opinions. For a preliminary analysis, I've simply operationalized **extremism** as the proportion of opinions in the population with an absolute value above 0.9, meaning they are much closer to an extreme than to the center. I've chosen this threshold somewhat arbitrarily; you are encouraged to see how the results change with different thresholds for extremism. We can keep track of extremism with the following code, which loops over each opinion of each agent, adds a count if the opinion is considered extreme, and then divides the count by the total number of opinions, *NK*.

NetLogo code 5.14

```
to-report extremism
  let total 0
  let count-hits 0
  ask turtles [
    let t 0
    while [t < num-opinions] [
      if (abs (item t opinions)) > 0.9 [set count-hits (count-hits + 1)]
      set total total + 1
      set t t + 1
    ]
  ]
  report (count-hits / total)
end
```

5.5.4 Analyzing the Model

As always, playing around with the model allows us to get a better sense of its dynamics and how they respond to changes in parameter values. The high dimensionality of opinion space makes this challenging, but our visualization scheme helps by allowing us to see the distribution of one opinion in the spatial organization of agents, and our scatterplot allows us to track how the first two opinions correlate and move toward either consensus or polarization. Batch runs that record long-run values for polarization and extremism over a wide range of parameter values will then allow us to characterize the behavior of the model more precisely.

Let's begin by considering the non-spatial version of the model, where interaction partners are chosen at random. When the number of opinions, K, is small, the population often becomes highly polarized with all opinions at extreme values, reminiscent of the single-opinion model with negative influence. When K is large, however, the system tends toward consensus. Why does the inclusion of more opinions cause the system to move from polarization to consensus? When there are more opinions, random pairs of agents are more likely to be similar overall—more on this later. Batch runs show that the strongest form of polarization, where equal numbers of opinions are at each extreme, is actually common only when $K = 1$ (Figure 5.10, left). As the number of opinions under consideration increases to $K = 2$, polarization declines. This is partly because there are more "corners" in opinion space (four in this case), which reduces the overall variance as opinions are spread among them. Extremism remains high, however, because all agents still hold extreme positions. When $K \geq 3$, we are much more likely to get convergence to the center. For a discussion of why this happens, see Box 5.1: Examining Model Assumptions.

The real world also has network structure, so let's now examine the spatial case in which agents interact only with their four closest spatial neighbors. As in the single-opinion scenario, the spatial model takes much longer to converge than the non-spatial model does, because opinions must percolate slowly through the network. Importantly, the spatial structure also allows a wider range of opinions to persist, and the population reaches neither full consensus nor full polarization. The right side of Figure 5.10 illustrates this. We still see a decrease in polarization and extremism as the number of opinions, K, increases, but the slope of the relation is less severe.

Why does the spatial model allow a wider range of opinions to persist? Figure 5.11 illustrates the dynamics of an individual model run with $K = 4$, in which the spatial and non-spatial constraints are contrasted in a single model run. For the first 1000 time steps,

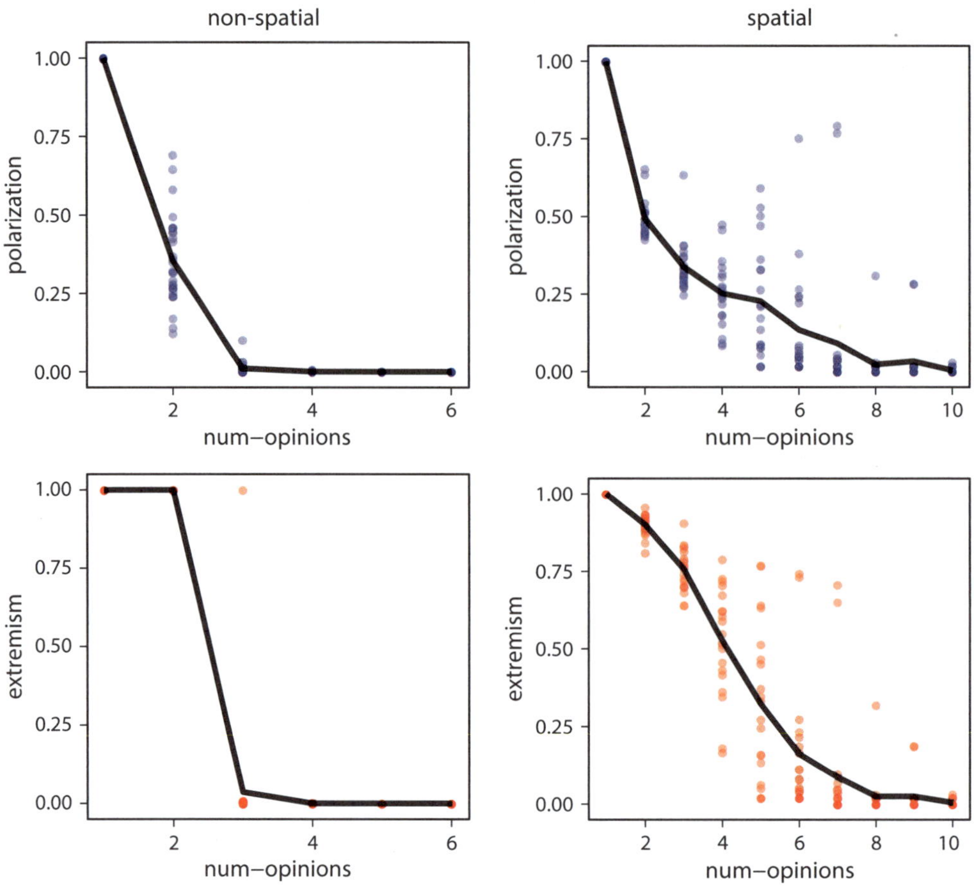

Figure 5.10 Polarization and extremism as a function of the number of opinions, K, for both non-spatial and spatial versions of the model. Circles are results from 30 individual simulations for each parameter combination, lines connect the means across runs. Simulations run to $t = 1000$, $N = 441$.

agents' interactions are restricted to their four closest spatial neighbors. The visualization of the first opinion shows what looks like the slow emergence of demographic zones of near consensus, though this is a flawed representation because only one of four opinions is depicted. Now consider the scatterplots showing the relationship between agents' first two opinions. These illustrate that a diverse range of opinions persists, because the spatial separation allows for a sort of dynamic equilibrium of push and pull to exist in the network. At $t = 1000$, the spatial constraint was removed, meaning that `spatial-interactions?` was flipped from true to false. At this point, the population rapidly organized into an equilibrium state with maximal extremism and high polarization (Figure 5.12). This might be viewed as illustrating the hypothesis that a population structured into small, tight-knit communities can enable the persistence of diversity in the population at large, while long-range communication of the type facilitated by the internet can foment polarization and extremism.[18]

[18]This result has been emphasized in other modeling work using more highly clustered network structures (Flache and Macy, 2011; Turner and Smaldino, 2018).

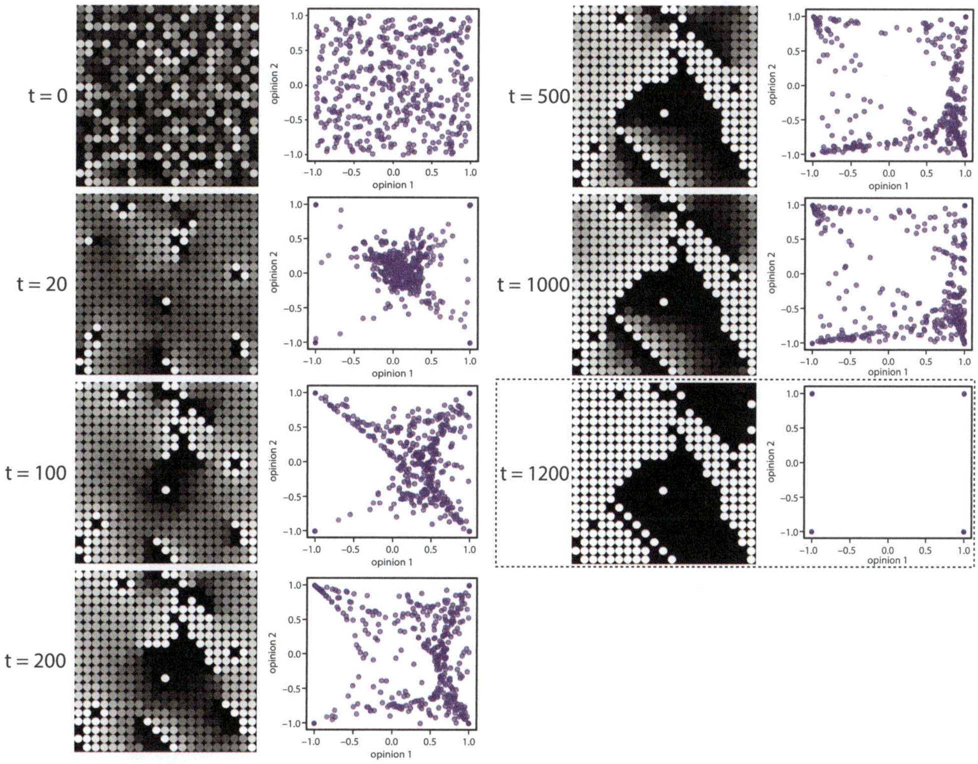

Figure 5.11 Example dynamics of the model with multiple opinions, in which interactions were spatially constrained until $t = 1000$, and then removed so that agents interacted with randomly chosen partners thereafter. Snapshots over time each depict two images. The left images show the spatial organization, with color indicating the value of agents' first opinion (black for -1, white for 1). The right images show a scatterplot of agents' first two opinions. Under spatial constraints, a diversity of opinions persist. Once interactions become randomly mixed, the population rapidly goes to an equilibrium with maximal extremism and high polarization. For these runs $N = 441$, $K = 4$.

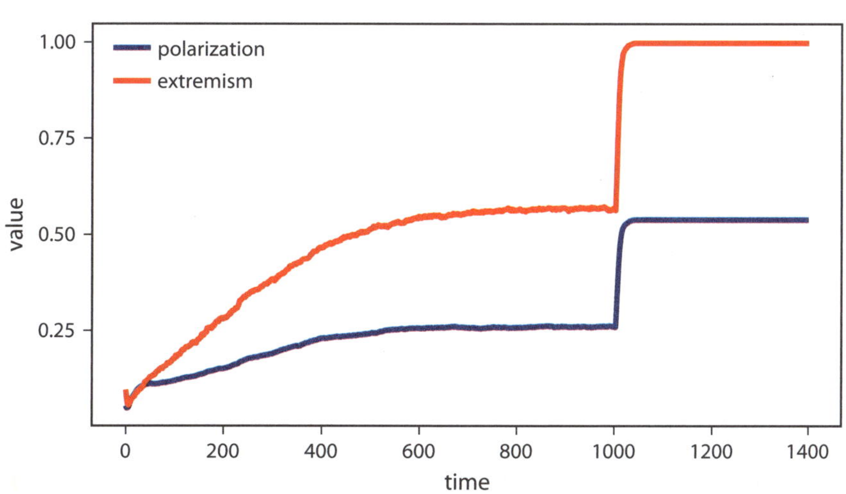

Figure 5.12 The dynamics of extremism and polarization for the model run depicted in Figure 5.11. Interactions go from spatially constrained to random at $t = 1000$.

BOX 5.1: Examining Model Assumptions

The conclusions one draws from analyzing a model follow wholly from the model's assumptions. When a model yields conclusions that appear interesting, it is our task as modelers not just to report those findings, but to explain them. When studying the world directly, as with experimental or observational research, it is often sufficient to report the mere existence of some phenomenon or relationship. With models, this sort of reporting is almost never sufficient.[19] The whole point of modeling is to examine the consequences of one's assumptions, so if a model generates an interesting result, we need to understand why it does so. Failure to explain one's results is an abdication of one's role as a modeler and a scientist. So let's examine one of *our* results in a little more detail in order to better explain it.

Analysis of the multi-opinion model presented above demonstrated that the magnitudes of both polarization and extremism decreased precipitously as the number of opinions, K, increased. Why? Polarization relies on negative influence—individuals' opinions growing further apart. In the model, this occurs only when the opinion distance between two individuals—defined as the average difference among all opinions—is greater than one. We assumed that opinions are initially independent and uniformly distributed in $[-1, 1]$. Two important consequences follow from this assumption that may not have been obvious.

First, let us ask: What is the expected distance at initialization between two agents when $K = 1$? You might think the answer would be one, but that isn't correct. This is because the opinions have an upper and a lower bound. For an agent with an opinion at ± 1, the expected distance from other agents *is* one, because all distances are equally likely when opinions are uniformly distributed. However, an agent with an opinion at zero cannot be more than one unit distant from any other agent, and has an expected distance from other agents of one-half. In fact, the expected distance between any two points in a line segment of length L is $L/3$. The proof of this is somewhat complicated, but you can get an intuition by considering that any two distint points on the line segment, excluding endpoints, divide the initial segment into three shorter line segments (see Figure 5.6; segment b is the distance we are calculating). Given that the combined lengths of these segments must equal L and that there are three segments, the expected length of segment b must be $L/3$ when averaging across all possible divisions. Because the space of possible opinion values in our model is a line segment of length $L = 2$ (from -1 to 1), the expected distance between any two opinions when $K = 1$ will be $2/3$, or roughly 0.667. In other words, 0.667 defines the median distance between any two opinions in our model. And therefore, because the model assumes that any distance less than *one* involves positive influence, the majority of initial influence will be positive.

Now, let us consider what happens when K increases. All opinions are still in $[-1, 1]$, and because the distance between two agents is just the averaged difference between each of their opinions, the expected distance between two agents remains the same: $2/3$. However, the variance in those distances does not stay constant. Another

[19] A possible exception is in the case of computer science experiments in which the ability of an algorithm to generate a result is often valuable regardless of whether one understands why it works.

way to look at distance is that it is the (normalized) sum of differences between uniformly distributed values. We have seen this sort of thing before, in chapter 2. The central limit theorem applies: as K increases, the distribution of distances will approach a normal distribution with a mean of $2/3$. And as more and more values are averaged, the standard deviation of the distribution will shrink. This in turn means that the expected distance between two random individuals is increasingly likely to be less than one as K increases, leading to a positive influence weight. Thus, when there are more opinions, more influence will be positive, resulting in more moderate, less polarized populations. This process is illustrated in Figure 5.13, which shows the distribution of initial distances for different values of K.

The question now becomes: Are our assumptions about the distributions of opinions and the calculation of distance reasonable? Maybe so. If individuals have more opinions to talk about, they are more likely to find common ground. Indeed, some political scientists have provided evidence suggesting that political polarization often tracks the increased correlations between opinions, which destroys the assumption of independence between opinions and makes larger distances in opinion space more likely (Mason, 2018). However, our assumptions may also lead to consequences that are less readily justified. For this reason, it is worth exploring as many consequences of a model's assumptions as possible, especially when they may influence outcome measures.

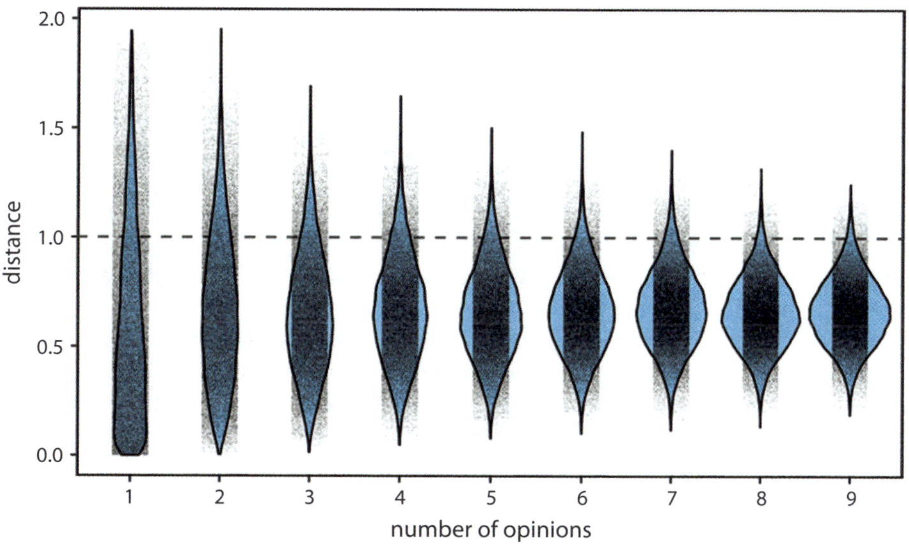

Figure 5.13 The distribution of distances between pairs of agents in the population when the model is initialized with $N = 200$, for different values of the number of opinions, K. As K increases, the number of dyads experiencing negative influence decreases, leading to more consensus and less polarization. For each value of K in this plot, $N(N-1)/2$ distances were calculated, represented here both by violin density plots and by the individual distances overlaid atop the violins.

5.6 Reflections

Among the approaches for modeling social influence and interaction that are considered in this book, the study of opinion dynamics is among the youngest and the least integrated with related theories of cognition, social behavior, and evolution. Deep considerations of the cognitive aspects of opinion representation and social influence in the models are still relatively rare, making integration of theory and empirical research difficult. For example, factors like reason, persuasion, or coalitional thinking are not well captured by the models we have looked at. Much of this chapter's value may lie in its efforts to help you examine the consequences of formal assumptions more generally.

I happen to think that the opinion models included here yield interesting dynamics and provide a good baseline upon which to construct richer models of opinions, beliefs, and social influence. There is a lot of open territory for the industrious scientist who wants to combine the cognitive science and sociology of opinions and beliefs. The dynamics of opinions is a very challenging system to model well because of the inherent difficulties in modeling both complex social interactions and complex cognitive phenomena. Moreover, real dynamics of opinions play out over multiple time scales, such that coherent or polarized populations interact with other populations and evolve over generations. Understanding these dynamics is extremely important for shaping and improving our world. To do this right, we need to know a lot more about the sociology of communicative interactions, the cultural evolution of beliefs, and the cognitive science of influence as it interacts with factors like language, categorization, bias, and identity. As of this writing, a formal understanding of this interplay is still nascent. That said, ever-increasing communication between social scientists, cognitive scientists, and cultural evolutionists is encouraging.

5.7 Going Deeper

The models examined in this chapter make very simple assumptions concerning how opinions are represented, communicated, and constructed. Although this modeling area is not as empirically grounded as some of the others addressed in this book, there are nevertheless several related traditions worth considering.

In the models we considered in this chapter, all opinion positions were equally likely and were influenced only through social interaction, perhaps representing exposure to ideas or even persuasion. However, not all opinions are equal. Some may be more useful or truer than others. Like, sometimes we believe things based on the things we see or think, rather than just based on what others tell us (I know, it's crazy, but I've heard that it happens). On the other hand, it is also clear that *misinformation* can spread perniciously on social networks. Some models of social influence have sought to address these factors (e.g., Zollman, 2013; Friedkin et al., 2016; Weatherall et al., 2020). A related consideration is the fact that an individual's true opinions may not be identical to what they communicate publicly. Although the interplay between public and private opinions has not been modeled extensively, some preliminary work suggests that complex dynamics can arise (e.g., Duggins, 2017).

We considered the case where people are motivated to become more similar to those to whom they are already similar. However, there are also cases in which individuals may want to actively differentiate themselves from others, whether to stand out from the crowd or to more easily find well-matched social partners (Brewer, 1991; Berger and Heath, 2008). Several models have considered distinctiveness-seeking or anticonformist incentives in group formation and opinion dynamics, showing that individual preferences are not always well aligned with group-level outcomes (Smaldino et al., 2012), that consensus can emerge without any preferences for conformity (Smaldino and Epstein, 2015; Touboul, 2019), and even that such preferences can stabilize the influence of extremists (Weisbuch, 2015). People also live in complex, multidimensional social worlds. They interact with different people in different contexts. They use their identity to navigate their social relationships and to decide who to learn from and who to communicate with. These factors influence communication and opinion dynamics in interesting ways (e.g., Battiston et al., 2017).

Finally, the models explored in this chapter all assume that beliefs and opinions can be represented by a single number. This discounts the role of uncertainty and, conversely, confidence. We may assign different weights to different beliefs without either buying into them completely or discounting them entirely. As such, beliefs might be better represented as a probability distribution rather than as a point estimate. Updating can be performed by the accumulation of evidence using **Bayes' Theorem**, a topic we will explore in chapter 8. This approach also allows us to incorporate evidence from multiple sources, including direct experience as well as social interaction. This approach has been utilized by social epistemologists (philosophers interested in how information arises and spreads within and across communities) studying populations of agents solving a *multi-armed bandit* problem. This is a classic problem in machine learning, in which individuals can choose among various options, each of which yields a reward with a unique probability, and the task is to figure out which option is the best (Zollman, 2013; Wu et al., 2023). In these models, agents can sample their options through experience but can also learn from the choices of others and so accumulate evidence both directly and indirectly.

5.8 Exploration

1. **Space takes longer.** Experiment with adding spatial interactions to the one-opinion model with positive influence only.

   (a) Add code to define consensus as the condition when the difference between the maximum and minimum opinion values in the population is less than some threshold, $s = 0.05$. Confirm that constraining interactions in local space means the system takes longer to reach consensus than when the population is well mixed. Use a learning rate of $\gamma = 0.5$.
   (b) What happens if you vary γ? Use BehaviorSpace to run simulations that allow you to plot the learning rate against the time to consensus for $\gamma = \{0.1, 0.3, 0.5, 0.7, 1\}$. What does the relationship between γ and consensus look like? Does something fishy happy when $\gamma = 1$? Why?

2. **Look what we have here.** Modify the Interface for the BC model to better observe the model outcomes dynamically.

   (a) Add a monitor that displays the number of cliques at the current time, and a corresponding plot that charts this number over time.
   (b) Add a scatterplot that plots initial position against current position. Describe the dynamics of how the model converges for large and small values of d.

3. **All at once.** Modify the BC model code to add a global Boolean switch called `synchronous?`. When true, every agent will initiate an interaction to potentially

update its opinion on each time step. Otherwise, only one agent initiates an interaction on each time step. Plot the output from example runs when this variable is true and false. How do the runs differ?

4. **Bounded space confidence.** Consider the BC model with $d = 0.18$. How many cliques emerge when space is not considered (i.e., `spatial-interactions?` = false)? What about when space is considered? Describe the differences in results both qualitatively and quantitatively. How important do you think network structure is in the way that opinions coalesce in social networks?

5. **Negative space.** Study the non-spatial version of the one-opinion negative influence model to consider the value of d that leads to polarization. Perform batch runs of the model for $d \in [0.8, 1.0]$ in intervals of 0.02, running each simulation for 10,000 time steps and performing at least 1 run per condition.

   (a) Plot the distribution of opinions as a function of d. There are many ways to do this. For example, you could plot the minimum and maximum opinion values or the standard deviation of values. Describe these results in terms of how the confidence threshold allows for either consensus or polarization.

   (b) For the following values of d, run at least one simulation of the spatial version of the model: $d \in \{0.8, 0.9, 1.0, 1.1\}$. Report plots of the spatial distributions of opinions. Describe how spatial organization affects the model dynamics with respect to the results from part (a).

6. **Law of the excluded middle.** Modify the code for the one-opinion negative influence model by introducing a new Boolean switch called `ignore-middle?`. If false, the model remains unchanged. If true, then an agent i will experience positive influence from an agent j if $|x_1 - x_2| < d$, negative influence if $|x_1 - x_2| > (1 - d)$, and no influence (ignoring agent 2) otherwise. Play around with several values of d (`confidence-threshold`). How do the results differ when `ignore-middle?` is false versus when it is true? How do you rate the realism of either assumption? Can you think of another option that better represents how individuals are influenced by others' opinions based on similarity?

7. **Fooled by randomness.** Start with the multiple-opinions model with spatial structure. Use $K = 4$ opinions, though you may wish to experiment with other values.

   (a) Add a global variable called `prob-random` that will be the probability that, on any given time step, an agent chooses their interaction partner not from among their closest spatial neighbors but at random (as they do in the non-spatial model).

   (b) Conduct batch runs to measure the polarization and extremism that emerge from different levels of randomness. Use values of `prob-random` in $\{0, 0.05, 0.1, \ldots, 0.3\}$. What is the effect of randomness on polarization and extremism?

   (c) Now consider that the previous runs all started with random distributions of opinions. Instead, write new code to automatically run the model with no added randomness for 100 time steps, and then increase the randomness by 0.01 every 20 time steps. Plot the output from one or more example runs. What happens? Is this the same or different from starting the simulation with added randomness to the structure of interactions?

8. **Beginning with agreement.** In all the models discussed in this chapter, the initial distribution of opinions is uniform across the entire range of possible opinions. This means that extreme opinions are just as likely as moderate ones at initialization. However, it's also reasonable to assume that extremists will often be rare, and we can consider a moderate population in which extreme views are rare or absent. Implement this in the multiple-opinions model by changing the range of the uniform distribution for initial opinions to $[-a, a]$, where a is a global parameter in (0, 1). Use $K = 2$ (though you may wish to experiment with larger K). You may use either the spatial or non-spatial versions of the model. How does the value of a alter the emergence of polarization and extremism? Illustrate your answer by plotting data from multiple runs of the model.

9. **Come on feel the noise.** In the models described in this chapter, we assume that individuals communicate and perceive their opinions with perfect accuracy. However, this is rarely the case.

   (a) Modify the multi-opinions model so that noise is added to each opinion when distances are calculated. Operationalize noise so that every time noise is added, it is a new random draw from a normal distribution with a mean of zero and a standard deviation s (in NetLogo, you may wish to use the primitive `random-normal`).
   (b) Consider how noise affects polarization and extremism for different assumptions about the initial distribution of opinions (using the parameter a in Problem 7), using $K = 2$ (though you may wish to experiment with larger K). Using batch runs with at least 5 repetitions of each set of parameter values, plot polarization and extremism as a function of $s \in [0, 0.2]$ for both $a = 1$ and $a = 0.5$. You may use either the spatial or non-spatial versions of the model.
   (c) What do you observe? What conclusions, if any, can you draw from this analysis?

10. **What do you think?** How do you rate the value of the opinion dynamics models we've studied in this chapter? What are some ways you might make a more interesting opinion dynamics model? What sorts of questions would you use the model to ask? What modifications or different approaches would asking those questions require?

6 Cooperation

> Sociability is as much a law of nature as mutual struggle.
> —Peter Kropotkin, *Mutual Aid: A Factor of Evolution* (1902)

Cooperation is everywhere. Your friend helps you move to a new apartment. A teacher spends her free time tutoring a struggling student. A group of hunters fan out to circle their prey. Soldiers risk their lives to defend their nation. Aunts and uncles babysit their nieces and nephews. Friends and family share food, shelter, and stories. Cooperative behaviors exemplified by these examples are foundational to the fabric of human existence. We are a cooperative species.

Despite its apparent ubiquity, cooperation is a special thing. All the major transitions in the evolution of life on Earth have involved new cooperative structures (Maynard Smith and Szathmary, 1997). Multicellularity is possible only when individual cells limit their own reproductive success in support of the reproductive success of the whole organism. And yet, although the first single-celled organisms appeared on Earth about 3.5 billion years ago, it was not until almost 3 billion years later that the first multicellular animals evolved (Pennisi, 2018). We cannot take cooperation for granted. Sociality is possible only when individual organisms can limit ruthless competition and help one another. Indeed, cancer can be seen as a sort of backslide in which cells cease behaving cooperatively and act only in their own (short-term) reproductive interest.

Overcoming the forces of pure competition is nontrivial, because evolution tends to favor strategies that allow individuals to outcompete their neighbors in the struggle for survival and reproduction. Among the most amazing examples of cooperative systems found in nature are symbolic communication systems, notably human language. Language is possible only when people can focus on a joint task for persistent coordination. Language requires cooperation at multiple stages, including the willingness to teach on the part of adults and to learn on the part of children, and a social structure in which individuals are continuously incentivized to communicate what they know to others (Kirby, 2017; Brand et al., 2021). More generally, humans are remarkable in the extent to which they are cooperative with others. Language, culture, technology, and our spread to every continent and nearly every ecosystem on Earth are testament to the power of cooperation.

The benefits of a cooperative society are obvious. However, it's *not* obvious how cooperation can be sustainable. If some individuals work to benefit others, those that are able to free ride will receive the benefit without paying the cost. If free riders outcompete cooperators, how can cooperation ever emerge? Once present, how can it be maintained?

In this chapter, we tackle these questions.[1] In doing so, we will also introduce two extremely important modeling approaches with broad applications. The first approach is **game theory**, a formal framework for analyzing interactions among behavioral strategies when the benefits and costs of one individual's behavior are affected by the behaviors of others. We will focus on one game in particular: the prisoner's dilemma. The second approach is **evolutionary dynamics**, a general framework for understanding how replicative systems change over time. Replication might refer to the transmission of genetic information across generations, but it can also refer to cultural transmission by social learning. We will use these approaches in both this chapter and the subsequent two chapters to understand how social behaviors might evolve.

6.1 The Prisoner's Dilemma

"Cooperation" is a word in common use, and it can colloquially refer to a lot of different things. To understand it better, we will need a model of cooperation that captures something essential about it, and, more importantly, formalizes the concept so that it is perfectly clear what sorts of cooperation are and aren't the targets of our analyses. In this chapter, we will focus on the sort of cooperative behaviors that provide a benefit to others but are costly to the individuals performing the cooperative acts. This is usually called **altruistic cooperation**, or simply altruism. Behaviors that directly benefit oneself but also benefit others as a by-product are often described as mutualisms (see Box 6.2: Other Models of Cooperation). For simplicity, I will often use the word "cooperation" in this chapter when I really mean altruistic cooperation.[2]

The archetypal model for altruistic cooperation is a two-player game called the **prisoner's dilemma** (PD). The name comes from a parable about two criminals in jail cells, each deciding whether to rat out his partner to the cops. If each cooperates by staying quiet, they will both be convicted of a minor offense. If each defects by ratting on his partner, they will share a much harsher sentence for their felony. If one stays silent while his partner defects, he will incur the full punishment alone while his partner goes free on a plea deal. In all cases, an individual does better (gets a shorter sentence or goes free) by defecting rather than cooperating. However, mutual cooperation is better than mutual defection—hence the dilemma.

Altruistic cooperation is costly. If we both cooperate with one another, we reap the benefits. However, if you choose not to cooperate, I end up paying a cost while you get all the benefit. The prisoner's dilemma game can be formally described by a **payoff matrix**, which defines the payoffs a player receives based on their behavior and the behavior of their co-player. The quantity in each cell of the matrix below indicates the payoff to the row player. Each player can either cooperate or defect. Cooperation incurs a cost, c, on the cooperator, and confers a benefit, b, on the recipient, where $b > c$ and both terms are nonnegative.

[1] For a nontechnical introduction to the vast literature on cooperation in humans and other organisms, see Raihani (2021).

[2] For a rich discussion of the semantic application of many of these terms in evolutionary biology, see West et al. (2007).

Table 6.1. Payoffs in the generalized prisoner's dilemma game. Values in each cell are fitness consequences to the row player.

	Cooperate	**Defect**
Cooperate	$b - c$	$-c$
Defect	b	0

From examining the payoffs in Table 6.1, two things should become clear in a manner that is more generalizable than our parable of the prisoners. First, mutual cooperation is better than mutual defection as long as $b - c > 0$. In other words, cooperation is the desired state of affairs as long as it provides a net benefit to the dyad. The second thing to notice is that, regardless of what the other player does, it is individually advantageous to defect. If the other player cooperates, you can exploit them, because $b > b - c$. If they defect, you should avoid being exploited, because $0 > -c$. Mutual defection is therefore the only stable equilibrium of the game when it is played only once.

In the language of classical game theory, mutual defection is the sole **Nash equilibrium** of the game. A Nash equilibrium, named for the mathematician John Nash, is an arrangement of player strategies in which no player can improve their payoff by unilaterally changing their strategy (some games have no Nash equilibria, while some have several). Yet it seems impossible that mutual defection is really the only equilibrium. Many social animals cooperate. Humans cooperate spectacularly, producing culture, art, teaching, science, etc. All of multicellular life is itself a form of cooperation among cells that halt their own reproduction in the service of the whole organism. How is cooperation stabilized when it appears to be advantageous to refrain from cooperating?

We can reframe the problem of cooperation in terms of the prisoner's dilemma game: Under what circumstances can we observe the emergence and maintenance of cooperation in a population of individuals playing the prisoner's dilemma? We will answer this question by examining our assumptions about just how the game is played. However, to do so, we first need to talk a little bit more about evolution.

6.2 Evolutionary Dynamics

One of the most powerful tools for understanding the world is the theory of evolution by natural selection, first introduced by Charles Darwin and Alfred Russell Wallace in the mid-nineteenth century. The theory provides a framework for asking questions about how things get to be the way they are, and what might happen if the world were a little different. I actually think that evolution is the single most important theoretical framework you can learn. It is a general theory about how populations change over time in response to features of the environment—and of course the environment can include other individuals. Learning about evolutionary dynamics obviously helps us think about how genes and traits change over generational time, but it can also help us understand human culture and even individual cognition on much shorter time scales.[3]

[3]This insight has been described many times. For noteworthy examples, see Campbell (1965), Skinner (1981), Richerson and Boyd (2005), and Laland (2017).

Understanding the evolution of any particular species, trait, or behavior is complicated and involves wrestling with many interlocking details. However, the core theory of evolution by natural selection is quite simple. Any system of individuals and traits that meets the following requirements will evolve by something like natural selection:

- *Variation.* There must be variation among traits within a reproducing population of individuals.
- *Heritability.* Those traits must be transmitted from generation to generation.
- *Selection.* Individuals' traits must have consequences for the transmission of their traits to others.

With these factors—variation, heritability, and selection—in place, any trait variant that provides a net advantage in the transmissibility of a trait (such as by its effect on survival or reproduction) will spread in a population, while any variant that provides a net disadvantage will fail to spread. We can think about behavioral strategies as traits in these terms. Let's assume that individuals possess behavioral strategies associated with how they play the prisoner's dilemma game, meaning that game strategies are traits with fitness consequences. Let's further assume that those strategies vary between individuals and are heritable. For example, one individual may be predisposed to cooperation, another to defection. If individuals using one strategy reliably outperform individuals using another strategy, the frequency of the first strategy in the population will increase while the frequency of the second strategy will decrease.

At this point, it's worth saying a little more about selection and heritability. Transmission of traits from one organism to another can occur via genes or through social transmission—that is, by learning.[4] In the case of human cultural traits, we are often talking about social transmission. Luckily, several transmission mechanisms are roughly equivalent in our model. If individuals copy their parents, then the individuals who have the most offspring will propagate their traits *as if* they were genetically determined. This sort of **vertical transmission** of social information can include active teaching as well as less conscious processes. Alternatively, consider the case where individuals actively assess potential targets for social learning. If certain behaviors lead to the accumulation of desirable resources, and if individuals can assess differences in resources and imitate the strategies of successful individuals, then once again the most successful traits will propagate in the population. This is sometimes called **success-biased transmission**. Either way, the dynamics of the model work out the same: successful individuals will preferentially transmit their strategies.[5]

The models we considered in the previous chapters of this book asserted particular distributions of agent characteristics, which allowed us to examine the social and structural consequences of those assumptions. Evolutionary modeling allows for another, complementary sort of analysis. In this case, we use what we know or suspect about the distribution of *possible* behaviors and about the social forces that select for certain outcomes to consider how particular behaviors are likely to become more or less common over time. Ultimately, the two approaches can be combined to explore the coevolution of individual behavior and social forces.

---

[4]There are other transmission mechanisms, including inheritance of environmental constraints on development, but I won't discuss these here. See Odling-Smee et al. (2003) and Kendal et al. (2011) for more on the inheritance of environmental constraints.

[5]Other potential social learning strategies are possible, and they can affect evolutionary dynamics. See Boyd and Richerson (1985), Laland (2004), and Kendal et al. (2018) for more on the role of social learning strategies in cultural evolution.

We can use evolutionary modeling to explore how two or more strategies change (or fail to change) in frequency when they interact in a population (for more on mathematical tools for modeling evolutionary dynamics, see Box 6.1: Replicator Dynamics). In particular, we will look at how a strategy fares both when it is already common in the population and when it is rare. The latter question is particularly important if we are interested in social change, because we want to know if a strategy can spread when only a few individuals initially employ it. The fitness of a strategy is often **frequency dependent**—it depends on its own frequency and the frequency of other strategies in the population—because the outcome of a social interaction often depends on the behavior of all parties involved. The rate at which an individual using one strategy encounters others using various strategies therefore matters. In the case of the prisoner's dilemma game, the utility of being a cooperator depends on the prevalence of other cooperators. After all, mutual cooperation can outperform mutual defection, but defectors can succeed by exploiting cooperators. A mechanism which makes cooperators more likely to interact with one another could help to make cooperation sustainable.

BOX 6.1: Replicator Dynamics

A simple way to get an intuition for the evolutionary dynamics of competing strategies is to mathematically model them using discrete-time **replicator dynamics** (there is also a continuous-time equivalent that I will not cover here). This is a way to consider how the relative frequency of a trait (such as a behavioral strategy) changes over time in a large, well-mixed population.

Let us consider two competing strategies, A and B, in a population of size N, where p is the frequency of A individuals in the population and $1-p$ is the frequency of B individuals. Let $V(A)$ and $V(B)$ represent the **fitness** of strategies A and B, respectively, where fitness connotes the extent to which a strategy will successfully be transmitted to the next generation (throughout this chapter and the next, I will use V to connote payoffs, which in evolutionary models are measured in some unit of fitness). For example, we might imagine that $V(A)$ is the probability that an individual employing strategy A will survive to become a reproductively viable adult. In this case, the number of reproducing adults using strategy A is:

$$\text{number of type } A \text{ individuals} = pNV(A) \tag{6.1}$$

If fitness differences translate to differences in the probability of survival, we can assume for simplicity that all reproducing individuals produce z offspring. The total number of type A offspring produced in the next generation is $pNV(A)z$. We can use this information to derive an equation for the total frequency of strategy A in the next generation, denoted p':

$$p' = \frac{\text{number of new } A \text{ individuals}}{\text{number of new } A \text{ individuals} + \text{number of new } B \text{ individuals}} \tag{6.2}$$

This works out to

$$p' = \frac{pNV(A)z}{pNV(A)z + (1-p)NV(B)z} \tag{6.3}$$

Notice that each term is multiplied by Nz, so these all cancel out, and we are left with

$$p' = p\frac{V(A)}{pV(A) + (1-p)V(B)} \tag{6.4}$$

The denominator of this fraction is the mean fitness in the population, which is sometimes written as $\bar{w}$ for convenience. The replicator equation can actually be generalized to an arbitrary number of competing strategies, such that the recursion for the frequency of strategy i, p_i, is given by

$$p_i' = p_i \frac{V(i)}{\bar{w}} \tag{6.5}$$

Just as we saw in our exploration of contagion dynamics, we can translate this recursion into a difference equation to focus on how trait frequencies change as a function of their relative fitnesses:

$$\Delta p_i = p_i' - p_i = p_i \left(\frac{V(i)}{\bar{w}} - 1 \right) \tag{6.6}$$

This equation makes it clear that a trait that confers a fitness above the mean fitness in the population will increase in frequency, while a trait that conveys a fitness below the population mean will decrease in fitness, and that the magnitude of that generational change will be proportional to the extent to which the fitness associated with a trait differs from the population's mean fitness.

These replicator dynamics make several simplifying assumptions, not least of which is that the population is well mixed. Once population structure comes into play, things get a bit more complicated.

6.3 Cooperation and Assortment

To study the evolution of cooperation, let's first consider a very simple model in which all individuals are either cooperators or defectors. These are **pure strategies**: cooperators always cooperate, and defectors always defect. We will denote these strategies *ALLC* and *ALLD*, respectively, though I will often continue to refer to them as cooperators and defectors. Imagine a large population in which individuals encounter each other at random. A proportion p of the population are cooperators; everyone else is a defector. Under these assumptions, we can consider the **expected payoff** for an individual using each of the two pure strategies. You can think of this in one of two ways. First, if an individual were to have random encounters many, many times, the average payoff would come out to this expectation. Alternatively, think of many, many individuals having one interaction. Considering all the individuals using a particular strategy, the average payoff among them would be the expectation. Expected payoffs are useful metrics for the sort of **evolutionary game theory** analyses we are doing throughout this chapter.

We use the notation $V(A)$ to denote the expected payoff to an individual of strategy A. In a large population, the probability that a randomly chosen interaction partner will be a

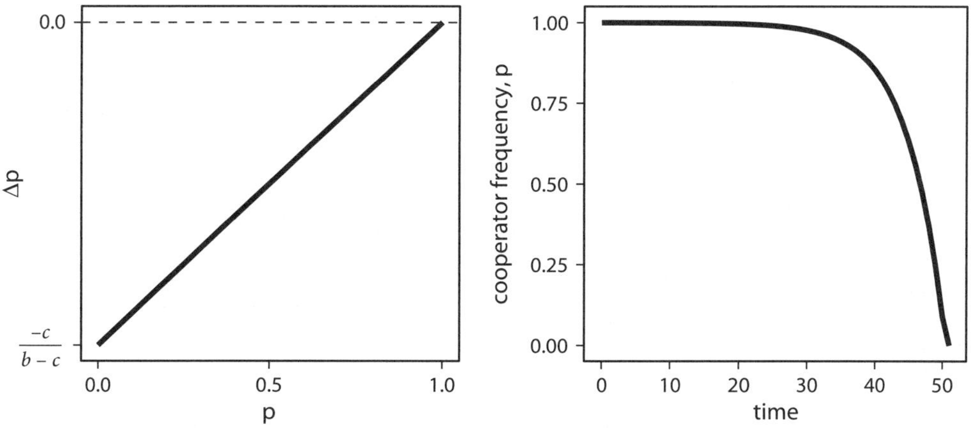

Figure 6.1 Replicator dynamics for the well-mixed PD game with pure strategies. Left: plotting the change in the cooperator frequency, Δp, as a function of p shows that the cooperator frequency always decreases. Right: An illustration of the system's evolutionary dynamics. Starting with $p_0 = 0.9999$, the frequency of cooperators decreases to zero (for this example, $b = 3$, $c = 1/2$).

cooperator is p. The expected payoff to a cooperator in such a population is therefore

$$\begin{aligned} V(ALLC) &= p(b-c) + (1-p)(-c) \\ &= pb - c \end{aligned} \tag{6.7}$$

A cooperator receives the benefit b when it encounters other cooperators, and always pays the cost c. The expected payoff for a defector is

$$\begin{aligned} V(ALLD) &= pb + (1-p)(0) \\ &= pb \end{aligned} \tag{6.8}$$

The payoffs are presumed to be directly related to evolutionary fitness—the ability of an individual to pass on its phenotype to members of the subsequent generation. This might reflect the ability to acquire calories or mates, to avoid death, or to acquire prestige or other measures of success and so become a more attractive target for social learning. Scenarios in which payoffs are not translatable into evolutionary fitness are not well captured by these models.

A defector receives the benefit b when it encounters cooperators, but never pays a cost. As long as there is some cost to cooperation (as long as $c > 0$), defectors will always outperform cooperators under the assumption of random interactions. Thus, defectors will always increase in frequency and any rare cooperators will fail to spread their strategies. We can see this by plugging the above payoffs into the replicator dynamics equation we derived in Box 6.1, plotted in Figure 6.1. The situation where $p = 1$ is an unstable equilibrium: any invading defectors will have higher fitness than the prevalent cooperators, and so they will increase in frequency until there are no cooperators left.

Let's now consider a situation in which interactions are *non*random. We'll use our tried-and-true model for structured populations: the square lattice. In particular, we'll study an agent-based model in which individuals are situated on a square lattice and interact with their nearest neighbors by playing a prisoner's dilemma game. Individuals with lower payoffs are replaced by their neighbors with higher payoffs. In this way, we can observe the evolution of cooperative or noncooperative strategies.

6.3.1 Cooperation in a Structured Environment

Consider a population of N agents, each situated on a unique cell of an $L \times L$ fully occupied square lattice with toroidal boundaries, so that $N = L^2$. Each agent is defined by a pure strategy of cooperate or defect. A newly created agent is initialized as a cooperator with probability p_0, otherwise it is a defector. The only other global parameters are the payoff variables b and c. Each agent i will keep track of its strategy, s_i, and its payoff, V_i.

The model proceeds in discrete time steps (as usual), each of which consists of two stages: game play and evolution. In the game play stage, each agent plays a prisoner's dilemma game with each of its four nearest neighbors (its von Neumann neighborhood). An agent's total payoff is therefore the sum of its payoffs from these four games. To calculate this sum, an agent considers its neighbors and counts the number of cooperators and defectors. Let these equal n_C and n_D, respectively. The payoff to a cooperator is then

$$V(C) = n_C(b - c) - n_D c, \tag{6.9}$$

and the payoff to a defector is

$$V(D) = n_C b. \tag{6.10}$$

In the evolution stage, each agent compares its own payoff with the payoffs of its neighbors. If the payoff held by any of these neighbors is higher than its own, the agent adopts the strategy of the neighbor with the highest payoff. These dynamics then repeat.

This model represents an extremely simplistic view of social behavior, structure, and evolution (be it cultural or genetic). Starting with simple models is a good thing. A baseline model like this will often be extremely simplistic. But simple models can give us insight, including insight into the sort of additional complexity we might or might not need to make sense of our systems. And, as I hope this book continues to demonstrate, it's not always obvious how an apparently simple system will behave.

6.3.2 Coding the Model

The NetLogo code for this model is **PD_simple.nlogo**. We will need three global parameters. First, we need the initial frequency of cooperators in the population, p_0, which we can call `init-coop-freq`. We also need the payoff parameters b and c, which we can call `payoff-benefit` and `payoff-cost`, respectively. I have used a 31×31 lattice with toroidal boundaries, implying a population size of $N = 961$.

Each agent needs to keep track of its strategy and its payoff, so each of those become agent-level parameters. The NetLogo code for this is:

NetLogo code 6.1

```
turtles-own [
  strategy ;;0=defector, 1=cooperator
  payoff
]
```

Based on the comment next to `strategy`, you'll notice that I have represented agent strategies as integers rather than as Boolean values. This allows us to add additional strategies (denoted by larger integers) to later versions of this model, which will come in handy.

The creation of agents upon the grid at initialization works similarly to the agent-based models of opinion dynamics in the previous chapter, controlled here by the procedure `make-agents`. Each cell of the grid "sprouts" an agent, which is then assigned to be a cooperator (with `strategy` equal to 1) with probability `init-coopfreq`; otherwise, it becomes a defector (with `strategy` equal to 0). Each agent's payoff is initialized to zero.

NetLogo code 6.2

```
to make-agents
  ask patches [
    sprout 1 [
      ifelse random-float 1 < init-coop-freq
      [ set strategy 1 ]
      [ set strategy 0 ]
      set shape "circle"
      set payoff 0
    ]
  ]
end
```

For the purposes of visualization, I have made defectors red and cooperators blue. For some reason, this color scheme is very common in evolutionary models of cooperation.[6] The procedure `recolor` ensures that agents' colors match their strategies.

NetLogo code 6.3

```
to recolor
  ask turtles[
    if strategy = 0
    [ set color red]
    if strategy = 1
    [ set color blue]
  ]
end
```

The model dynamics are controlled by the `go` procedure, which first checks to see whether either of the two pure strategies have completely dominated the population. If it has, the model stops running. This assumption is not strictly necessary, and may even be detrimental if rare strategies can return through mutation. However, in the absence of mutation it is useful to ensure that time steps in which no dynamics occur are not needlessly simulated, reducing the computing time needed for batch runs. If both cooperators and defectors are present, procedures are called for the game play and evolution stages of the model dynamics, followed by a call to `recolor`.

[6] It is likely that this convention was established by a paper by Nowak and May (1992), which presented one of the first spatially explicit agent-based models of cooperation to be visualized and to have a wide audience.

NetLogo code 6.4

```
to go
  if (all? turtles [strategy = 0] or all? turtles [strategy = 1]) [stop]
  play-game
  evolve
  recolor
  tick
end
```

In the `play-game` procedure, each agent counts the number of its neighbors playing either strategy and then calculates its payoff according to Equations 6.9 and 6.10.

NetLogo code 6.5

```
to play-game
  ask turtles [
    set payoff 0;;reset payoff for this time step
    ;;count neighbors with each strategy
    let neighbors-C count (turtles-on neighbors4) with [strategy = 1]
    let neighbors-D count (turtles-on neighbors4) with [strategy = 0]
    ;;if I'm a cooperator
    if (strategy = 1)
    [ set payoff (neighbors-C * (payoff-benefit - payoff-cost) -
       neighbors-D * payoff-cost) ]
    ;;if I'm a defector
    if (strategy = 0)
    [ set payoff (neighbors-C * payoff-benefit) ]
  ]
end
```

Next, each agent compares its payoff from the current round with the payoffs of each of its neighbors. If the neighbor with the highest payoff has a different strategy, the agent switches to that strategy. The code below uses several valuable NetLogo primitives, particularly `max-one-of`, which allows an agent to identify the neighboring agent with the highest payoff; this neighbor is denoted by the local variable `best-neighbor`. If this neighbor's payoff is higher than its own, the agent adopts the neighbor's strategy.

NetLogo code 6.6

```
to evolve
  ask turtles[
  let best-neighbor max-one-of (turtles-on neighbors4) [payoff]
  if ([payoff] of best-neighbor) > payoff
    [set strategy [strategy] of best-neighbor]
  ]
end
```

There's a minor flaw in the last step of the code I've presented here. Each agent copies the strategy of its highest-performing neighbor. However, because each agent is updating its strategy in turn, there may be some cases in which the strategy being copied is not the strategy that helped the agent get their high payoff. Such cases are probably rare, but this choice does add some stochasticity to the model that was not part of the original model description. It is left as an exercise to the reader to remove this source of potential error (hint: you can use the fact that an agent's color and strategy carry redundant information).

We can now study this model by visual inspection, observing the spatial patterns of red and blue agents on the grid. We can also plot the frequency of cooperators in the population over time in order to get a better sense of the dynamics—I have included such a plot in the NetLogo code. As usual, I encourage you to play around with this model and observe what sorts of patterns emerge under different parameter values. Some of the outcomes may surprise you.

6.3.3 Analyzing the Model

For all of our analyses, we can fix the benefit of receiving cooperation at $b = 1$, and only vary the cost, c, since it is really the *relative* benefit–cost ratio that matters. To get a sense of the model's behavior, let's run a few preliminary computational experiments. We'll start with an arbitrary value for the cost of cooperation, $c = 0.2$, and an initial population of 50% cooperators. Again, remember the dilemma: in the two-player game, mutual cooperation is better than mutual defection (here yielding a payoff of 0.8 "fitness points" per interaction instead of zero), but all things equal, one can always increase their payoff by switching from cooperation to defection. When interactions are random, cooperators always lose. Interactions here are not random, but the initial arrangement of agent strategies are. So what happens? Do cooperators go extinct? On the contrary. When we run the simulation, cooperators dominate, with only a few defectors remaining! Perhaps this isn't such a dilemma for cooperation after all. What's going on?

Let's take a closer look. Figure 6.2 shows several snapshots of the population at different times during a single simulation run, using the parameter values just described. Notice that the frequency of cooperation first goes *down* before it goes up. Defectors can increase rapidly when cooperators are scattered—that is, randomly distributed. A lone cooperator surrounded by defectors receives no benefits and pays large costs. However, by chance, some cooperators will start out next to a few other cooperators. In these cases, the benefits they receive can outweigh the costs they pay. And so, locations where cooperators happen to assort initially can survive and grow.

Now slowly increase the cost of cooperation in your simulation. You should observe that something quite dramatic occurs around $c = 0.25$. When the cost of cooperation increases above this value, cooperation suddenly collapses, with the population being overrun by defectors. Indeed, if we start with a mixed population of 50% initial cooperation, cooperators increase in frequency if and only if $c < b/4$. If $c \geq b/4$, defectors increase in frequency. To illustrate how stark this result is, I ran 10 simulations, each for 100 time steps (plenty of time for the model to reach equilibrium), for values of c between 0.01 and 0.5 in increments of 0.01 (Figure 6.3). In this plot, the dots are the individual runs, and the line is their average. We see that there's a sudden change in outcomes—a phase transition, as the physicists say—right at $c = 0.25$. Why does this happen?

In our model, each agent plays with exactly four neighbors. We already know that one act of mutual cooperation cannot outcompete one act of exploitation. What about *two* acts of mutual cooperation? Consider a cooperator with two other cooperators as neighbors (Figure 6.4A). Its payoff is $2b - 4c$. A nearby defector who interacts with one cooperator has a payoff of b. Cooperators with at least two cooperative neighbors will therefore spread when $2b - 4c > b$, or when $c < b/4$. Our simulations reflect this. When the cost c is above this threshold, cooperation declines, but does not vanish entirely. Some cooperators usually

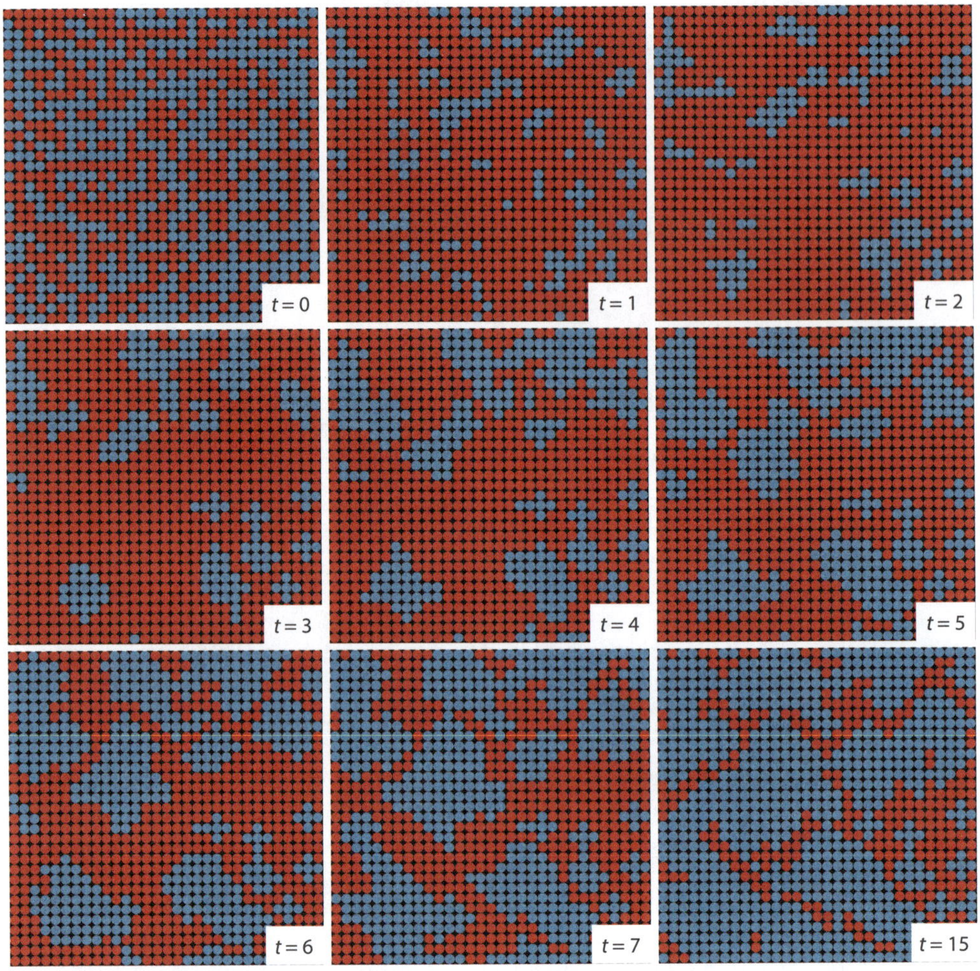

Figure 6.2 Spatiotemporal dynamics for the simple cooperation model with assortment. Each square shows the spatial distribution of cooperators (blue) and defectors (red) at a different moment in time. The final image shows the model at $t = 15$, at which point the population has reached a stable equilibrium.

stick around as long as $c < b/2$, but above that threshold cooperators go to zero always. Why? In our model, a cooperator can receive at most four acts of mutual cooperation, as it does when all of its neighbors are cooperators. When is even this not enough? Consider a cooperator surrounded by four other cooperators (Figure 6.4B). The central cooperator in this formation gets a payoff of $4b - 4c$. A nearby defector can interact with two of those cooperators and get a payoff of $2b$. So, the cooperator formation is stable as long as $4b - 4c > 2b$, or as long as $c < b/2$.

The insight here is that cooperation can do quite well if the cost of cooperating isn't too high relative to the benefit and, critically, if there is sufficient assortment. We can see the importance of assortment if we consider simulations modeling the invasion of just a few scattered cooperators (say, 5%) into a population of mostly defectors. You can verify that cooperation rarely increases in these cases, because there is not enough initial assortment.

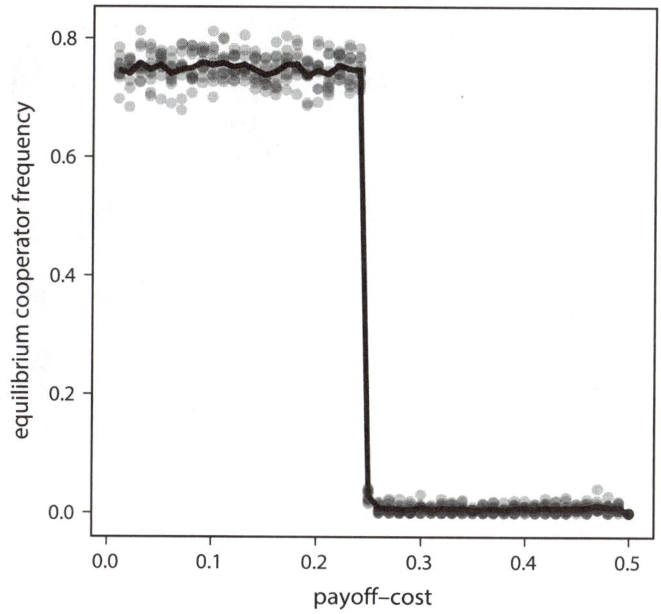

Figure 6.3 The equilibrium frequency of cooperators plotted against the cost of cooperation (c). Open circles are individual runs, the black line connects the means across 10 runs for each condition. The benefit of cooperation was $b = 1$ in all cases.

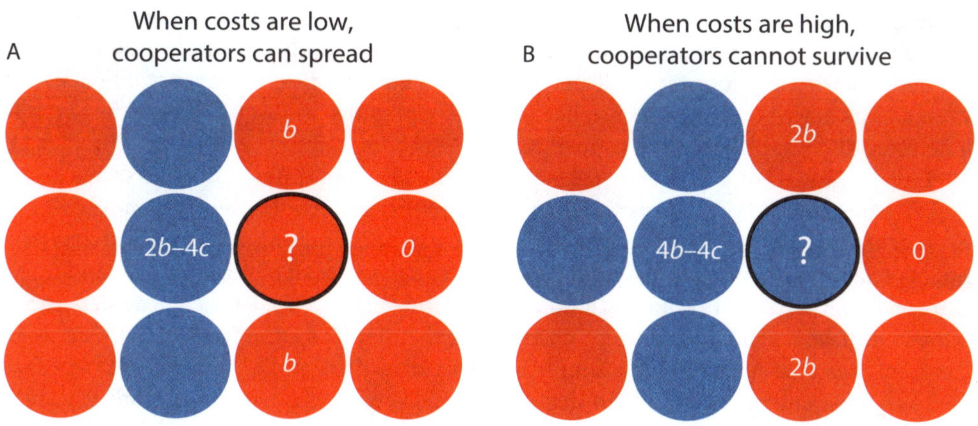

Figure 6.4 Considering the payoffs of agents in the spatial prisoner's dilemma model. (A) A cooperator with two neighboring cooperators will outperform a neighboring defector when $c < b/4$. (B) A cooperator surrounded entirely by cooperators will still be outperformed by a defector when $c \geq b/2$.

In general, our assumption of assortment is quite strong in this model. We have assumed that social networks are very rigid and never change, or equivalently that offspring will inherit the social connections of their parents. This provides cooperators a mechanism for maintaining contact with other cooperators, which aids cooperation because mutual cooperation (clusters of cooperators) outperforms mutual defection (clusters of defectors). In the next section, we will see that our reliance on strong assortment makes the sort of pure cooperation used in this model rather fragile.

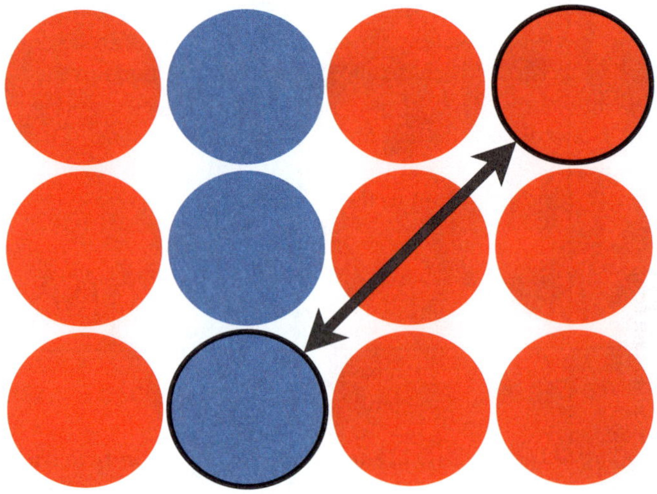

Figure 6.5 Probabilistic randomization. With some probability, each agent may swap locations with another randomly chosen agent.

6.4 Reducing Assortment

In the previous model, agents interacted with their neighbors in a fixed network structure. An agent's neighbors at the start of a simulation are its neighbors forever (or, alternatively, its neighbors' offspring are its offspring's neighbors—biologists call the phenomenon of offspring remaining close to home **limited dispersal**). Consider why limited dispersal is good for cooperation. If a cooperator disperses to a neighboring patch, it creates a benefit for itself and its neighbors, who are also likely to be cooperators and so will return the favor. If a defector disperses to replace a cooperator, its neighbors suffer.

The assumption of a fixed network structure may be appropriate for some scenarios in which groups boundaries are very strong and persist over time. However, in most cases—even in highly structured populations—there's at least some mixing around of network ties. So let's relax the assumption of a rigid network structure, at least a little bit. We'll do this by introducing what I call **probabilistic randomization** (Figure 6.5). At each time step, every agent will have a small but nonzero chance of switching its spatial position with another randomly selected agent. This is a simple mechanism to disrupt the spatial assortment that can otherwise emerge from limited dispersal, while keeping the overall network structure of the population fixed. Probabilistic randomization isn't intended to model any specific behavioral or social process. Rather, it is a generic process that is intended to reflect the extent to which social networks are fixed or change over time. When the probability of randomization is close to zero, most of the structure will be maintained and assortment can persist. As the probability increases, assortment is increasingly disrupted. The aim here is to examine the effect this disruption has on the evolution of cooperation in our spatial model.

6.4.1 Coding the Model

The NetLogo code for this model is **PD_randomized.nlogo**. To our previous model we need to add one new global parameter: the probability of randomization, `randomization-prob`. The initialization of this model is exactly the same as the previous model with simple assortment, and the dynamics differ only in that before agents play the prisoner's dilemma game, they each consider whether or not to move to a new random location (swapping positions with the agent at their new position). For this, we introduce a new procedure, `randomize-locations`, which is called before `play-game` in the `go` procedure.

This new procedure allows each agent to probabilistically select another agent at random and swap *x*- and *y*-coordinates. When an agent decides to move, it first records its own position. It then selects another agent to swap with and records the other agent's position. Each agent then sets its new position to the other's recorded position.

NetLogo code 6.7

```
to randomize-locations
  ask turtles [
    if random-float 1 < randomization-prob [
      let my-x xcor
      let my-y ycor
      let partner one-of other turtles
      let their-x [xcor] of partner
      let their-y [ycor] of partner
      set xcor their-x
      set ycor their-y
      ask partner [
        set xcor my-x
        set ycor my-y
      ]
    ]
  ]
end
```

6.4.2 Analyzing the Model

If you observe the model's behavior while playing around with the new parameter `randomization-prob`, you should notice that disrupting spatial assortment decreases the amount of cooperation that can be maintained in the population. It also makes the relative frequencies of cooperators and defectors in the population more variable over time—the system never settles to a perfectly stable mixed equilibrium of cooperators and defectors.

If we disrupt assortment enough, we drive cooperation to extinction. I've run batches of the model over a range of values for `randomization-prob`, with 20 runs per value, holding the payoffs fixed at $b=1$ and $c=0.2$. Each simulation was run for 200 time steps, which was generally enough time for the population to settle into a stable dynamic. Recall that without any randomization, a high level of cooperation was stabilized in the population under these payoffs. Figure 6.6 shows the long-run frequency of cooperators for these simulations. It is easy to see that when noise is added to the population's network structure, cooperation suffers. Past a certain point, it can't be maintained at all.

These simulations illustrate the importance of **positive assortment**. Positive assortment is the tendency for individuals that share some trait to interact at a higher rate than would be predicted by chance.[7] Our simulations indicate that positive assortment is a key factor for the evolution and maintenance of cooperation. In fact, much of the work documenting

[7] Positive assortment can be quantified by measuring the *excess assortment* of a population, which is the extent above chance that interactions are between similar individuals (Pepper, 2007; Smaldino and Lubell, 2011).

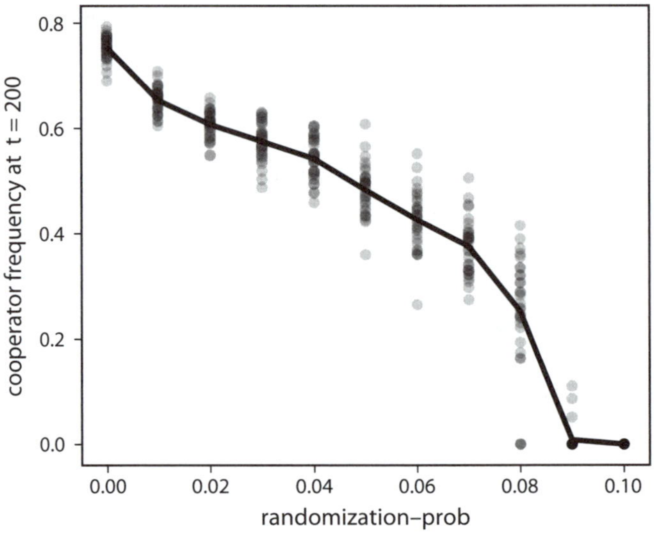

Figure 6.6 The frequency of cooperators at $t = 200$ plotted against the probability of randomization at each time step, starting with 50% cooperators initially. Open circles are individual runs, the black line connects the means across 20 runs for each condition. Payoffs were $b = 1$, $c = 0.2$ in all cases.

mechanisms for the evolution of cooperation is in reality the documentation of mechanisms that can generate positive assortment (Apicella and Silk, 2019).

These agent-based simulations are useful for gaining intuition about the need for assortment in the evolution of cooperation. However, the simulation model is also quite idiosyncratic. Each agent interacts with exactly four others in a rigid lattice network structure. It would be nice if we could derive more general principles about assortment and the evolution of cooperation. To do this, we return to mathematical modeling.

6.5 Positive Assortment and Hamilton's Rule

The modeling work we've done so far indicates that evolving cooperation is possible, but only when cooperators can effectively maintain relationships with other cooperators instead of being exploited by defectors. A rigid spatial structure allows this, because once a cluster of cooperators is established, they can maintain an advantage through mutual cooperation and expand locally, increasing their advantage. We also saw that when that structure was disrupted, cooperation decreased. In this section, we will derive a more general mathematical principle of assortment for the evolution of cooperation that doesn't rely on any particular details of spatial or network structure.

As before, we will assume a large population of individuals who pair up with others for a potential cooperative interaction, which is modeled as a prisoner's dilemma game. Individuals are again defined by pure strategies of cooperate and defect, and they can transmit their behavioral traits to their offspring either through genetics or social learning. Suppose the frequency of the cooperative trait in the population is p. If interactions occur at random, the probability that a cooperator encounters another cooperator will simply be p, and the probability that a defector encounters another defector will be $1 - p$. However, let us now assume that there is some mechanism in place that facilitates nonrandom interactions, allowing individuals to interact preferentially with others using the same behavioral strategy. We are talking about a mechanism for positive assortment. There are many possible ways this could work. For example, there could be a rigid spatial or social network structure, as in our first agent-based model, or a cognitive-behavioral mechanism that allows individuals to interact preferentially with close kin or others with similar behavioral traits.

Let r be the probability *above chance* that an individual encounters another individual of the same type. Positive assortment therefore occurs when $r > 0$. If we imagine that individuals attempt to employ a search strategy to find interaction partners of their same type, then r is the probability that they are successful in their attempt. When they are unsuccessful, they accept an interaction partner at random, who could still be of the same type purely by chance! The probability that a cooperator will encounter another cooperator is therefore:

$$\Pr(ALLC|ALLC) = r + (1-r)p \tag{6.11}$$

The probability that a defector will encounter another defector is similarly:

$$\Pr(ALLD|ALLD) = r + (1-r)(1-p) \tag{6.12}$$

The expected payoff to a cooperator is then the sum of its payoffs against both cooperators and defectors, weighted by the probability of encountering each type of partner:

$$V(ALLC) = \left[r + (1-r)p\right](b-c) - (1-r)(1-p)c \tag{6.13}$$

And similarly for a defector:

$$\begin{aligned} V(ALLD) &= \left[r + (1-r)(1-p)\right](0) + (1-r)pb \\ &= (1-r)pb \end{aligned} \tag{6.14}$$

Now that we have our expected payoffs, we can ask when cooperators can expect to outperform defectors. That is, when is $V(ALLC) > V(ALLD)$? Recall that without positive assortment, this inequality is never true. Our simulations suggest that positive assortment makes the difference. We are now in a position to say exactly what sort of difference it makes, though doing so requires us to flex our algebra muscles a bit. The analysis is mathematically simple, but it can nevertheless be difficult to keep track of all the terms, so I recommend trying to work it out on your own in addition to following along below.

The equation for $V(ALLC)$ is the gnarlier of the two payoffs, so let's see if we can reduce it a bit. Multiplying out the first term, we get:

$$V(ALLC) = r(b-c) + (1-r)p(b-c) - (1-r)(1-p)c \tag{6.15}$$

Multiplying out the second and third terms gives us:

$$V(ALLC) = r(b-c) - rp(b-c) + p(b-c) - (1-p)c + r(1-p)c \tag{6.16}$$

Now, let's rearrange the first two terms, and multiply out the third and fifth terms:

$$V(ALLC) = r(1-p)b - r(1-p)c + pb - pc - c + pc + r(1-p)c \tag{6.17}$$

Notice that now we've got a few terms that cancel out. This allows us to simplify the expression quite a bit:

$$\begin{aligned} V(ALLC) &= r(1-p)b - \cancel{r(1-p)c} + pb - \cancel{pc} - c + \cancel{pc} + \cancel{r(1-p)c} \\ &= r(1-p)b + pb - c \end{aligned} \tag{6.18}$$

Our goal here is to derive the amount of assortment, r, needed for $V(ALLC) > V(ALLD)$. Let's bring back the full inequality, multiplying out the expression for $V(ALLD)$ as well as our new expression for $V(ALLD)$:

$$rb - rpb + pb - c > pb - rpb \tag{6.19}$$

Notice the the term $pb - rpb$ appears on both sides of the inequality, so we can subtract this term from both sides, leaving us with

$$rb - c > 0 \tag{6.20}$$

Adding c to both sides yields a rather famous inequality:

$$\boxed{rb > c} \tag{6.21}$$

This inequality says that cooperators will outperform defectors when positive assortment is sufficiently strong so that the benefits conferred to cooperators will compensate for the costs of altruistic cooperation. In other words, altruistically cooperative behaviors increase in frequency when the benefits are preferentially bestowed on individuals who are themselves cooperative. This allows cooperators to recover the costs of being cooperative and to avoid giving an unfair advantage to free riders. To reiterate the main lesson: cooperation can evolve if there is a mechanism that allows cooperators to positively assort.

Equation 6.21 is widely known as **Hamilton's rule**, after William D. Hamilton, who first derived it in 1964 (Hamilton, 1964). Hamilton focused on the evolution of genetically transmitted prosocial traits, and discussed how they could evolve if individuals preferentially assorted with close relatives, who would be likely to also possess the same prosocial trait. In this framing, r is sometimes called the "coefficient of relatedness." Hamilton's analysis gave rise to the concept of **inclusive fitness**, which is the idea that the evolutionary fitness associated with a trait or behavior is related to the total number of future offspring who share that trait, whether or not they inherited it from the focal individual. Thus, we can make sense of Hamilton's remark that, at least in the world represented by his model, "we expect to find that no one is prepared to sacrifice his life for any single person but that everyone will sacrifice it when he can thereby save more than two brothers, or four half-brothers, or eight first cousins" (Hamilton, 1964, p. 16), because on average the probability above chance that an individual shares any particular gene is $1/2$ for a full sibling, $1/4$ for a half sibling, and $1/8$ for a first cousin. This focus on relatedness stems from a gene-centered view of evolution, in which the units of selection are not individuals, but the genetic vehicles by which they transmit their traits (Dawkins, 1976). A gene that leads to more copies of itself in the population will be successful, even if it decreases the number of offspring in a specific host. Inclusive fitness, also called the theory of **kin selection** (Maynard Smith, 1964), indicates that prosocial interactions should be expected most often between close relatives. However, our derivation shows that the mechanism for assortment need not be kin-based.

Any mechanism that promotes positive assortment among individuals with prosocial traits can facilitate the evolution of altruistic cooperation.

Our spatial simulations from the previous section illustrate this principle. When cooperators clustered together, they produced positive assortment, which allowed cooperation to spread. The mathematical analysis presented in this section allows us to make a more general statement about the relationship between assortment and cooperation than was possible using the more idiosyncratic spatial model. However, the spatial model also illustrates something that is not obvious given only the mathematical formulation. This is that assortment need not be constant over time. In the spatial model, cooperators and defectors were initialized at random locations, and so assortment was low and cooperation therefore *decreased*. Positive assortment occurred among some cooperators who *by chance* happened to be located near other cooperators. Once the non-assorting cooperators were replaced by neighboring defectors who could exploit them, the only remaining cooperators were those who were assorting with other cooperators. This drove up the population level of positive assortment. Once the population self-organized into assortative clusters of cooperators, cooperation was able to increase.

6.6 Reciprocity

Naïve cooperators, who cooperate indiscriminately, can spread into a population of defectors and be maintained at high rates if the costs of cooperation are sufficiently low relative to the benefits and if there is sufficiently strong assortment across generations. Strong assortment may suffice to explain cooperation among close kin or in situations where social relations can be maintained over generational time. However, the requirement of strong assortment seems like a problem for explaining other types of cooperation, including cooperation between members of different species (such as that observed between cleaner fish and their hosts) as well as most human cooperation (Bshary and Raihani, 2017; Apicella and Silk, 2019). In many real-world cases, cooperative partnerships are not quite so rigidly determined. We interact with lots of people, many of whom could easily take advantage of us if we are just blindly altruistic, giving our time and resources to anyone we meet. Without assuming that cooperative clusters can be maintained indefinitely, how can we explain the evolution and maintenance of cooperation? If assortment is present, can we explain how it might arise?

So far, we have considered only pure strategies. In particular, we have considered only cooperators that always cooperate with everyone. That's not particularly savvy. Surely it would be better to cooperate only with fellow cooperators, thereby reaping the synergistic benefits of cooperation while avoiding exploitation by free-riding defectors. This sort of strategic behavior is challenging if we assume that opportunities for cooperation are accurately modeled as one-shot games in which an interaction lasts for only one round. In one-shot games, a player's payoff is fully determined by their opening move. However, what if interactions are instead extended in time, so that individuals have multiple opportunities to interact and could even learn from their partners' past behavior? This consideration allows us to move from pure strategies to **contingent strategies**, in which an individual's behavior is based on the previous behavior of their co-player. We call this game scenario the **iterated prisoner's dilemma game** (IPD), which just means that pairs of players play the game repeatedly for multiple rounds, over which they accumulate payoffs.

Table 6.2. Payoffs for the iterated prisoner's dilemma game when games always last for exactly k rounds. Values in each cell are fitness consequences to the row player.

	ALLC	TFT	ALLD
ALLC	$k(b-c)$	$k(b-c)$	$-kc$
TFT	$k(b-c)$	$k(b-c)$	$-c$
ALLD	kb	b	0

A new game calls for a new strategy, and here we will focus on a contingent strategy called **tit-for-tat (TFT)**. Tit-for-tat agents are cooperative but responsive. They don't like being exploited. TFT always starts out cooperating, but thereafter copies its co-player's previous move. So it will happily continue to cooperate with another cooperator for the duration of the interaction, but will only cooperate with a defector once (unless the defector switches tactics and starts to cooperate). In other words, TFT adheres to a principle of **reciprocity**.

Let's return to our agent-based model and add reciprocity in order to try and get some intuitions from our simulations. We will tackle reciprocity analytically in the following section. We need to update the model to account for multiple rounds of game play; this can be accomplished by specifying a fixed number of rounds for each game. We'll continue to use the abbreviation TFT for tit-for-tat and ALLC and ALLD for pure cooperators and defectors, respectively.

Let k be the number of iterations played with each neighbor on each time step. Before we tackle how the agent-based model works—with agents playing four neighbors simultaneously—let us consider an updated payoff matrix with one-on-one play (Table 6.2). Notice that when the two cooperative strategies play each other, there is no difference in payoffs, because each cooperates fully with the other. The differences come from their interactions with ALLD. ALLC continues to be exploited on each round. TFT is also exploited, but only once, after which it switches to defection. Our task is now to see if this responsiveness can suffice to maintain cooperative behavior in a population.

To keep things simple, we will focus on competitions between ALLD and one cooperative strategy—either ALLC or TFT. The model will be initialized just as before, except that we will create a Boolean switch that, when true, will replace all the ALLC agents with TFT agents. During the game play stage of the model dynamics, each agent will play the iterated PD game with each of its four neighbors for k rounds. As before, let n_C denote the number of an agent's cooperative neighbors, and n_D the number of defector neighbors. Recall that payoffs are denoted with the letter V. The payoff to an ALLC agent is therefore

$$V(ALLC) = kn_C(b-c) - kn_Dc, \tag{6.22}$$

while the payoff to a TFT agent is

$$V(TFT) = kn_C(b-c) - n_Dc. \tag{6.23}$$

If the game lasts for only one round ($k = 1$), TFT is exactly equivalent to a pure cooperator, since by the time it's ready to respond, the game is over. The difference is in the extent to which the two strategies pay the cost of cooperating with defectors over two or more rounds. The ALLC agent is exploited by defectors for every round of the interaction, while the TFT agent is only exploited once. Similarly, the payoff to defectors is also affected, because ALLD can exploit ALLC for *k* rounds but TFT for only one round. Reciprocity, insofar as it is captured by TFT, decreases the extent to which cooperators will be exploited by free riders. We can now code up the model to explore the consequences of reciprocity a bit more precisely.

6.6.1 Coding the Model

The NetLogo code for this model is **PD_reciprocity.nlogo**. To our previous model parameters, we need to add a global parameter that controls the number of rounds for each iterated game, `num-iterations`. We will also add a Boolean switch called `TFT?`. When this is false, cooperative agents will play a strategy of ALLC; when true, they will play TFT. This allows us a direct comparison of ALLC and TFT strategies against the exploitative ALLD in the iterated prisoner's dilemma game. For the curious, we will briefly investigate interactions between all three strategies later in this chapter.

The initialization of this model is the same as the previous models. However, I have modified the `recolor` procedure to make TFT agents green instead of blue so that it is readily apparent which cooperative strategy is being used.

NetLogo code 6.8

```
to recolor
  ask turtles[
    if strategy = 0
    [ set color red] ;;ALLD = red

    if strategy = 1
    [ ifelse TFT?
      [set color green] ;;TFT = green
      [set color blue] ;;ALLC = blue
    ]
  ]
end
```

The dynamics of this model differ slightly from the previous model in the way game payoffs are calculated, because we have to account for the iterated game and whether cooperative agents are ALLC or TFT. The `play-game` procedure must be updated accordingly. The following updated code is used in this procedure to compute payoffs for each of the three agent strategies:

NetLogo code 6.9

```
;if I'm a cooperator
if (strategy = 1 and not TFT?)
[set payoff num-iterations * (neighbors-C * (payoff-benefit -
        payoff-cost) - neighbors-D * payoff-cost)]

;;if I'm a TFT
if (strategy = 1 and TFT?)
[set payoff num-iterations * (neighbors-C * (payoff-benefit -
        payoff-cost)) - (neighbors-D * payoff-cost)]

;if I'm a defector
if (strategy = 0) [
```

```
    ifelse TFT?
    [set payoff (neighbors-C * payoff-benefit)]
    [set payoff num-iterations * (neighbors-C * payoff-benefit)]
  ]
```

Once payoffs are accumulated, no change in the `evolve` procedure is needed, as agents continue to imitate the strategy of their highest-earning neighbor.

6.6.2 Analyzing the Model

This analysis focuses on scenarios where `TFT?` = **true**. If the number of iterations is one, the model is exactly equivalent to the previous model with pure strategies only. You can confirm this by setting `num-iterations` to 1 and recovering similar results. Let's see what happens when we increase the number of iterations. Try it with $k=4$. Start with strong spatial assortment (`randomization-prob` $=0$), 50% initial cooperators, $b=1$, and $c=0.25$. Recall that this parameter combination is where pure cooperators got into trouble. Yet TFT not only outperforms defectors, it totally dominates. When we used only pure strategies, there were usually still some ALLD agents remaining even when ALLC dominated. In this case, however, the population evolves to 100% TFT agents. To get a sense of the power of reciprocity, try cranking the cost way up to $c=0.6$. Remember, at this point, ALLC is toast. Not TFT! It dominates again.

So far we've assumed very strong assortment, which allows cooperative clusters to persist. We know now that assortment aids the evolution of cooperation. Let's get rid of assortment altogether and set `randomization-prob` all the way up to 1, so there is no longer *any* persistent spatial assortment across generations. In other words, we've got totally random interactions. Remarkably, TFT still usually wins out, because it benefits from interactions with other TFT agents (and confers benefits on them as well), but still avoids getting exploited too badly during its unavoidable interactions with defectors. Thus, TFT can handle much higher amounts of randomization than ALLC. This is positive assortment of a different kind: when mutually cooperative partnerships arise, they can last a while, so that most interactions involving cooperation are between pairs of cooperators.

If we continue to increase the cost of being exploited, we eventually reach a point where TFT does better with less randomization—there are still benefits to assortment. Consider $c=0.7$. TFT persists in a mixed equilibrium for `num-iterations` $=4$ and no randomization, but loses to ALLD as the randomization increases (Figure 6.7). TFT can also invade when rare with less initial assortment than pure cooperators, because it resists being exploited. Try running simulations where `num-iterations` $=4$, $c=0.3$, `init-coop-freq` $=0.05$, and `randomization-prob` $=0.1$. Under these conditions, ALLC would get creamed (you can verify this). However, even though TFT agents rarely find each other, as long as they do occasionally, the benefits of their repeated interactions still outweigh the costs of being occasionally exploited. Figure 6.8 shows a more rigorous exploration of the simulation model, examining the frequency of TFT agents at $t=100$ for multiple values of `num-iterations`, `payoff-cost`, and `randomization-prob`. TFT can outperform ALLD across a wide range of conditions. The figure also shows that, unlike the case of pure strategies, we rarely get mixed equilibrium. One strategy tends to completely dominate the other.

In 1980, the political scientist Robert Axelrod held a pair of tournaments in which participants could enter strategies to compete against the other entries in the iterated prisoner's dilemma. TFT, submitted by the game theorist Anatol Rapoport, won both times (Axelrod,

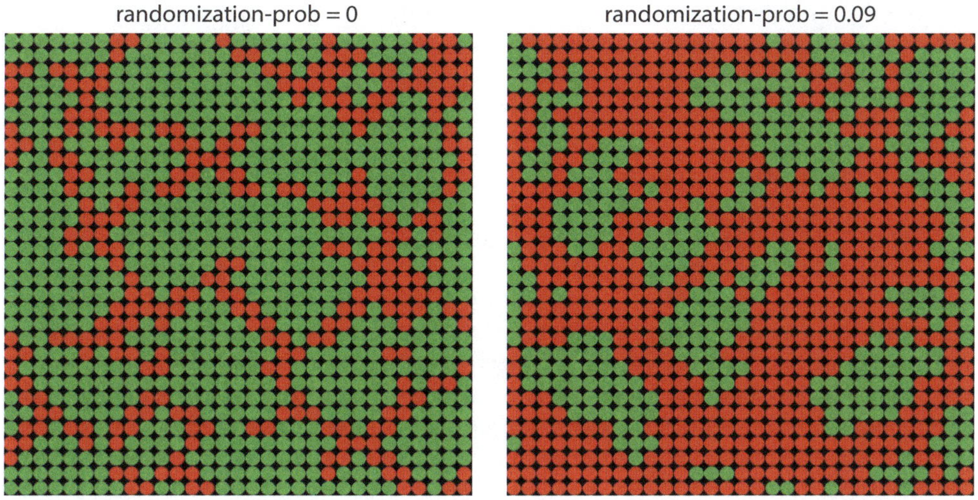

Figure 6.7 Spatial distributions of TFT (green) and ALLD (red) agents for different levels of randomization. The left figure is a stable equilibrium, while the right figure is a momentary profile of a dynamic population. ALLD completely dominates when `randomization-prob` increases to 0.1. For these runs, $p_0 = 0.5$, $b = 1$, $c = 0.7$, and $k = 4$.

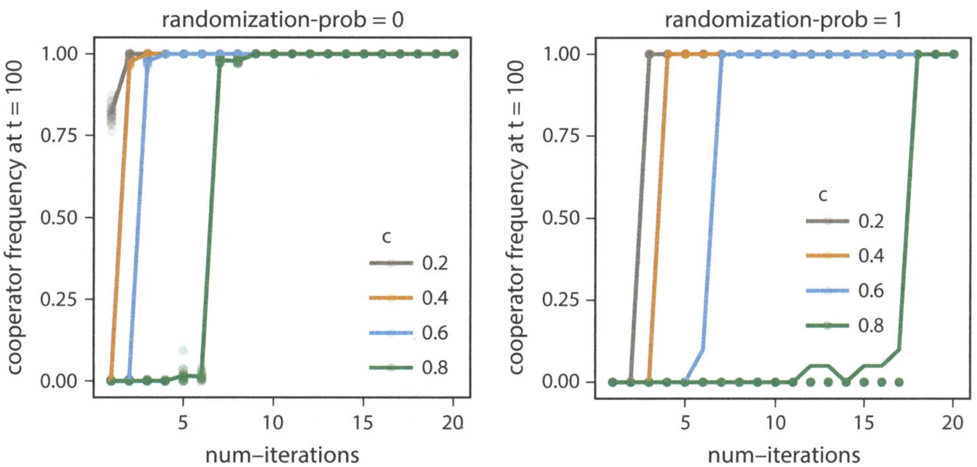

Figure 6.8 The frequency of TFT agents at $t = 100$ plotted against the number of iterations for each round of game play, for several values of the cost of cooperation, c. Simulations started with 20% TFT agents initially, and used $b = 1$. Reciprocal cooperation requires fewer iterations when aided by spatial assortment (left) compared with when interactions are randomized (right). Open circles are individual runs, solid lines connect the means across 20 runs for each parameter combination.

1984), which is one reason the strategy is noteworthy. TFT also has several qualities that are often analogized to human characteristics. It is *nice*, because it assumes the best of a new co-player and starts out by cooperating. It is *retaliatory*, because it returns defection with defection. And it is *forgiving*, because it will return to cooperating if its co-player does the same. TFT can permit the persistence of cooperation under greater costs and lower assortment than can ALLC, as long as interactions persist long enough so that there is ample opportunity for reciprocal cooperation. This is indicated by our computational simulations but, once again, it would be nice to have a more precise mathematical articulation of the power of iterative interactions. So let's do that.

6.7 The Evolutionary Stability of Reciprocity

The motivating idea behind the iterated prisoner's dilemma game is that over the course of their lives, individuals may have many opportunities to interact, and their behavior during an interaction may be informed by their partner's behavior during their past interactions. Our simulations indicate that the longer a relationship persists, the more effective a strategy of reciprocity can be at promoting cooperation. Here, we'll make this argument more precise. In doing so, we'll also learn about evolutionary stability analysis, which is an important modeling technique with broad applications. But first, we need to address a conceptual problem with our simulation model: the fact that the number of interactions was fixed.

6.7.1 The Problem with a Fixed Number of Iterations

The agent-based model we studied in the previous section assumed that all interactions lasted for a constant, known number of iterations. This is a convenient modeling assumption, but it also has a conceptual problem. If an individual knows exactly how long the game will last, they might be incentivized to always defect when the end was in sight in order to briefly exploit their partner. Let me explain what I mean by this.

Classical game theory (of the sort typically employed by economists and political scientists) often involves the assumption that individuals are perfectly "rational" agents with complete information, who can calculate the outcomes of all possible scenarios and thereby decide on an optimal, payoff-maximizing strategy. Although we've seen that TFT can outcompete ALLD in our simulations, this occurs because our agents are not really making rational decisions. Instead, they merely execute simple strategies, and the model allows those strategies to compete. The assumptions of perfect rationality and complete information are often limiting, because most organisms—humans very much included—are *bounded* in their rationality. This means they are limited in terms of the time, access to information, and computational power required to actually calculate optimal strategies. Moreover, a strategy that is optimal in one context may fail catastrophically in another context, and so part of the calculation must be to infer what context one is in—a calculation which often involves some error. As a number of scholars have convincingly argued, humans and other animals are likely to rely on generally successful "good-enough" heuristics in their decision-making (Simon, 1972; Gigerenzer and Selten, 2002; Todd et al., 2012). Nevertheless, the assumptions of rationality and perfect information are often worth engaging with to establish a baseline of what one *would* expect from so-called rational agents.

Recall that in the one-shot prisoner's dilemma game, the only Nash equilibrium is mutual defection. A cooperating player can increase their payoff by switching to defect, regardless of what their co-player does. Recall also that mutual cooperation is better than mutual defection. Our analyses have shown that to maintain cooperation, some mechanism is required that allows cooperation to be directed toward cooperative individuals, and our simulations above suggest that iterated games may provide the conditions for reciprocal cooperation to flourish. However, this may not be the case if the number of rounds in the iterated game is fixed and known to all players. The logic works like this. Imagine that we are playing an iterated game that we know will last for exactly k rounds. On the kth round, both players know that it's the last round, and so the game is equivalent to the one-shot game. The optimal move is to defect, and so rational players will do so. On the $(k-1)$th round, both players know that the next round is the end and that both will defect. As such, they know that their behavior in the current round will not be used by their co-player to choose future behaviors, and so once again the scenario is equivalent to a one-shot game in which both players will defect. This continues until we are back at the first round. Because perfectly rational players know that their co-player's moves in all future rounds are already determined, they will defect from the start.

The assumption of a fixed, known number of rounds is a strong one, and it isn't a great characterization of how repeat interactions tend to work. The fact is, we often do *not* know when or whether we will interact with someone again, and so repeat interactions may be better captured as probabilistic rather than deterministic. And indeed, this is the way the iterated prisoner's dilemma game is often modeled: with additional rounds occurring probabilistically. In classical game theory, the **folk theorem** (so called because it was widely known among game theorists decades before anyone had published a proof) establishes that *any* set of interactions, including mutual cooperation, can be maintained as long as players don't know when the interaction will end and place sufficient value on the future outcomes. The classical game theory approach[8] is valuable, but it requires very strong assumptions about the information individuals have and the way they weigh proximate versus distal outcomes. The evolutionary approach that this chapter focuses on relies only on competition between strategies and allows us to examine dynamics at the population level.

6.7.2 The Probabilistic IPD

We will now instantiate the iterated prisoner's dilemma as follows. Players always play the game for at least one round. After each round of play, the interaction will persist and the pair will play an additional round with probability w. As before, we'll consider the following three game strategies: ALLD, ALLC, and TFT.

Iterated games are now probabilistic, but our analyses for the IPD above required us to be able to predict agent payoffs, which relied on knowing the precise number of rounds players would play. This seems like a problem. However, if we want to calculate *expected* payoffs for our agents, we simply need to know the *expected* number of rounds they will play, under the assumption that variation will average out in the long run. Thankfully, we can determine the expected number of rounds due to our assumption that the probability of playing another round is a constant, w. The game will last for at least one round with certainty. The probability of the second round is w. A third round occurs with probability w conditional on the second round occurring, so their joint probability is w^2—in other words, the probability of a third round is w^2, and so on. The expected number of rounds—the length of the entire interaction—is the sum of the probabilities of each round occurring. Let this be represented by K, capitalized to distinguish it from the fixed number of rounds used in the previous section:

$$K = 1 + w + w^2 + w^3 + \dots \tag{6.24}$$

This is an infinite series, which might look daunting to analyze. We can't add up infinite numbers, can we? We can, actually, because math is nifty. Although the series extends to infinity, we can show that it converges to a finite quantity. The fact that the series approaches

[8] Spaniel (2014) provides a gentle introduction of classical game theory for the novice, including an accessible discussion of the folk theorem.

a finite sum shouldn't be too surprising when you consider that $w < 1$, and so each term w^n will be smaller than the last, approaching zero as $n \to \infty$. You can simulate some of these series and see for yourself that the series does indeed converge for all $w \in (0, 1)$. Now let's figure out exactly what it will converge to. Let's start by subtracting one from each side of the equation:

$$K - 1 = w + w^2 + w^3 + \ldots$$

Every term on the right side has a w, so we can use the distributive property of multiplication to pull one of those out. You'll notice that what we're left with is our original number K:

$$K - 1 = w\underbrace{(1 + w + w^2 + w^3 + \ldots)}_{K}$$

Substituting K for the infinite series, we get

$$K - 1 = wK$$

Adding one and subtracting wK on both sides gets us

$$K - wK = 1$$

Finally, we solve for K to get

$$\boxed{K = \frac{1}{1 - w}} \tag{6.25}$$

Mathemagic! A quick sanity check shows that if $w = 0$, the expected number of rounds is one. As w increases, so does K, reaching a limit of infinity when $w = 1$, which makes sense for the case where there is always another round.

Now that we know how many rounds each game should last on average, we can consider how the strategies fare against one another. To do so, we'll need a way to figure out when strategies should increase or decrease in frequency. We could use replicator dynamics to simulate the evolution of each strategy, and this can indeed be useful. However, there is another way that doesn't require any simulation at all.

6.7.3 ESS Analysis

Can a cooperative population of reciprocating TFT agents resist being outcompeted by invading ALLD agents? Could a population of free-riding, uncooperative ALLD agents be invaded and outcompeted by a small band of TFT agents? To answer these questions, we can consider whether either strategy is an **evolutionarily stable strategy**, or ESS. What we will do in this section is sometimes called ESS analysis.[9] Given two competing strategies A and B, we can ask two questions about the evolution of A. First, will a rare A individual be competitive in a population dominated by Bs and therefore increase in number? In other words, can A **invade** B? Second, will a population of mostly As be able to resist being invaded by a small number of Bs? In other words, is A an **ESS** against B? The answers to these two questions are not always identical; a strategy may perform well when common but poorly when

[9]For a deeper dive into the sort of analytic evolutionary game theory approaches explored in this chapter, see McElreath and Boyd (2007).

rare (recall our earlier discussion of frequency dependence). ESS analysis was introduced by Maynard Smith and Price (1973) and is used extensively to study important topics related to social behavior and social cognition. It is important to note that a strategy can only be an ESS in relation to another particular strategy or set of strategies. It has been proven that in the IPD, there is always some condition (such as a combination of simultaneous competing strategies) under which *any* strategy can be invaded (Boyd and Lorberbaum, 1987).

First, let's use ESS analysis to confirm that, without population structure, a naive cooperative strategy like ALLC cannot compete with the ruthless defector ALLD in the IPD. We will use the notation $V(A|B)$ to denote the expected payoff to an agent using strategy A against an agent using strategy B. In a large population where nearly everyone is a defector, an ALLD agent always plays other ALLD agents and receives no payoff, so its expected payoff is

$$V(ALLD|ALLD) = 0 \tag{6.26}$$

An invading ALLC cannot find other ALLC agents, and so must play ALLD. Its expected payoff is

$$V(ALLC|ALLD) = -c - wc - w^2c - \cdots = \frac{-c}{1-w} \tag{6.27}$$

Since $-c < 0$, ALLC can never invade. Being indiscriminately nice in a population where everyone is selfish does not pay.

Next, we can show that, should there somehow arise a community of gentle souls who cooperate warmly with everyone without thought, it could be rapidly and catastrophically invaded by those who would exploit their generosity. If nearly everyone plays ALLC, the expected payoff to a cooperator is

$$V(ALLC|ALLC) = b - c + (b-c)w + (b-c)w^2 + \cdots = \frac{b-c}{1-w} \tag{6.28}$$

This seems pretty good, and it is definitely better than the zero payoff everyone gets in a world of defectors. In other words, a population of cooperators does better than a population of defectors.[10] Can it last? An invading defector receives the benefit of cooperation but doesn't pay the cost. Its payoff is

$$V(ALLD|ALLC) = b + bw + bw^2 + \cdots = \frac{b}{1-w} \tag{6.29}$$

Since $c > 0$, ALLD can always invade, and indeed will increase to fixation, as we saw in Figure 6.1.

All is not lost, however. This section is about reciprocity, so instead of naïve cooperators, let us now assume a cooperative population of savvy TFT individuals. From the outside, this population looks just like ALLC individuals, because they all happily cooperate with one another. Their expected payoff when common is the same as the population of ALLC:

$$V(TFT|TFT) = b - c + (b-c)w + (b-c)w^2 + \cdots = \frac{b-c}{1-w} \tag{6.30}$$

[10] Indeed, there are arguments for the evolution of cooperation based on theories of *group selection*, which posit that strong group structure may facilitate the assortment needed for cooperative behaviors to spread (Wilson and Wilson, 2007).

When ALLD agents attempt to invade, however, they find something quite different from ALLC. After being exploited on the first round of play, TFT returns defection with defection. An invading defector's expected payoff is therefore:

$$V(ALLD|TFT) = b + (0)w + (0)w^2 + \cdots = b \tag{6.31}$$

TFT is an ESS against ALLD whenever the expected payoff to invading defectors is less than the expected payoff to the dominant TFT individuals. This happens whenever

$$\frac{b-c}{1-w} > b \tag{6.32}$$

Take a moment and solve this inequality for c. We will see that TFT can resist invasion by ALLD whenever

$$\boxed{wb > c} \tag{6.33}$$

TFT will be able to resist invasion by ALLD as long as interactions last long enough so that repeated acts of mutual cooperation are sufficiently beneficial that they outweigh the benefit of one-shot exploitation. Furthermore, the length of interactions (w) needed to stabilize cooperation increases when the cost of cooperation (c) is larger and decreases when the benefit of receiving cooperation (b) is smaller. You should notice that this inequality looks very similar to Hamilton's rule, and in fact it conveys the same basic idea. Here, positive assortment stems not from the extent to which cooperative *individuals* assort with one another, but the degree to which *acts* of cooperation are directed toward cooperative individuals.

OK, so a savvy, reciprocating cooperator like TFT can resist invasion by ALLD. Reciprocity can stabilize cooperation. But can such a strategy make inroads when ALLD is dominant? The short answer is no. This becomes clear when you consider that TFT cooperates once with each defector, so $V(TFT|ALLD) = -c$ while $V(ALLD|ALLD) = 0$. Because $-c < 0$, ALLD is an ESS against TFT. This seems problematic. However, we should also recall that cooperators do best when they can assort. Our ESS analysis has so far assumed that all interactions are random. It should be no surprise at this point that if a small community of invading TFT individuals can interact with each other at a rate higher than chance, then TFT can indeed invade and spread in a population of ALLD.

6.7.4 TFT Can Invade ALLD with Positive Assortment

Here we will perform an analysis similar to our derivation of Hamilton's rule. Let the frequency of TFT in the population be denoted as p, and let r be the probability *above chance* that an agent interacts with another agent of the same type. Let $\Pr(B|A)$ denote the probability of encountering an agent of strategy B given that one's own strategy is A. The total probability that a TFT interacts with another TFT agent is thus:

$$\Pr(TFT|TFT) = r + (1-r)p,$$

where the first term on the right side of the equation is the probability of a nonrandom assortative interaction, and the second term is the probability that an assortative interaction occurs by chance. Purely random mixing occurs when $r = 0$. In a population with both TFT and ALLD individuals, the expected payoff to a TFT agent is

$$V(TFT) = \Pr(TFT|TFT)V(TFT|TFT) + \Pr(ALLD|TFT)V(TFT|ALLD) \tag{6.34}$$
$$= \left(r + (1-r)p\right)\frac{b-c}{1-w} - (1-r)(1-p)c$$

and the expected payoff to an ALLD agent is

$$V(ALLD) = \Pr(TFT|ALLD)V(ALLD|TFT) + \Pr(ALLD|ALLD)V(ALLD|ALLD) \tag{6.35}$$
$$= (1-r)pb$$

recalling that $V(ALLD|ALLD) = 0$. When TFT is rare in the population, the probability of encountering a TFT agent by chance is effectively zero, so we can use the approximation $p \approx 0$. TFT will invade a population of ALLD when V(TFT) > V(ALLD), or when

$$r\frac{b-c}{1-w} - (1-r)c > 0 \tag{6.36}$$

We can first confirm that when there is no positive assortment and $r = 0$, the condition for invasion reduces to $-c > 0$, which is never true. Solving for r, we can obtain the minimum amount of positive assortment for TFT to invade ALLD:

$$r > \frac{1-w}{b/c - w} \tag{6.37}$$

This inequality indicates that if there is positive assortment among cooperators, then TFT agents *can* invade a population of pure defectors. Moreover, the amount of assortment needed for TFT to invade decreases when interactions last longer (larger w) and when the benefit of cooperation is higher relative to its cost. There is another noteworthy aspect to this expression. When games only last for one round ($w = 0$), the expression reduces to $rb > c$, which is Hamilton's rule for the evolution of cooperation. This indicates that the amount of assortment needed for a reciprocating strategy like TFT to spread is substantially less than the assortment needed by a pure cooperating strategy when games are not iterated.

6.7.5 Three's Company

The analyses in this chapter have focused on competitions between two strategies. Adding a third strategy into the mix can make things more complicated, because the frequency dependences interact. Let us consider the case where ALLC, ALLD, and TFT are all present in a population, and TFT is dominant. TFT can resist invasion by ALLD. However, ALLC can invade by **neutral drift**, because it has the same expected fitness as TFT in the absence of ALLD. Once ALLC is sufficiently common, ALLD may be able to invade by exploiting ALLC. However, if TFT is sufficiently common, ALLD will be driven out once the ALLC population has diminished, leading to cycles of cooperation and defection.

This system can be solved analytically using ESS analysis and replicator dynamics. However, it is also instructive to observe this dynamic in simulation. We can alter the code for our reciprocity model so that all three strategies can coexist. In order to allow repeated invasion of new strategies, we can add a mutation parameter, μ, which is the probability that an agent will adopt a random strategy instead of the strategy of its highest-performing neighbor. I have not provided the code for this model extension, and have instead left its creation as an exercise. An example of the cyclical dynamics that can emerge is shown in Figure 6.9.

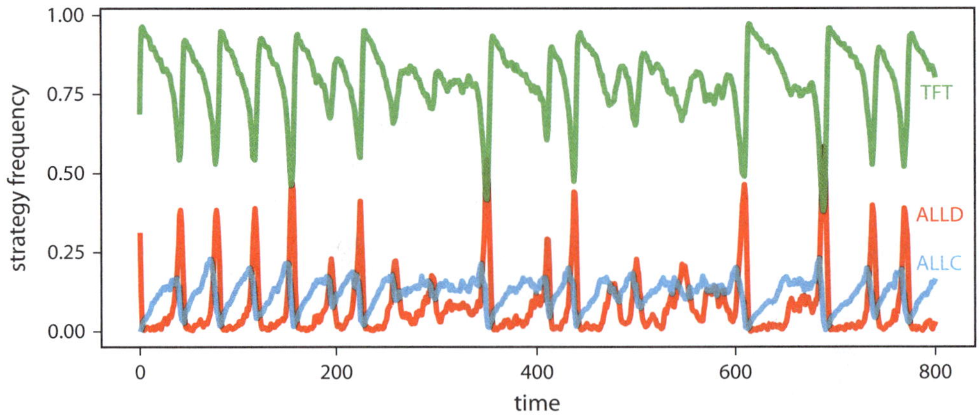

Figure 6.9 Temporal dynamics of the frequencies of ALLC (blue), ALLD (red), and TFT (green) in an agent-based model where all three strategies compete in a well-mixed population. Here $b = 1$, $c = 0.5$, $k = 4$, and $\mu = 0.02$.

6.8 Reflections

Altruistic cooperation can evolve and be maintained by mechanisms that preferentially direct the benefits of cooperation toward other cooperators. This statement is about as strong as general principles of social evolution get, and it has profound implications for understanding the nature of our psychology, our norms and institutions, and our cultures. It not only helps us understand how cooperation arises, but also how it can break down, as when free riders are able to target and exploit cooperators. The analyses in this chapter also hopefully further highlight the complementary nature of analytical and agent-based approaches.

We should ask whether our models for the evolution of strategies in the prisoner's dilemma game are actually good representations of the types of cooperation we're interested in knowing about. For example, not all cooperation is equivalent—two different cooperative partners may yield different benefits due to differences in their intrinsic abilities, their suitability to the specific task, and their compatibility with the focal agent concerning norms, goals, and expectations. Exactly how the benefits of cooperation are actually generated is an important consideration, and one that has been less widely studied (Calcott, 2008; Smaldino, 2014, 2019). In addition, cooperation is not always a good thing, as when companies collude to raise prices, bullies coordinate to antagonize their victims, or conquering armies exterminate their foes. Cooperation can go hand in hand with parochialism and xenophobia. Further, universal cooperation is never a stable equilibrium, as there will always be some force that can disrupt it (Boyd and Lorberbaum, 1987). It is possible, for example, that widespread cooperation could lead to sufficient levels of tolerance and naïve generosity that would create the conditions for the invasion of free riders.

I'd like to end this section by reflecting on what has become known as experimental or behavioral game theory (Camerer, 2003). Behavioral scientists have used the predictions from economic and evolutionary game theory to generate hypotheses that are then tested on human (and occasionally non-human) participants. Individuals are asked to play games like the prisoner's dilemma with (usually anonymous) partners, where the payoffs typically take the form of cash payments. The results of these experiments are often quite interesting, particularly when actual human behavior deviates from the predictions of the models, or when there is substantial cultural variation in participant behavior (Henrich et al., 2005; Gerkey, 2013). A problem arises, however, when the experimental games are then used to draw overly strong conclusions about the classes of behavior the mathematical games were developed to model. The prisoner's dilemma is a model of a cooperative dilemma involving the benefits and costs of providing aid to others. In reality, cooperative behaviors take many forms and often involve factors like shared identity, joint attention, common ground, and both verbal and non-verbal communication. The model abstracts all of these factors away to focus on a core, coarse-grained aspect of the behavior. When studying real humans (and other animals), we should also be interested in their real, consequential behavior. Experiments are themselves models—they represent an abstraction of a larger class of behaviors and scenarios. Experimental games then can be seen as models of models, in which the target representation is not real-world behavior per se, but the abstraction of the formal models. Whether this is a problem is open to debate.

6.9 Going Deeper

There is a vast, vast amount of research on models of cooperation, much of it focused on solving the prisoner's dilemma (though not always—see Box 6.2: Other Models of Cooperation). In general, a lot of the research on cooperation explores various mechanisms for assortment, which also includes work on group structure, network structure, movement strategies, environmental harshness, and reputation management, sometimes called indirect reciprocity. There is also a lot of empirical work on cooperation, in both humans and non-human organisms (not just animals, but plants and bacteria too!). It's a very rich literature. The research presented here on cooperation and assortment dates to Hamilton's (1964) work on altruism and inclusive fitness. The work on reciprocity stems from two seminal papers, one by Trivers (1971) and the other by Axelrod and Hamilton (1981). It has been shown, however, that tit-for-tat performs substantially less well when agents can make mistakes. For example, if an agent's well-intentioned act of cooperation is misperceived as defection, TFT can become destabilized into a cycle of alternating cooperation and defection.

BOX 6.2: Other Models of Cooperation

The prisoner's dilemma game is the archetypal model for altruistic cooperation, and the most widely studied of all cooperation models. However, it is not the only model of cooperation. Here I will mention three others. The first two are also two-player games like the prisoner's dilemma, each with two possible moves, and can be defined in terms of the relationships between the values in the payoff matrix. Consider a generic payoff matrix (Table 6.3). The game is a prisoner's dilemma when $B > A > D > C$ (and, strictly speaking, when $A > (B + C)/2$, so that players cannot just take turns exploiting one another). Mutual cooperation is better than mutual defection, but exploiting a cooperator is the best of all, while being exploited is the worst. The form of the prisoner's dilemma game used in this chapter is a special case in which the cost and benefit of cooperation are modeled explicitly.

Table 6.3. Generic payoff matrix for a two-player game with two possible moves.

	Cooperate	**Defect**
Cooperate	A	C
Defect	B	D

A related game is the **snowdrift game**, in which $B > A > C > D$. Note that the last two terms have flipped. The story that gives the game its name goes something like this. Two friends are driving somewhere during a blizzard. The car veers off the road and gets trapped in a snowbank. There are shovels in the car, which the friends can use in order to dig a path back to the road. If both players cooperate and shovel together, they do much better than if they both refuse to help. If your friend cooperates and shovels, you do better by defecting and staying toasty warm while your friend does all the work. However, mutual defection is *not* a Nash equilibrium here. If your friend refuses to help, you do better by shoveling the snow yourself, because being cold and tired (and probably a little resentful) is still better than being trapped indefinitely in a snowbank. The best solution is in fact anticoordination—doing the opposite of whatever your co-player does—although the defector comes out on top in this situation. The snowdrift game is sometimes used to model situations involving mutualisms, in which an individually beneficial action produces a benefit to others as a by-product.[11] Cooperation here is not altruistic, because a cooperative act benefits the cooperator directly (Sugden, 1986; Doebeli and Hauert, 2005).

Another sort of cooperation is captured by the **stag hunt game**, in which $A > B > D > C$. The story here concerns two hunters, who can choose to cooperate and hunt collaboratively for stag, or defect and hunt individually for hare. Successfully hunting stag is presumed to require cooperation, whereas hare can be successfully hunted solo. This game is actually a type of coordination problem, which we will discuss more in the next chapter. Not only is mutual cooperation better than mutual defection, but there is no incentive to defect from mutual cooperation, because working together is better than working alone. The problem comes from the fact that both mutual cooperation and mutual defection are Nash equilibria. Although cooperation is the best possible outcome, the difficulty is to convince your partner to switch their behavior when doing so unilaterally is costly. This game is sometimes used to model problems of social cohesion, including how the benefits of cooperation can be generated (Skyrms, 2003; Calcott, 2008).

All of the games discussed so far are for two players. This captures many cooperative interactions, but also omits the fact that cooperation often involves larger groups. A resource may be sought or shared by many individuals, who can contribute to the group interest or attempt to free ride on the contributions of others. There are many n-person cooperation games (Ostrom et al., 1994), but for the sake of brevity I will discuss only the most widely studied one. The **public goods game** is a sort of n-person prisoner's dilemma game. There are a few variations on this game, but the simplest works as follows. Each of n players can contribute to a public good at a cost c. Their contribution provides a benefit $b > c$ to the group that is divided evenly among its members, so that each individual gets $b/n < c$ for every cooperative group member.

The dilemma is similar: the group does best when all individuals contribute, but individuals gain a relative advantage by defecting. The game provides a formalization of what has been called the "tragedy of the commons," in which a group destroys commonly held resources by overexploiting them. An important result is that reciprocity generally fails to maintain cooperation in the public goods game, because other cooperators end up suffering as a result of retaliatory defection (McElreath and Boyd, 2007). The public goods game is often used to study how institutions such as punishment and policing can enforce cooperation even when reciprocity fails (e.g., Boyd and Richerson, 1992; Hooper et al., 2010).

[11] The hawk-dove game is a reimagining of the snowdrift game that is often used to model conflict. See Maynard Smith (1982).

When errors like this occur, a more ruthless strategy called Pavlov can dominate (Nowak and Sigmund, 1993). Pavlov cooperates on the first move and repeats its previous move if its opponent cooperates but switches its move (to defect if it had been cooperating, to cooperate if it had been defecting) if its opponent defects. In this way, Pavlov can exploit suckers who will cooperate naïvely while simultaneously being able to recover from errors. Starting with cooperation isn't even always the best move. Axelrod (1997a) used genetic algorithms (see chapter 10) to evolve potentially complex strategies under a range of conditions, where each strategy played the IPD against many others. Reciprocal strategies like TFT often evolved, but many successful strategies also defected rather than cooperated on the first move.

The assumption that cooperation is a discrete behavior, and that the only alternative is to defect, is quite a strong one that raises new questions. What are the logistic mechanisms that generate the benefits of cooperation? What cognitive mechanisms are required to recognize cooperators and effectively coordinate with them? Moreover, why play the game at all? The prisoner's dilemma game framework typically assumes that all individuals have the same number of interactions, and must simply choose to cooperate or defect. But perhaps sociality is itself a variable strategy. The ability to leave the game or seek out multiple games is surely an important consideration. For example, simple models have shown how assortment (and hence cooperation) can be enhanced if individuals simply walk away from partners who defect (Aktipis, 2004; Santos et al., 2006a; Pepper, 2007). On the other hand, cooperation can also be harmed if it is defectors who are mobile. If defectors can widely explore the space of possible partners, they can avoid being outcompeted by cooperators (Koella, 2000; Smaldino and Schank, 2012b) and even overcome reciprocal strategies like TFT (Dugatkin and Wilson, 1991; Enquist and Leimar, 1993). Observable markers or tags can evolve to facilitate cooperation with the in-group, and cognitive mechanisms may also have evolved to navigate markets for better partners (Noë and Hammerstein, 1994; Tooby and Cosmides, 1996; Hammond and Axelrod, 2006; Smaldino, 2019).

Conditions that are initially bad for cooperation can give rise to conditions that are good for it. For example, harsh environments without assortment can lead cooperators to be exploited to extinction, but this may set the stage for greater cooperation in the future when most survivors are those who managed to work together (Smaldino et al., 2013). However, conditions that are initially good for cooperation can also give rise to conditions that are bad for it, such as when expanding network ties create more opportunities for cooperation, and hence for exploitation (Akçay, 2018). There are also problems involving cooperation beyond merely avoiding or punishing free riders. Group boundaries and markers can help individuals identify who they should cooperate with or avoid, for better or worse. Cooperation is part and parcel with intergroup conflict, and because more cooperative groups will be better able to compete, conflict may actually drive cooperation (Choi and Bowles, 2007; Makowsky and Smaldino, 2016; Turchin, 2015; Zefferman and Mathew, 2015).

6.10 Exploration

1. **You've changed.** Consider the ABM of the simple PD game with assortment. The `evolve` procedure has a problem. Each agent copies the strategy of its highest performing neighbor. All agents accumulate payoffs before anyone copies their neighbors. However, due to the way the procedure is coded, there may be some cases in which the strategy being copied is not the strategy that helped the agent get their high payoff. Fix the code so that agents always copy the strategy that was used by their neighbor to acquire their payoff (hint: you can use the fact that an agent's color and strategy carry redundant information). Include a switch for a Boolean variable called `synchronous` that allows you to toggle between the original version of the update rule and your new version. Is there any difference between the two conditions in terms of either the proportion of cooperators at equilibrium or the spatial patterns that emerge?

2. **Everyone's a winner.** Consider the simple ABM with assortment.
   (a) Write equations for the payoffs to a cooperator and a defector as a function of the number of cooperative neighbors each has (this can vary between 0 and 4).
   (b) Using the plotting software of your choice, plot $V(C) - V(D)$ as a heatmap (or equivalent) where the x and y axes are n_C for the cooperator and defector, respectively. If the color changes at zero, it will become clear how strongly cooperation or defection is favored. Keeping $b = 1$, make three plots for $c = \{0.2, 0.4, 0.8\}$.
   (c) How do you reconcile that there appear to be scenarios in which cooperation is favored even for high c, while defection usually wins in the corresponding model simulations?

3. **Random invasion.** Consider the ABM with pure strategies and randomization. We saw that without randomization, cooperation was readily maintained in the population as long as the cost of cooperation was low. The long-run frequency of cooperation decreased as randomization increased, until sufficient randomization made it impossible for cooperation to persist. However, the analyses in the chapter assumed that cooperators were common in the initial population, and therefore that the chance of a cluster of cooperators forming at random was high. Now, instead of assuming that cooperators are initially common, assume that they are initially rare. In order to spread, they must "invade" a noncooperative population. Perform a batch run with the model to recreate Figure 6.6 with one major difference: you will start your simulations with only 5% cooperators in the population instead of 50%. Are the results similar to when we started with 50% cooperators? If not, why not?

4. **Reciprocal invasion.** How easy is it for TFT to invade ALLD? Run batches of the ABM with TFT with only 5% initial cooperators, $b = 1$, and $c = 0.2$. Vary the probability of randomization from 0 to 0.2, using several values for the number of

iterations, {1, 2, 3, 4}. Run 20 simulations per batch. What do you find? Plot your results.

5. **Three's company.** Extend the agent-based model of reciprocal cooperation to allow all three strategies (ALLC, ALLD, and TFT) to coexist simultaneously. You will need two additional changes. First, remove the stopping condition so that the model does not stop when one strategy dominates. Second, add a parameter to control mutation, which is the probability that an agent will adopt a random strategy instead of the strategy of its highest-performing neighbor.
   (a) Report the code for this model.
   (b) Write a formal description of the updated model, such as would appear in an academic paper or technical report.
   (c) Describe the behavior of this model when all three strategies are initially present in comparison to models with only ALLC or only TFT. Do you observe dynamics as in Figure 6.9? Support your description with graphs or screenshots, exploring several combinations of parameter values.
6. **The new mutants.** Repeat Problem 5, but now add another global parameter called `mutation-rate`. This is the probability that, rather than copying the strategy of their highest-performing neighbor, an agent adopts a strategy at random. *This can include strategies that have gone extinct.* Plot the frequencies of all three strategies over time for several different parameter values. What do you observe?
7. **The even newer mutants.** Modify the model from Problem 6 so that the number of iterated rounds of the PD game played by each group of agents is stochastic. In the ABM provided for the IPD, the number of rounds was constant, controlled by the parameter `num-iterations`. However, the analytic model assumed that the number of rounds was probabilistic. Update the model code so that the number of rounds each agent plays with its neighbors is a random draw from a distribution of your choice (e.g., a uniform or a normal distribution), with a mean of `num-iterations`. Add a Boolean switch that allows you to toggle between a deterministic and probabilistic number of rounds. How does the additional stochasticity change the overall population dynamics?
8. **Sticking together.** Show that TFT can invade a population of ALLD when agents can preferentially assort. Let r be the probability above chance that similar strategies interact, and let p be the frequency of TFT in the population.
   (a) Write equations for the expected payoffs of TFT and ALLD.
   (b) Show that when cooperators are initially rare (so $p \approx 0$), TFT can invade ALLD if

$$r > \frac{1-w}{b/c-w}$$

   How do you interpret this inequality? What happens when play is not iterated ($w=0$)? What happens when there is no positive assortment ($r=0$)?
9. **Sticking together in simulation.** Demonstrate, using agent-based modeling, that TFT can invade a population of ALLD when agents can preferentially assort. Develop an algorithm that initializes TFT under various levels of assortment using an assortment parameter we can call α. When $\alpha = 0$, all agents will be initially placed randomly on the grid. As α increases, TFT will be increasingly likely to be placed

next to other TFT agents. Show that as α increases, TFT can invade ALLD with increasingly small initial numbers of TFT agents.

10. **Evolutionary dynamics of TFT.** Both TFT and ALLD are ESSes against each other when the population is well mixed. That is, TFT can resist invasion by ALLD, and vice versa. That implies that there is a threshold frequency of TFT above which TFT will increase in frequency, and below which it will decrease. Find this threshold. Assume a well-mixed population where all individuals are either ALLD or TFT, and let p be the frequency of TFT.

    (a) Write out the equations for the expected payoffs to both TFT and ALLD as a function of p, b, c, and w.

    (b) Show that the threshold frequency at which the two strategies have equal fitness is

    $$\hat{p} = \frac{(1-w)c}{w(b-c)}$$

    (c) One can track the evolutionary dynamics of multiple strategies by considering their discrete-time *replicator dynamics* (see Box 6.1). For two strategies A and B, where p is the frequency of A in the current generation, the replicator dynamics are represented by the following difference equation, so that the change in the frequency of a strategy is a function of its relative fitness—its fitness relative to the fitness of other strategies in the population:

    $$\Delta p = p(1-p)\frac{V(A) - V(B)}{pV(A) + (1-p)V(B)}$$

    Plot the change in the frequency of TFT, Δp, as a function of the current frequency, p. Use values $b = 1$, $c = 0.5$, and $w = 0.8$. Show that this function crosses from positive to negative at $\hat{p}$.

11. **Making mistakes.** Sometimes someone might intend to cooperate, but nevertheless fail to do so. For example, you might intend to drive your friend to the airport, but you mark the wrong date on your calendar and oversleep. Consider the evolution of reciprocal cooperation when there is this sort of error.

    (a) Consider the following table of payoffs for an IPD game with $k = 8$ iterations.

	Round	1	2	3	4	5	6	7	8
TFT 1	Payoff	$b-c$	$b-c$	$b-c$	$b-c$	$b-c$	$b-c$	$b-c$	$b-c$
	Behavior	C	C	C	C	C	C	C	C
TFT 2	Behavior	C	C	C	C	C	C	C	C
	Payoff	$b-c$	$b-c$	$b-c$	$b-c$	$b-c$	$b-c$	$b-c$	$b-c$

    The total payoff for each TFT player is $k(b-c)$. However, consider what happens if player 1 makes a mistake in round 3, and accidentally defects. Rewrite this table with that error accounted for. Assuming no further errors, what is the total payoff for each player after k rounds? Why might errors of this type create conditions for ALLD to invade?

(b) How could you implement errors of this sort in the agent-based model of reciprocity? Consider the difficulties in doing so given how payoffs are calculated as described in the chapter (and observed in the accompanying NetLogo code). Explain the problem and then write a revised formal description of the model dynamics in which errors occur with probability ϵ. An error occurs when an agent who intended to cooperate accidentally defects. It is up to you whether the agent accidentally defects with all four of its neighbors or only one.

7

Coordination

> But the Lord came down to see the city and the tower the people were building. The Lord said, "If as one people speaking the same language they have begun to do this, then nothing they plan to do will be impossible for them. Come, let us go down and confuse their language so they will not understand each other."
>
> —*Genesis* 11:5-7, New International Version

Cooperation involves more than simply avoiding free riders. Once we've solved the free rider problem and managed to assort with other cooperators, we shouldn't settle for just *any* cooperative partner, because all cooperators are not the same. Cooperation involves more than acting generously or providing others with a benefit; it's also about finding ways to most effectively generate that benefit and avoid conflict. If cooperative partners are misaligned on their expectations or goals, can't communicate well, or just don't do things in the same way, cooperation can be impeded despite best intentions. A functioning society thus requires not only cooperation, but **coordination**.

Language is a prominent example of an activity that requires coordination. Conversation partners must agree on a lexicon of words and how those words are mapped to meanings for communication to be effective. If we can't coordinate on a set of words to denote particular situations, we are liable to have a misunderstanding. Of course, misunderstandings can occur regardless, but they are likely to be less severe among people who speak the same language than between those who do not. In the Old Testament book of Genesis, God exacts a severe punishment on the people for their hubris of attempting to build a tower to Heaven. He confounds their speech so that they no longer understand one another, and therefore can no longer effectively work together to build the Tower of Babel. It is telling that this lesson about the importance of coordination was so deeply ingrained in a widespread and long-lasting cultural mythology. It speaks to the critical importance of being able to anticipate one another's behaviors and motivations in order to work together.

One way that societies solve the problem of coordination is by instituting and enforcing **norms**, which dictate how people in a society will or should behave in a particular situation. Norms help people to reduce uncertainty and manage their expectations about how others will behave. Norms can come in different flavors. For example, some scholars have distinguished between *descriptive* and *injunctive* norms (Cialdini et al., 1990; Bicchieri, 2005; Gavrilets, 2020), which can be thought of as the distinction between the "is" and "ought" meaning of norms. Descriptive norms describe what is typical or, as it were, *norm*al. They reflect an expectation of how people will behave. They are what people *usually* do. Injunctive norms reflect the morals and institutions of a society. They are what people ought to do, such that people will receive disapproval or even more punitive sanctions if they fail to adhere to the norm. Norms often differ from place to place. For example, in Berlin, people crossing a street against a traffic light (jaywalking) will meet with staunch disapproval, while in Manhattan, *failure* to jaywalk will be met with derision.

The distinction between descriptive and injunctive norms is not as cut-and-dried as it may appear, however. Some behaviors are normative merely because they are the most effective means of doing something, or because they happen to be popular or trendy at a given time. But even completely arbitrary norms can serve an important function, which is to facilitate coordination. As such, individuals may be penalized for failing to adhere to a norm not only through injunctive judgment, but also due to their failure to effectively coordinate. Consider that in most of the world, cars and trucks are driven on the right side of the road, but in the UK, Australia, and Japan, cars are driven on the left. It doesn't really matter which side of the road you drive on, only that everyone does it the same way. You will get into a lot of trouble if you try to drive on the wrong side of the road—not only because it is illegal, but because it creates a failure of coordination that can lead to accidents. In this chapter, I focus on a view of norms as *strategies for coordinating behavior.*

Although norms can be arbitrary, one norm isn't always just as good as any other norm. Some, like sharing a cigarette after certain vigorous activities, can be directly harmful. Other norms can be prosocial, improving the collective good. For example, norms proscribing murder and assault encourage civil order, and make us feel relatively secure. Property rights encourage productive effort and innovation. Well-managed taxation provides roads, schools, and other public goods. Institutionalized norms regarding product standards, building codes, and rules of professional conduct allow more efficient commerce and protect citizens from harm. Norms governing the filling of political offices reduce the chances of a civil war over political disputes.[1]

There are often many possibilities for exactly which norms will be established in a particular domain.[2] Drive on the right or the left? Speak English or German? Wear a business suit or yoga pants and flip-flops? Allow disputes to escalate or prohibit violence? Some norms may be more in line with basic human psychology and practicality than others (for example, we can probably take driving in swirly spirals off the table as a potential traffic norm), but it's still the case that many options are often possible. Human cultural diversity is a testament to this fact.

In this chapter, we'll consider how norms might spread or fail to spread in a population. Understanding coordination, as with cooperation, requires a perspective in which the utility

[1] Hopefully.

[2] As a stark illustration of varying norms across similar domains, consider my first two postdoctoral fellowships, which involved very similar work. For my first postdoc at the Johns Hopkins School of Medicine, "professional" attire was mandated, and I was forbidden from wearing jeans or sneakers at the office. To my horror, I was even required to wear a tie every Friday. On my first day at my second postdoc, in the much more laid-back Department of Anthropology at UC Davis, my supervisor met with me in his office while wearing a loose-fitting T-shirt, shorts, and flip-flops.

of a behavior depends on the behavior of others.[3] Most of the chapter will focus on games in which players must align on the same behavior. Later, we will also consider games in which players coordinate by adopting complementary behaviors. We'll once again use the framework of evolutionary game theory, in which strategies leading to higher payoffs are preferentially transmitted. In other words, we'll assume that individuals interact with others using particular strategies, and that those interactions lead to payoffs. Individuals will then have the opportunity to observe others and switch to a strategy that is more successful. To begin, we'll consider a scenario in which two norms compete, and neither has any intrinsic a priori advantage.

7.1 Norms with Symmetric Payoffs

The real world is quite complicated, and involves interactions with many people with many different options for how to behave. To help us make sense of this, we'll once again reduce the situation to something very simple: two-player games with two possible options. Let's imagine a population in which everyone uses one of two behaviors in their social interactions. We'll creatively refer to these behaviors[4] as norm 1 and norm 2. This is not a cooperative dilemma; it is always better to coordinate than not. Even noncoordination is presumed to be cooperative (that is, nonexploitative) and therefore yields a positive payoff to both players. In other words, we can assume that in this scenario, the dilemma of cooperation has been solved, and now the issue is how to maximize the benefit between potential cooperators.

We'll start with a **symmetric coordination game**, in which coordination always yields an advantage over noncoordination. Imagine two people walking toward each other down the center of a long hallway, each carrying dinner platters full of precariously stacked dishes. They both need to move to get out of the other's way. They can each veer to their right or to their left. If they both move in the same direction (egocentrically, so that they end up on different sides of the hallway), they avoid crashing and continue on their way. If they move in opposite directions, they end up colliding and all the dishes fall and shatter. It doesn't matter which way they veer, they just need to veer the same way. It would help if there were a norm for which way to move!

Without any loss of generality, we can set the baseline payoff to 1, and let δ be the additional payoff that each player gets by using the same norm as the other player. This yields the payoff matrix below.

A game of matching pennies. Moving to one side of a walkway. Driving on a fixed side of a two-way street. Shaking hands versus bowing when greeting someone. It doesn't matter which behavior you choose as long as it's the same one as your co-player. So, how does a population converge on a norm?

Let's assume that individuals have interactions with many other members of their community, during which they accumulate payoffs. We assume many interactions because norms involve populations that are already highly social, and individual events may be relatively less consequential than acts involving altruistic cooperation. We can therefore calculate the expected payoff to individuals using either of the two norms. Let p denote

[3] An early and influential analysis of coordination games to study norms and conventions is Lewis (1969), which contains useful philosophical discussion of many of the topics tackled in this chapter.

[4] Philosophers who study norms and conventions may feel comfortable referring to a behavior as a "norm" only once it is widely adopted. This is a reasonable objection, but for simplicity I'm going to ignore it.

Table 7.1. Payoff matrix for a simple, symmetric coordination game, indicating payoffs to row player. Playing the same strategy as one's opponent creates a benefit $\delta \geq 0$.

	Norm 1	Norm 2
Norm 1	$1+\delta$	1
Norm 2	1	$1+\delta$

the proportion of individuals using norm 1, so the frequency of norm 2 in the population is $1-p$. If interactions are random, the payoff to an individual using norm 1 is

$$V_1 = p(1+\delta) + (1-p)(1) = 1 + p\delta \tag{7.1}$$

This makes sense, because all interactions yield at least a payoff of 1, while only interactions with others using norm 1 will yield the additional benefit δ. Similarly, the expected payoff to an individual using norm 2 is

$$V_2 = (1-p)(1+\delta) + p(1) = 1 + (1-p)\delta. \tag{7.2}$$

Norm 1 should increase in frequency when $V_1 > V_2$. This will occur whenever

$$1 + p\delta > 1 + (1-p)\delta, \tag{7.3}$$

or whenever

$$p\delta > (1-p)\delta \tag{7.4}$$

Note that when $\delta = 0$, there is no benefit to coordination, and neither norm will have an advantage. If we assume that there *is* a coordination benefit and $\delta > 0$, we can divide each side of the inequality by δ. Some minor algebraic manipulation reveals that norm 1 will increase in frequency whenever

$$p > \frac{1}{2}. \tag{7.5}$$

In other words, whichever norm is initially more popular will spread, because individuals have the best chance of receiving the benefit of coordination if they adopt the more popular norm. Moreover, each norm is an ESS against the other, because the rare norm can never increase in frequency (as long as the assumption of a well-mixed population holds).

We can explore this model further with the replicator dynamics we discussed in chapter 6 (see Box 6.1: Replicator Dynamics). The discrete-time dynamic is given by the following recursion:

$$p' = p\frac{V_1}{pV_1 + (1-p)V_2} \tag{7.6}$$

Plugging in our values for V_1 and V_2 from Equations 7.1 and 7.2 yields the following:

$$p' = p\frac{1+p\delta}{1+\delta(1-2p+2p^2)} \tag{7.7}$$

We can also use this to derive a difference equation by recalling that

$$\Delta p = p' - p. \tag{7.8}$$

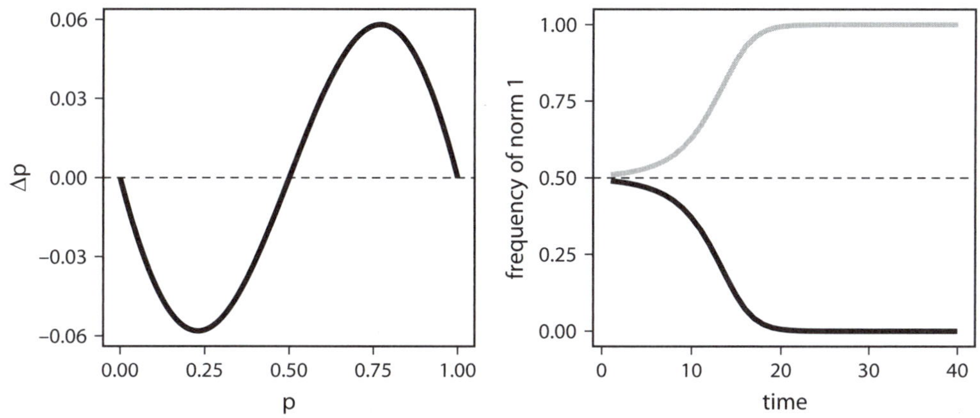

Figure 7.1 Equilibrium dynamics of the symmetric coordination model. Left: the difference equation for Δp plotted against p, the frequency of norm 1 in the population. Right: Replicator dynamics of norm 1 starting at either $p_0 = 0.49$ or $p_0 = 0.51$. In all cases, $\delta = 1$.

Plotting Δp against p shows us that the pace of evolution will increase and then decrease as p deviates from 0.5 (Figure 7.1, left). Importantly, Δp is negative for all $p < 0.5$ and positive for all $p > 0.5$. In the language of dynamical systems, $p = 0.5$ is therefore an **unstable equilibrium**. At this value, users of either norm receive exactly the same payoff, meaning that neither norm has an advantage, and so the system is at an *equilibrium*. The equilibrium is *unstable* because the slightest perturbation will send the system toward one of two extremes—complete saturation by either norm. We can see this illustrated by the right side of Figure 7.1, in which the temporal dynamics for p are plotted with p_0 either barely above or barely below 0.5.

Given the simplicity of the model and the thoroughness of our mathematical analysis, we probably don't need a simulation model to understand how norms spread under the assumptions of symmetric payoffs and random interactions. But we're going to build one anyway, both because it will be instructive and because it will allow us to build up more complex models whose behaviors are perhaps less intuitive. To proceed, we need to say a little about how transmission will work in the agent-based model.

7.1.1 Probabilistic Copying

In the agent-based models of cooperation we explored in the previous chapter, individuals interacted with a small set of neighbors, which introduced heterogeneity in payoffs among agents using any particular strategy. Agents then used a deterministic copying rule: copy the neighbor with the highest payoff. Here, we'll try something different in order to sample some of the different methods used in evolutionary modeling. Recall our assumption that individuals have *many* interactions with a variety of social partners. This means that, in the absence of a particular population structure, an individual's payoff will be well approximated by the expected payoffs we calculated above (we will consider the spread of norms in structured populations later in this chapter). If we were to assume deterministic copying here, then all agents would immediately copy the most popular strategy and the simulation would end very quickly. However, one of the strengths of agent-based modeling is its ability to capture heterogeneity and stochastic, even low-probability events. We *will* add stochasticity to our model of norms, but with regards to transmission rather than social interaction.

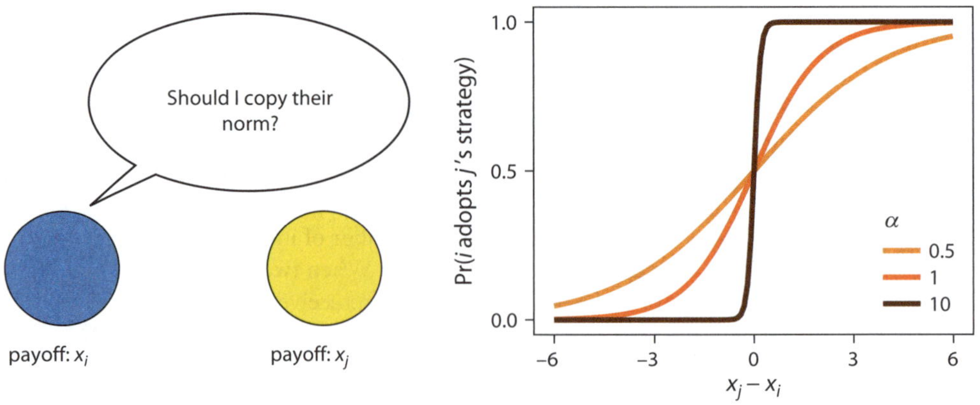

Figure 7.2 A sigmoid function describing the probability of copying an observed individual as a function of the difference in payoffs.

Success-biased copying produces an evolutionary dynamic in which the most successful individuals (the most "fit") are most likely to transmit their traits, and so is analogous to evolution by natural selection. However, there are usually *many* factors that lead an individual to be more or less fit, above and beyond their successes and failures at coordination on whatever norms we are modeling. This is why I think probabilistic copying makes sense. If your payoff is much higher than mine, I should be very likely to copy you. If your payoff is only a little higher than mine, I should still probably copy you, but I should also be more likely to stick with my current strategy (or try something different altogether) than I would be if your advantage was greater. After all, small differences in the real world are often the result of random chance. Similarly, if your payoff is worse than mine, I should generally stick with my current strategy, but every now and then I might make an error in judgment or want to explore, so I will occasionally copy an apparently worse strategy. Conveniently, there is a functional form that fits this description nicely: the **sigmoid** function (Figure 7.2).

Sigmoid functions are used often in models of social behavior and evolution (as well as in cognitive and decision models) because of how well they capture a sort of "soft threshold" approach. Sigmoids are also generated by a number of social processes, including contagion with social influence (chapter 4), as well as by the pattern of adoption under conformist learning strategies (Boyd and Richerson, 1985; Smaldino et al., 2018a). We'll use a sigmoid here to model the probability that individual i with payoff x_i adopts the behavior of observed individual j with payoff x_j,

$$\Pr(i \text{ adopts } j\text{'s strategy}) = \frac{1}{1 + e^{-\alpha(x_j - x_i)}}, \tag{7.9}$$

where α is a parameter that controls the slope of the increase from a probability near zero to a probability near one. It can be interpreted as the extent to which small advantages matter, and is sometimes called the "strength of selection" on the behavior in question. For simplicity, I've used $\alpha = 1$ for all the simulations presented in this chapter.

7.1.2 An Agent-Based Model of the Evolution of Norms

Now that we have a decent sense of how game play and evolution will work, we can build a simple agent-based model of norm adoption. We will study a population of N individuals, each of which is characterized by its use of one of two possible norms (1 and 2). Agents will be situated on a (now-familiar) fully occupied square lattice of size $L \times L$, so that $N = L^2$, though for now we will ignore spatial position and the effects of local neighborhoods (minimally, the spatial organization will help us to visualize the distribution of norms in the population). At initialization, each agent is assigned norm 1 with probability p_0 and norm 2 otherwise. Each agent will keep track of its own norm and its payoff.

The dynamics of the model can be subdivided into two stages: interaction and evolution. During the *interaction* stage, each agent has a large number of interactions with randomly selected partners, from which they accumulate payoffs. When two partners have the same norm, they each receive a payoff of $1 + \delta$, otherwise they receive a payoff of 1. Agents will receive their expected payoff based on the distribution of norms in the population. An agent using norm i will receive a total payoff of

$$V_i = 1 + p_i \delta, \tag{7.10}$$

where p_i is the frequency of norm i in the population. Technically, this isn't quite right when the population is finite and relatively small, because an agent can't interact with itself. The true probability of encountering an agent of one's own type is therefore $(n_i - 1)/(N - 1)$, where n_i is the number of agents using norm i. However, for reasonably large N, $(n_i - 1)/(N - 1)$ is extremely well approximated by $n_i/N = p_i$, and so we will continue to use the latter term. It is worth being aware of this approximation, however, because it could matter when modeling very small populations. During the *evolution* phase, each agent chooses one other agent from the population as its "model." The agent compares its payoff to that of its model, and uses the sigmoid function defined in Equation 7.9 to probabilistically copy its model's norm—it is more likely to copy if the model has a substantially higher payoff than itself. The model dynamics continue until the entire population has adopted one of the two norms.

7.1.3 Coding the Model

The NetLogo code for the coordination model with symmetric norms is **coordination_simple.nlogo**. The model is very simple, and only has two important global parameters: the initial prevalence of norm 1, p_0 (`init-norm1`), and the benefit of coordination, δ (`coordination-benefit`). As noted above, I have represented agents as colored circles on a square grid as we have done previously—this will help give us an intuition about the population dynamics, and will make it easy to modify the model to account for spatial interactions. However, this population structure is not strictly necessary; one could readily code this model without any consideration of spatial structure. If the lattice is not strictly necessary, why include it? I've done so for two reasons: (1) it makes the model dynamics easy to visualize, and (2) it allows us to easily add more rigid structure to the population. As you should now be familiar with much of the coding techniques used for this model, I have only presented the code for new elements specific to this model.

Agents must keep track of two agent-level variables: their own norm and their most recent payoff (`payoff`). We code an agent's norm as a Boolean variable `norm1?`, which is true if the agent uses norm 1 and false if it uses norm 2. At initialization, an agent is created on each cell of the grid and assigned a norm. With probability `init-norm1`, the agent uses norm 1, otherwise it uses norm 2. For the purposes of visualization, I have made norm 1 agents yellow and norm 2 agents blue (Figure 7.3).

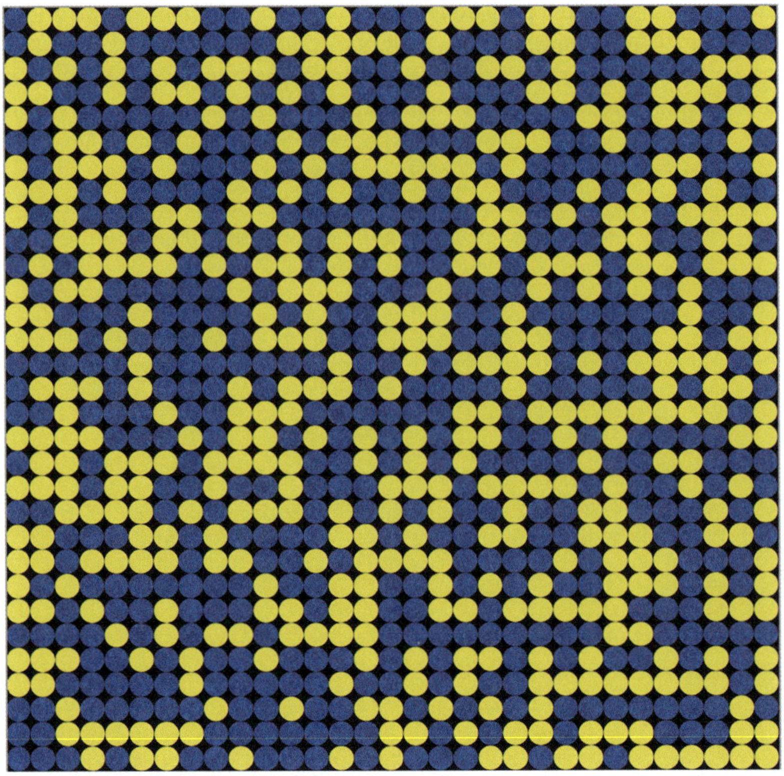

Figure 7.3 Spatial organization for the simple coordination model, with $p_0 = 0.49$ and $L = 31$. Norm 1 agents are yellow, norm 2 agents are blue.

The dynamics, controlled by the go procedure, are in many ways similar to the cooperation model from the previous chapter. First, the simulation checks to see if all agents are using the same norm, and stops if that is the case. Again, this is a useful assumption for minimizing the computational time needed for batch runs, but it can be problematic if the universal adoption of one norm is not a steady state, as in the case of mutation (in which case the stopping condition should be removed). If both norms persist, the dynamics proceed in two stages: interaction (`calculate-payoffs`) and evolution (`evolve`). Finally, agents are recolored based on their norms.

NetLogo code 7.1

```
to go
  if (count turtles with [norm1?]) = (count turtles) or
     (count turtles with [norm1?]) = 0
  [ stop ]
  calculate-payoffs
  evolve
  recolor
  tick
end
```

In the interaction stage, the procedure `calculate-payoffs` is called. The expected payoffs are calculated for an agent using either norm, using Equation 7.10. Agents are then assigned payoffs based on their norms. Because we assume a large number of random interactions, this expected value is the same for all agents using a given norm.

NetLogo code 7.2

```
to calculate-payoffs
  let p1 (count turtles with [norm1?]) / (count turtles)
  let norm1-payoff 1 + (p1 * coordination-benefit)
  let norm2-payoff 1 + ((1 - p1) * coordination-benefit)
  ask turtles with [norm1?][ set payoff norm1-payoff ]
  ask turtles with [not norm1?][ set payoff norm2-payoff ]
end
```

In the evolution stage, each agent chooses another agent at random to observe. The observer switches to the norm of the observed agent with a probability based on the sigmoid function described earlier (with $\alpha = 1$). Otherwise it keeps its own norm. On some occasions the agent will observe another agent that is already using the same norm as itself. On these occasions, the agent will always keep its current norm.

NetLogo code 7.3

```
to evolve
  ask turtles[
    let model one-of other turtles
    let other-payoff [payoff] of model
    let prob-copy (1 / (1 + exp (-1  *  (other-payoff - payoff))))
    if random-float 1 < prob-copy
    [set norm1? [norm1?] of model]
  ]
end
```

7.1.4 Analyzing the Model

Our mathematical analysis tells us that whichever norm is initially dominant should increase to fixation, and we will see that this is indeed what happens in our stochastic model. But first, let's examine the case where there is no benefit to coordination ($\delta = 0$). Both norms can, in theory, persist indefinitely in this case, as the frequency of each norm over time is essentially a random walk. In the long run, one norm *will* eventually go to fixation even without any selective advantage, just by chance. In evolutionary theory this is called **neutral drift**. Drift is known to be more rapid and its effects more substantial—traits can both enter and leave the population—in small populations where each individual represents a large proportion of the total (Wright, 1938; Kimura and Ohta, 1969). An example dynamic is illustrated in Figure 7.4. Notice that although the frequency of norm 2 exceeds 50% for quite a long time, it is norm 1 that eventually goes to fixation. It could have gone either way, of course. Accounting for random factors such as drift is important to keep in mind when evaluating dynamic processes. I have illustrated this further in Figure 7.5, which shows the dynamics of 10 runs of the model for varying population sizes. With no selective force acting, smaller populations go to fixation much faster than large populations.

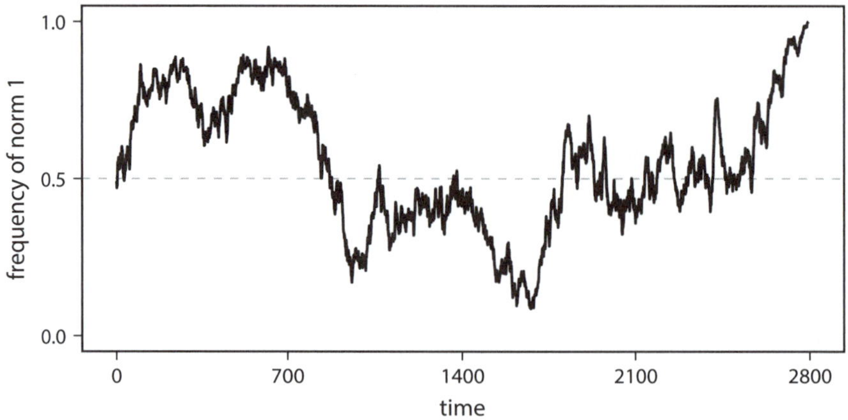

Figure 7.4 An example of neutral drift. In a population in which neither norm has a selective advantage ($\delta = 0$), one norm may nevertheless go to fixation through neutral drift.

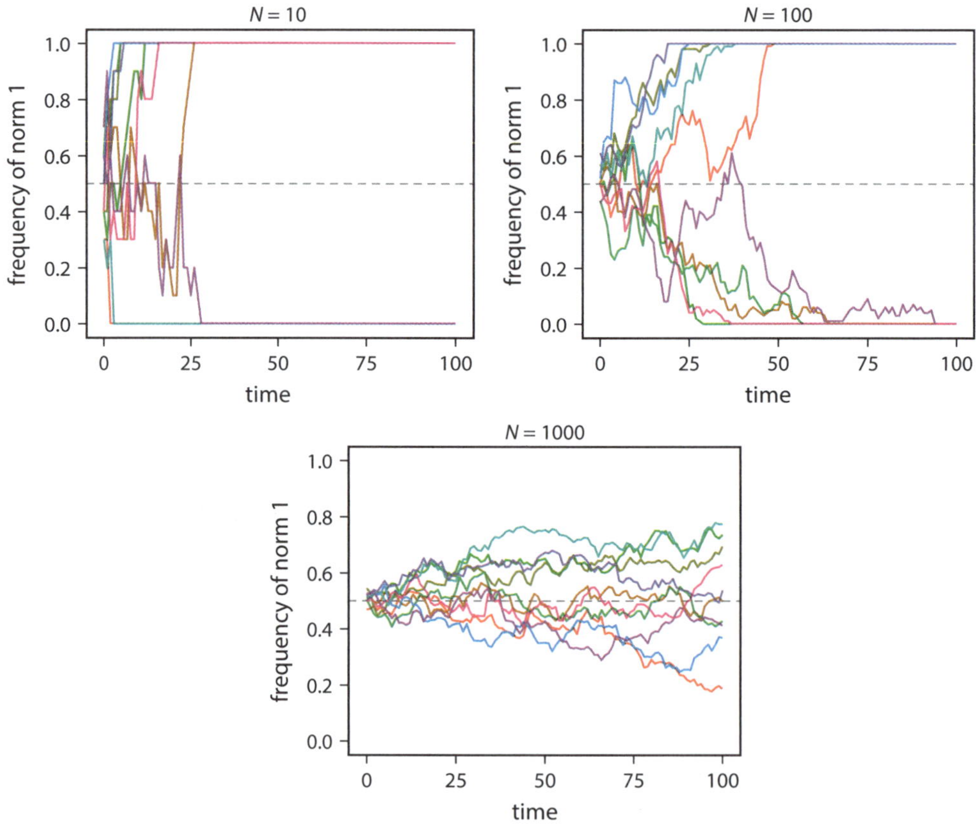

Figure 7.5 Neutral drift and population size. Each plot shows the dynamics of ten runs of the model for different population sizes. When the population is small, it can quickly drift to fixation.

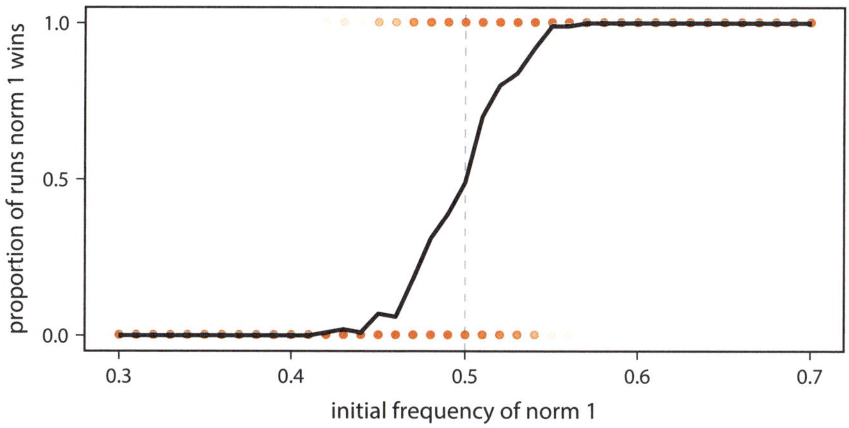

Figure 7.6 The proportion of simulation runs that ended with norm 1 dominating as a function of the initial frequency of norm 1, p_0. Orange circles are results for individual runs; the black line connects the means across 100 runs for each condition. For all simulations, $N = 961$, $\delta = 1$.

As soon as there is any benefit to coordination ($\delta > 0$), the more common norm almost always increases to fixation (Figure 7.6), as predicted by our mathematical analysis. However, there is still *some* uncertainty when the two norms start out at very similar frequencies ($p \approx 0.5$), due to the stochasticity in the copying rule. Noise can make the outcomes of dynamical social systems less predictable when selective forces are weak.

Our simple model appears to have major implications for the spread of norms in a population. It suggests that once a norm is established in a population, it should be extremely stable. And indeed, norm change is far from easy—it is often a great challenge to alter how things are done. However, even a cursory knowledge of cultural history tells us that, at least sometimes, norms *do* change. This implies that norms that are initially rare *can* sometimes increase in frequency and become dominant. If the outcomes of our model are not aligned with empirical observations, then a key assumption of our model must be wrong. One of our assumptions has been that the two competing norms are perfectly symmetrical—that is, that neither norm has an intrinsic advantage over the other. That assumption does not always hold. We will tackle asymmetric payoffs in the next section.

7.2 Group-Beneficial Norms

Let's consider two competing norms such that groups that adopt one norm tend to outperform groups that adopt the competing norm. Even group-beneficial norms still require coordination, however, and can be costly when rare. We can even consider norms that are cooperative dilemmas in this light. A norm of wearing your pants on your head instead of on your legs would be costly and impede coordination, though presumably if everyone did it, head-pants might be reasonably stable. More realistically, consider a normative behavior involving a contribution to a public good, like volunteering at local schools or paying dues to one's local workers union. If the common norm is to *not* contribute, then a maverick who bucks the trend and contributes will provide a benefit to others in the community but will also pay a net cost, because others are merely free riding on their generosity. However, if

Table 7.2. Payoff matrix for an asymmetric coordination game in which norm 1 is a group-beneficial norm, indicating payoffs to row player.

	Norm 1	**Norm 2**
Norm 1	$1+\delta+g$	$1-h$
Norm 2	$1+g$	1

everyone contributes, not only does everyone receive the benefit, but they also receive *additional synergistic benefits* from coordinating in a prosocial activity. This benefit could arise by several means. For example, sufficient volunteering at a school can give rise to a diversity of after-school enrichment programs, while sufficient engagement with workers unions can empower those unions for collective bargaining. To repeat: this is in addition to the benefit associated with being around others who employ the prosocial norm; it arises from the *interaction* between multiple individuals employing that norm.

We can formalize this sort of scenario as a coordination game with asymmetric payoffs, as in the following payoff matrix, where all variable terms are assumed to be nonnegative.

As before, coordination is always preferred to noncoordination, and both strategies are Nash equilibria. This makes the game different from the prisoner's dilemma, in which the only Nash equilibrium is the undesirable case of mutual defection. If you know your partner will use norm 1, you should also use norm 1, which carries an advantage of δ over using norm 2. If you know your partner will use norm 2, you should also use norm 2, which in this case carries an advantage h over using norm 1. Even though both norms are equilibria, they are not equivalent. Norm 1 is a prosocial, group-beneficial norm. Those employing norm 1 confer a benefit g on everyone, regardless of their partner's norm (this is the prosocial component). Partners also receive an additional benefit δ when they coordinate on norm 1 (this is the group-beneficial component). It is clear that a population using norm 1 will be better off compared with a population using norm 2 by a factor of $(g+\delta)$. If the current equilibrium is to use norm 2, everyone would be better off switching to norm 1.

Norm 1 carries a cost h when rare, however, and this presents a problem for would-be norm switchers. In the absence of such costs $(h=0)$, norm 1 could invade a population using norm 2 by neutral drift and then expand rapidly once sufficiently common. But norm 1 may be costly to implement, it may be burdensome to execute alone, or it may be actively punished by nonadopters. All these are real possibilities for uncommon behaviors. This quandary may be reminiscent of idealistic conversations you have had in which someone (perhaps you) points out some flaw in how society operates and suggests that everyone would be better off if only everyone would all do things differently. It may even be correct that things *would* be better if everyone did things in some other way.[5] The trouble is likely to be that you can't get there from here.[6] In the language of evolutionary game theory, norm 2 will resist invasion by rare norm 1 agents. How, then, can costly prosocial norms spread when rare?

If the frequency of norm 1 in the population is denoted by p, the expected payoff to an individual using norm 1 is the average of the two possible payoffs in the first row of Table 7.2,

[5] An additional problem for suggestions of this sort is that it is unlikely to be the case that *everyone* would be better off by switching. Incentives are rarely perfectly aligned in a society.

[6] If possible, this sentence is best imagined spoken in a strong New England accent.

weighted by the probability of encountering a co-player using either norm:

$$V_1 = p(1 + \delta + g) + (1 - p)(1 - h) \tag{7.11}$$

Similarly, the expected payoff to an individual using norm 2 is:

$$V_2 = p(1 + g) + (1 - p)(1). \tag{7.12}$$

Before going through further mathematical analysis, it can be valuable to study an agent-based model in order to get some intuition for how the model's dynamics operate. Luckily, this requires only a small extension to our previously coded model. It may seem counterintuitive to do the agent-based modeling *before* the mathematical analysis. In most academic papers that include both types of models, the math is usually presented first. The rationale behind this approach is that one should prove what one can, and then simulate whatever is too complicated to tackle analytically. This is a perfectly reasonable approach for communicating theoretical results. However, our task here is to develop an intuition for dynamical social systems. For many people, agent-based simulations are simply more intuitive than equations, perhaps because they involve rules that are followed rather than the mysterious hieroglyphics of mathematical notation. It can also be gratifying to first observe a pattern in a simulation and then use math to explain that pattern. This is how we will proceed.

7.2.1 Coding the Model

The NetLogo code for the coordination model with asymmetric payoffs is **coordination_asymmetric.nlogo**. Only two changes are needed to modify this from our model with symmetric payoffs. First, instead of a single global parameter for the benefit of coordination, we now have three payoff parameters (Table 7.2): the synergistic benefit to coordinating on norm 1, δ (`norm1-coord-benefit`), the prosocial benefit that users of norm 1 confer on all their interaction partners, g (`norm1-prosocial-benefit`), and the cost of using norm 1 when one's interaction partner uses norm 2, h (`norm1-rarity-cost`). The other update to the code is a minor revision to the `calculate-payoffs` procedure, reflecting the new payoff matrix.

NetLogo code 7.4

```
to calculate-payoffs
  let p1 (count turtles with [norm1?]) / (count turtles)
  let norm1-payoff ((p1 * (1 + norm1-prosocial-benefit +
      norm1-coord-benefit)) + ((1 - p1) * (1 - norm1-rarity-cost)))
  let norm2-payoff (p1 * (1 + norm1-prosocial-benefit) + (1 - p1) * 1)
  ask turtles with [norm1?][ set payoff norm1-payoff ]
  ask turtles with [not norm1?][ set payoff norm2-payoff ]
end
```

7.2.2 Analyzing the Model

When payoffs to the competing norms were symmetric and neither coordination outcome was intrinsically better, whichever norm was initially most common had the advantage (Figure 7.6). We can now ask how the asymmetry in the payoff matrix affects this dynamic.

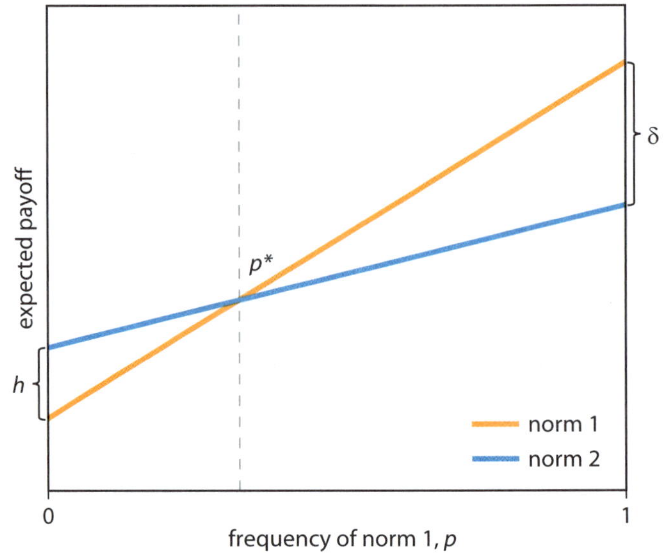

Figure 7.7 The expected payoffs for norms 1 and 2 in the asymmetric coordination game as a function of the frequency of norm 1, p. This plot also illustrates interpretations of the payoff values δ and h.

Let's start out with some arbitrary values and play around with our simulation. Let $\delta = g = 1$ and $h = 0.5$, and let the initial frequency of norm 1 be $p_0 = 0.5$. Try running the model a few times. If this was the symmetric model, each norm would come to dominate about half the time. Instead, you should observe that the group-beneficial norm 1 dominates *every* time. This happens even if we lower its initial frequency to $p_0 = 0.4$. Huzzah! The group-beneficial norm spreads even when rare! Let's keep it going. Will norm 1 always spread, no matter how rare it is initially? Keep lowering the initial frequency of norm 1. You'll find, dishearteningly, that as the initial frequency dips to around $p_0 \approx 0.35$, norm 2 begins to dominate occasionally. If $p_0 < 0.3$, the prosocial norm almost never spreads. It appears that there is some threshold value of p_0 below which norm 2 is the one to dominate. Let's dig deeper.

The fitness of each norm is frequency dependent. Each norm does best when common, worst when rare. Consider the expected payoffs we derived in Equations 7.11 and 7.12. We can plot these payoffs as a function of p, as illustrated in Figure 7.7. Notice that there is a critical value of p where the expected payoff of norm 1 exceeds the expected payoff of norm 2. Once norm 1 reaches this level of prevalence in the population, it should increase to fixation, because its fitness will exceed the population mean (recall replicator dynamics from the previous chapter). Similarly, if the frequency of norm 1 drops below this threshold, norm 2 will increase to fixation. This critical value is therefore an unstable equilibrium. We can derive an exact equation for the critical value, p^*, by setting the expected payoffs for the two norms equal to one another and solving for p. We begin by writing out the full equations we used in our agent-based model.

$$V_1 = V_2$$

$$p(1 + \delta + g) + (1 - p)(1 - h) = p(1 + g) + (1 - p)(1) \tag{7.13}$$

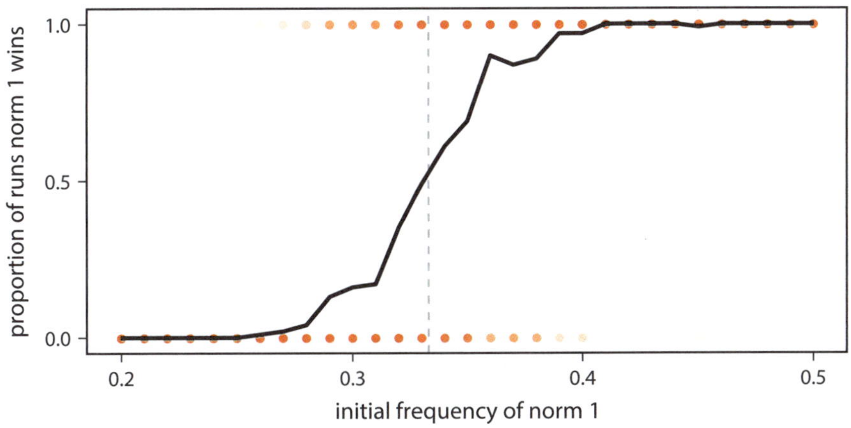

Figure 7.8 The proportion of simulation runs (out of 100) in which norm 1 dominated as a function of its initial frequency, p_0. Norm 1 wins more often when $p_0 > p^*$. For these simulations, $\delta = g = 1$, $h = 0.5$, $p^* = 1/3$.

We then multiply out several terms, and cross out terms that appear on each side of the equation.

$$\begin{aligned} p + p\delta + pg + (1-p) - (1-p)h &= p + pg + (1-p) \\ \cancel{p} + p\delta + \cancel{pg} + \cancel{(1-p)} - (1-p)h &= \cancel{p} + \cancel{pg} + \cancel{(1-p)} \\ p\delta - (1-p)h &= 0 \end{aligned} \tag{7.14}$$

We can then consolidate all the terms involving p.

$$\begin{aligned} p\delta - h + ph &= 0 \\ p(\delta + h) &= h \end{aligned} \tag{7.15}$$

Solving for p yields the critical value:

$$\boxed{p^* = \frac{h}{\delta + h}} \tag{7.16}$$

This is the value of p for which both norms have exactly the same expected payoff. Whenever the frequency of norm 1 increases above this threshold, its frequency will continue to increase to fixation. Whenever the frequency of norm 1 drops below this threshold, that frequency will continue to decline until norm 1 has been eliminated entirely. If you are using the provided NetLogo code, you'll notice that I've added a monitor that automatically calculates and displays this threshold value in the Interface tab.

The agent-based model parameter values we used earlier ($\delta = g = 1$, $h = 0.5$) yield a threshold of $p^* = 1/3 \approx 0.333$, which is consistent with our exploratory simulations. As with our simpler symmetric model, the mathematical analysis represents the expected outcome, but stochasticity can lead to either norm spreading when the initial frequency of norm 1 is close to the critical value. This is illustrated in Figure 7.8, which shows the proportion of runs (out of 100 for each value of p) in which norm 1 dominated as a function of p_0. The presence of stochasticity means that norm 1 dominates for some values of $p < p^*$ and norm 2 dominates for some values of $p > p^*$.

There are a few things to note concerning the calculation of p^*. First, the threshold frequency for norm 1 to spread is lower when the group benefit of coordinating on norm 1 (δ) is higher. In other words, group-beneficial norms spread more easily when coordinating on those norms provides a higher payoff. Second, the threshold p^* is higher when the cost of using norm 1 with a norm 2 agent (h) is higher. This makes sense. Even rare interactions may be attractive if they have a very high payoff, but those benefits must outweigh the costs of the noncoordinating interactions.

A third thing we learn is that the prosocial benefit generated by norm 1, g, is completely irrelevant to its spread! It can be very large, or it can be zero—its value will have absolutely no influence on the spread of norms in this model. This result is perhaps less intuitive. Surely it matters if a norm benefits everyone? The reason it *doesn't* matter becomes apparent when we compare the equations for the expected payoffs to users of the two norms. Norm 1 agents provide the benefit g to *everyone* in the population, regardless of which norm their partner is using. This means that g plays no role in differentiating the payoffs of the two norms. It may be better for the group to have more people using norm 1, but it doesn't help to differentiate those using the norm from those simply taking advantage of its benefits.

What we have modeled is a norm that is better when common but still costly when rare. It is group beneficial in that a group using norm 1 will be better off than one using norm 2, but its prosocial nature is apparently irrelevant to its ability to spread. Recall what I said above: *it may be better for the group to have more people using norm 1.* As long as we are considering competition purely within a group, this won't matter, because evolutionary dynamics don't optimize group-level fitness when competition occurs solely between individuals within the group. If we consider competition *between* groups, however, a norm's prosocial characteristic becomes relevant. We will tackle between-group competition in the next section.

Our analysis suggests that if a group-beneficial norm is not sufficiently common at the outset, it will not spread. Instead, it will disappear. This is a little depressing. Imagine a superior norm that, if we could only just adopt it, will make everyone better off. The asymmetric coordination model suggests that unless a sufficiently large faction can be rapidly coerced into adopting the new norm, thereby demonstrating its superiority, the superior norm will tragically fizzle and die. Is there any other hope for the spread of new beneficial norms?

All hope is not lost. There are, in fact, several mechanisms that allow for the spread of rare norms, one of which we will explore in detail. We've so far assumed that individuals are organized into a single population of interacting agents. It's true that in previous chapters we considered simple lattice networks, but these don't capture the fact that most real human populations are usually structured into groups in which members interact primarily (though not exclusively) with other group members. Might group structure play a role in the spread of norms?

7.3 Group-Beneficial Norms in a Structured Population

Humans are obligately social and inherently cultural. We are embedded in cultural communities that shape our interactions and hence our norms. Of course, people also interact with members of other communities from time to time via travel, commerce, etc. These extra-community interactions allow people to observe norms that may not be common in their own communities, and to thereby make comparisons in terms of the benefits and costs associated with various norms. For example, we might notice that members of another community tend to be doing much better than members of our own community, or much worse. In comparing these costs and benefits, it may not be apparent exactly which behaviors are causally related to differences in success between groups, or how particular behaviors generate those differences. We need only assume that people can observe the norms and payoffs of members of other groups and compare those to their own norms and payoffs. If members of another community seem to be doing better than members of one's own community, an individual may decide to emulate the behaviors associated with that other community.

This dynamic is probably quite common. During the Warring States period in ancient China (fifth through third century BCE), the wearing of men's pants spread rapidly throughout various states, replacing skirt- or toga-like clothes. This was because pants allowed for superior control of horses, which in turn was essential for military dominance. States that did not adopt pants lost to states that did.[7] Similarly, workers in modern corporations who are not unionized may observe unionized employees of other companies who enjoy higher salaries and more secure benefits. Crucially, it doesn't matter whether individuals understand why a norm is associated with a higher payoff for it to be the target of imitation. The causal opacity involved in norm adoption is exemplified by the "cargo cults" that sprung up on various Pacific Islands several times in the last couple of centuries (White, 1966; Dennett, 2006). In what is probably the best-known example, during World War II, Americans established a military base on the island of Efate and conscripted workers from among the islanders of nearby Tanna—both islands are part of what is now Vanuatu. The islanders observed that the Westerners were rich in "cargo," a blanket term denoting the Westerners' unimaginable material wealth. They came to believe that the Westerners' strange practices were rituals by which they appealed to their gods, who then blessed them with cargo. Soon the people of Tanna, eager to get some of the cargo for themselves, began marching in parades, marking "USA" on their chests, and carving bamboo figurines of American warplanes, helmets, and rifles for use as religious icons. During festivals, elders would perform a dance based on American military drills. Islanders even built plane runways, with some individuals stationed as flaggers and others up in towers wearing headphones made from coconuts.

The cargo cults are famous for their *mismatch* between behavior and its associated reward. Adoption of maladaptive behavior in this way is, of course, hardly limited to non-Westerners or even nonscientists. The use of inappropriate statistical tests and indicators of "significance," for example, can spread through scientific communities among community members who are untrained in their application and interpretation. Other community members, themselves underqualified to assess the relationship between these "statistical rituals" and the true state of the world, may simply offer congratulations on a new publication, thus contributing to crises of nonreplication in some fields (Feynman, 1974; Gigerenzer, 2018; Ritchie, 2020). We will return to this dynamic in chapter 8.

In many cases, of course, behaviors that are statistically associated with benefits are indeed causally related to those benefits. Pants really do help riders to control their horses. And there really is a direct relationship between membership in labor unions and the ability

[7] For a short and entertaining piece on the cultural evolution of pants, see: https://peterturchin.com/cliodynamica/cultural-evolution-of-pants/.

of union members to engage in collective bargaining. If an individual living in a community without labor unions observes that people who live in communities with strong unions enjoy a better quality of life, they may be motivated to advocate for union membership in their own community, even if doing so carries short-term costs.

We can formalize this idea by extending our coordination model with asymmetric payoffs so that individuals are assigned to one of two groups, each of size N. Let's creatively call these group A and group B. Individuals interact and attempt to coordinate primarily with members of their own groups, but occasionally observe the norms and associated payoffs of members of the other group. The model works similarly to the single-group model we studied in the previous section, with two exceptions. First, each group can start with a different frequency of agents using norm 1, so that it may be initially common in one group and rare in the other. Let p_{A0} be the initial frequency of norm 1 in group A, and p_{B0} be the initial frequency of norm 1 in group B. Second, during the evolution stage of the model dynamics, agents occasionally observe an agent from the out-group. Recall that each agent chooses another agent to observe at random, with whom the focal agent will compare norms and payoffs. With probability m, the observed agent is chosen from the out-group, otherwise the observed agent is chosen from the in-group. At either extreme, we replicate the single-group model: with a doubly large population size when $m = 1$, or twice with the original population size when $m = 0$. At small but nonzero values of m, the two groups remain distinct but may nevertheless exert influence over one another.[8]

I want to focus our attention on the particular scenario in which different norms are initially common in each of the two groups. Certainly it's reasonable to assume that two communities could converge on different norms. Recall from our earlier analyses that almost any norm can become entrenched in a population once it becomes sufficiently common, as long as it confers a coordination benefit. Let us assume that in group A, the group-beneficial norm 1 is dominant, while the inferior norm 2 dominates in group B. In both cases, the frequency of norm 1 is far from p^*, so that the nondominant norm in each group cannot increase in frequency by purely within-group processes. Nevertheless, members of group A receive consistently higher payoffs compared with members of group B—it is observably better to be a member of group A than a member of group B. If members of group B occasionally observe the superior payoffs common in group A, perhaps group B can come to adopt the group-beneficial norm as well. Is this plausible? Let's turn to our agent-based model to investigate.

7.3.1 Coding the Model

The NetLogo code for the two-group coordination model is **coordination_2groups.nlogo**. The extension to two groups does not require a great deal of additional coding, though it may be worth putting some effort into effectively visualizing the two groups. First, we will need a few new global parameters. We replace the old `init-norm1` with two parameters in order to control the initial frequencies of norm 1 in *both* groups: `init-norm1-groupA` and `init-norm1-groupB`. This is important if we want to model scenarios in which there are initial differences between the two groups regarding the dominant norm. We then add a new parameter for the rate of out-group observation, m (`prob-outgroup-observation`).

Agents must now keep track of their group identity, and will do so with a newly added variable, `groupID`. Although we have been using A and B to denote the groups in the text, in the code I will use 0 and 1 to denote those same groups. Indexes in programming

[8] This model is a simplified version of a model presented by Boyd and Richerson (2002).

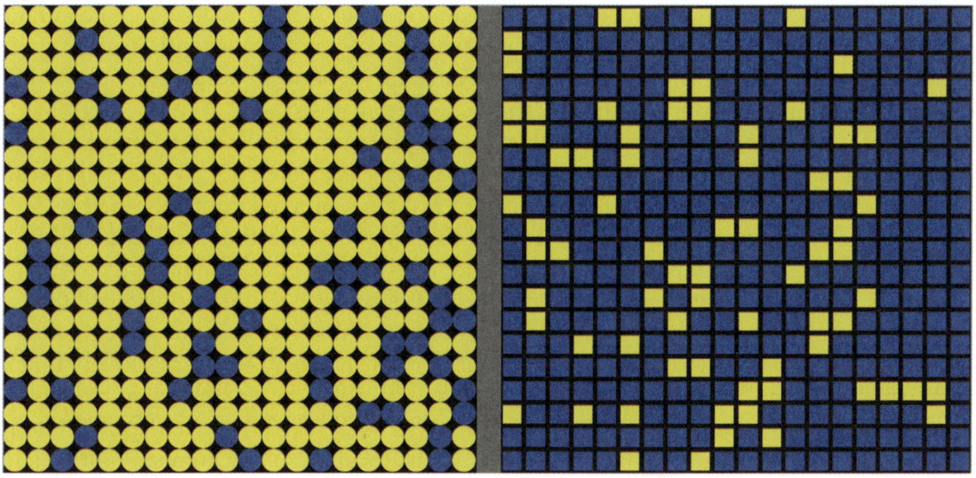

Figure 7.9 Visualization of agents arranged into two distinct groups. Norm 1 agents are yellow, norm 2 agents are blue. Group A on the left (circles) has 80% of agents using norm 1, while only 20% of agents in group B (squares) use norm 1. The center column of patches are grey to provide a visual separation between the groups.

languages often start at zero, and there is a practical reason to use these numbers to denote group identity, as we will see below. Using numbers rather than letters also makes it easier to extend the model to include more than two groups. We will assume that group identities are permanent, and that individuals and their offspring remain in their initial group.[9]

I mentioned that you may find it worthwhile to update the visualization of the model to reflect the new population structure. And indeed, the way I've used NetLogo's patch layout to organize agents almost demands that we do this, though it is possible to build versions of these coordination models without using patches or spatial structure at all. We are making two groups, so we can double the width of the grid in order to double the number of patches in which we will place agents. We can also leave a gap between the groups to more easily delineate the boundary, though this is an aesthetic choice. Our previous one-community models used an $L \times L$ grid, so this model will use a grid that is $L \times (2L + 1)$ in area, consisting of two $L \times L$ grids and a $1 \times L$ strip dividing them (Figure 7.9).

To initialize the model, we need to assign agents to groups. There are several possible ways to do this.[10] I have taken advantage of the fact that patches in NetLogo can be easily assigned colors, and that agents can quickly ascertain the color of the patch they sit on. The `setup` procedure therefore calls a procedure to color the grid patches before creating the agents.

NetLogo code 7.5

```
to setup
  clear-all
```

[9] It is possible to use payoff-based migration instead of mere observation in a model like this, with similar results (Boyd and Richerson, 2009).

[10] It is a truism of programming that there are almost always multiple ways to do almost anything. Figuring out better ways to do things is part of the process of becoming a solid coder.

```
  color-patches
  setup-turtles
  recolor
  reset-ticks
end
```

Most of this is familiar. However, there is a new procedure, `color-patches`. This will be used to divide up the patches on the grid into three categories: group A, group B, and the dividing line between them. It doesn't matter what colors we make them, as long as they're all different. I've made the left side, for group A, pure black, which NetLogo maps to the number 0. The right side, for group B, I've made very dark grey, using the number 1. The difference from pure black is almost imperceptible to the eye, but that doesn't matter, because the computer can tell the difference. The boundary between the groups will be assigned the color `grey`, corresponding to the color code of 5. Note that I've also gone into Settings to make sure the origin coordinate is at the lower left corner of the grid, so that the center line is defined by the x-coordinate of `max-pxcor / 2`.

NetLogo code 7.6

```
to color-patches
  ask patches [
   if pxcor < max-pxcor / 2 ;;left side
   [ set pcolor 0 ]
   if pxcor > max-pxcor / 2 ;;right side
   [ set pcolor 1 ]
   if pxcor = max-pxcor / 2 ;;center line
   [ set pcolor grey ]
  ]
end
```

The procedure `setup-turtles` uses the information about patch colors to create agents belonging to each of the two groups while keeping the dividing line free of agents. Colors, shapes, and norms are assigned based on group identity, though of course agents' colors and shapes are for visualization purposes only and do not affect the model dynamics.

NetLogo code 7.7

```
to setup-turtles
  ask patches with [pcolor = 0 or pcolor = 1][
    sprout 1 [
    set payoff 0
    set norm1? false
    ]
  ]
  ask turtles with [pcolor = 0] [
   set groupID 0
   set shape "circle"
   if(random-float 1 < init-norm1-groupA)
     [set norm1? true]
  ]
```

```
    ask turtles with [pcolor = 1] [
     set groupID 1
     set shape "square"
     if(random-float 1 < init-norm1-groupB)
       [set norm1? true]
    ]
  end
```

The dynamics, controlled by the `go` procedure, again involve the two stages of game play and evolution. Each of the associated procedures must be altered to accommodate the group structure. Let's start with `calculate-payoffs`, in which agents interact with others and calculate their expected payoffs. Recall that game interactions still occur only between members of the same group. Thus, agents in each group calculate their expected payoffs based on the number of agents using each norm in their own group. This works just as it did in the one-community models, but to calculate expected payoffs we need to use the frequency of each norm within the agent's own community only, and not across the entire population. One way to do this is to calculate the payoffs for agents in each group, one group at a time. A benefit of coding things this way is that if we use variables to stand in for groups, we can easily extend the model to more than two groups in the future. I have used the NetLogo primitive `foreach`, which is used to loop over a set of values for some variables. Here, the code loops over each group `k`, calculating payoffs for agents using either norm.

NetLogo code 7.8

```
to calculate-payoffs
  let num-agents (count turtles) / 2 ;;num agents in each group
  (foreach [0 1][
    [k] ->
    let p1 (count turtles with [groupID = k and norm1?]) / num-agents
    let norm1-payoff ((p1 * (1 + norm1-prosocial-benefit +
         norm1-coord-benefit)) + ((1 - p1) * (1 - norm1-rarity-cost)))
    let norm2-payoff (p1 * (1 + norm1-prosocial-benefit) + (1 - p1) * 1)
    ask turtles with [groupID = k and norm1?][set payoff norm1-payoff]
    ask turtles with [groupID = k and not norm1?][set payoff
      norm2-payoff]
  ])
end
```

Finally, we turn to the `evolve` procedure, in which each agent chooses an agent to observe and potentially copy. Recall that an agent chooses to observe an individual from the out-group with probability m (`prob-outgroup-observation`), otherwise they observe an individual from their own group. The agent then uses the sigmoid function to decide whether or not to copy the observed agent's norm. The agent first notes its own group, `myID`. It then establishes the group of the agent it will be observing, `otherID`, which by default is set to the focal agent's own group. If it is to make an out-group observation, however, the value of `otherID` has to change. Recall that we indexed group IDs starting at zero rather than one. The reason is that we can once again use modular arithmetic and the `modulo` operator to determine the group identity of the observed agent. Because `groupID` can be either 0 or 1, the code `(groupID + 1) mod 2` will always evaluate to either 1 or 0. The full code for the `evolve` procedure is given below.

NetLogo code 7.9

```
to evolve
  ask turtles[
    let myID groupID
    let otherID groupID
    if random-float 1 < prob-outgroup-observation
    [set otherID ((myID + 1) mod 2)]
    let model one-of other turtles with [groupID = otherID]
    let other-payoff [payoff] of model
    let prob-copy (1 / (1 + exp (-1 * (other-payoff - payoff))))
    if random-float 1 < prob-copy
    [set norm1? [norm1?] of model]
  ]
end
```

7.3.2 Analyzing the Model

Take a moment to think about how you expect this two-community model to work, given what we have learned about the single-community version in the previous section. Norm 1 is superior to norm 2 when common, but it is penalized by users of norm 2 when rare. Thus, norm 1 thrives only when it is sufficiently common to generate a superior payoff. However, when groups are distinct but not completely isolated, different norms have the opportunity to compete. If norm 1 is common in group A and rare in group B, members of group A will have higher payoffs on average than members of group B. If members of group B observe and copy the members of group A, norm 1 may gain a foothold in that group. Conceptually this makes sense. However, it is not clear just *how much* out-group observation is actually needed to make this happen. If it's a very large amount, say more than 50% of observations, then the model may indicate that intergroup observations are insufficient to facilitate the spread of group-beneficial norms. Adding precision to our intuitions is, of course, why we build models.

Let's start by running some simulations to get a sense of what is possible with our new model. Using the notation from the payoff matrix in Table 7.2, we can once again let $\delta = g = 1$ and $h = 0.5$. Recall that for these values, $p^* = 0.333$. Set the initial frequency of norm 1 to be 80% in group A and only 20% in group B. If the probability of out-group observation is $m = 0$, then what happens next is predicted by our prior analyses. Norm 1 will increase to fixation in group A and completely die out in group B. Next, increase m to 0.01. Now, norm 1 persists to a small degree in group 2. However, most observations (99%, in fact) are still within-group, and so the group-beneficial norm fails to spread. Now increase m to 0.02. You may observe that norm 1 persists in group B at low levels for a while until, all of a sudden, BOOM! Norm 1 crosses the threshold to become the dominant norm in group B. This is encouraging. It looks like even a small amount of between-group interactions can facilitate the rapid spread of group-beneficial norms from one group to another.

We can and should analyze this in more depth. Figure 7.10 shows the results of simulations run over a wide range of values for m and h, with 50 simulations per parameter combination. This figure illustrates that the amount of out-group observation needed for the spread of norm 1 is larger when the cost of using norm 1 when rare (h) is greater. This makes sense, since greater costs work against the evolution of the group-beneficial norm. Nevertheless, even when the costs are quite high—even equal to the group benefit

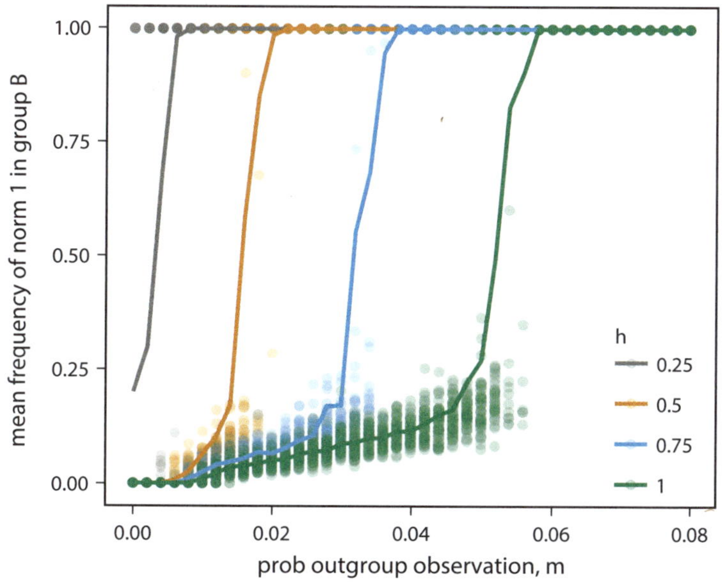

Figure 7.10 Mean proportion of agents in group B adopting norm 1 at $t = 500$, as a function of the probability of outgroup observation, m, for several different costs of using norm 1 when rare (h). Initial frequencies of norm 1 were 0.8 in group A and 0.15 in group B. Other parameters used are $\delta = g = 1$. Fifty simulations were run for each set of parameter values. Circles are results from individual runs, lines are means.

to coordination—norm 1 can still spread when a very small minority of observations are of out-group individuals.

When observations occur purely within one's community, we observed that the prosocial benefit of norm 1, g, was inconsequential to the spread of norm 1. This is because everyone receives this benefit regardless of which norm they use, and so g does not contribute to a distinction between users of norm 1 and norm 2 in terms of payoff. When group structure is taken into account, however, the prosocial benefit *can* help us to distinguish between members of a group where norm 1 is common and one in which it is rare. Larger prosocial benefits from users of norm 1 enhance the relative advantage of their group compared with other groups. This is illustrated in Figure 7.11, which plots the effect of the prosocial benefit, g, on the extent to which norm 1 spreads in group B. There is a consistently positive effect: the larger the prosocial benefit, the better the group-beneficial norm spreads.[11]

This analysis shows that group structure can help aid the spread of group-beneficial and prosocial norms. This happens because community structure allows groups to compete, adding a group level to the evolutionary dynamics (see Box 7.1: Cultural Group Selection). Recall from the previous chapter that explicit spatial or network structure could also facilitate the spread of prosocial behaviors—altruistic cooperation in that case. It seems possible that if interactions involving coordination are restricted to stable network ties, small clusters of individuals using group-beneficial norms could emerge and compete locally with

[11]I have not included a purely mathematical analysis of the group-structured model. Readers interested in how to tackle this model analytically should see Boyd and Richerson (2002).

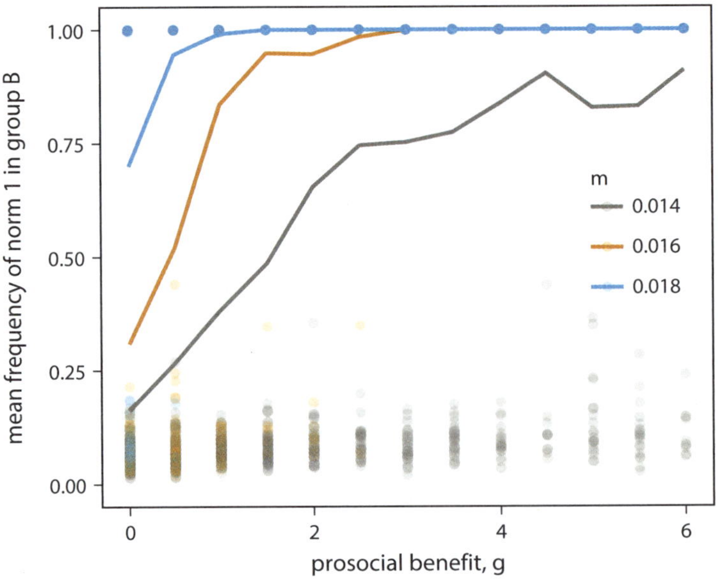

Figure 7.11 Mean proportion of agents in group B adopting norm 1 at $t = 1000$, as a function of the prosocial benefit, g, for several values of the probability of outgroup observation (m). Initial frequencies of norm 1 were 0.8 in group A and 0.15 in group B. Other parameters used are $\delta = 1$, $h = 0.5$. 100 simulations were run for each set of parameter values. Circles are results from individual runs, lines are means.

other norms, without the need for a superordinate community identity. In reality, local network effects probably interact with larger group identity effects. It is left as an exercise to incorporate local interaction effects into our group-structured model of norm evolution.

BOX 7.1: Cultural Group Selection

The model presented in this chapter involving group-beneficial norms spreading in a structured population is a variant of a model introduced by Boyd and Richerson (2002), and is a stylized illustration of a dynamic that is sometimes called **cultural group selection**. Cultural group selection is a theory of cultural evolution that concerns why social norms tend to be prosocial. It is based on the more general principles of **multilevel selection**, which is a framework for analyzing the evolution of traits that accounts for competition both within and between groups. If the variance of some trait distribution *between* groups is sufficiently high relative to the variance within groups (that is, if groups are sufficiently distinguished from each other) and members of different groups nevertheless compete for resources, then group membership can be a factor in the evolutionary dynamics of that trait. This aspect of selection is called **group selection**.

Group selection does not require the group to be the reproductive unit of selection; it merely indicates that group membership exerts a strong influence on the strength of selection for some traits. Because we need to understand how selection *between groups* and competition *between individuals* within those groups interact, multilevel selection theorists have devised methods to separate a trait's contributions to fitness into individual selection (contributions to fitness by virtue of possessing a trait in direct competition with others) and group selection (contributions to fitness by virtue of being part of a group with a particular trait distribution). For more details about the mathematics of multilevel selection, see Okasha (2006). The theory of multilevel selection has at times been controversial, because the evidence that group membership is important to selection is mixed for most species. The best examples regarding the genetic evolution of biological traits come from species sorted into competing kin groups. This has led many researchers to note the formal equivalence, under some conditions, between multilevel selection and inclusive fitness perspectives (Okasha, 2016).

In human cultural evolution, groups are a bit different, for two primary reasons. First, a variety of cultural and psychological mechanisms work to maintain between-group differences. Cultural transmission between groups is difficult when groups speak different languages and have different norms. Second, cultural traits can be transmitted even when individuals exhibiting those traits do not produce more offspring than others—consider how teachers and prestigious individuals can have an outsized influence on the knowledge and behaviors that spread in a cultural population. This can facilitate the cultural evolution of traits that are group-beneficial, prosocial, and even altruistic in nature.

There is occasionally some controversy among social scientists concerning the predictions made by theories of cultural group selection. Much of this is healthy debate among scholars. However, it is worth noting that the theory of cultural group selection is not explicitly about the spread of altruistic behavior, in the sense of costly cooperation. Rather, it is a theory about why social norms tend to be group-beneficial when groups compete. Groups lacking socially-beneficial norms—which may include norms of trust, conformity, or cooperation—may arise when individuals compete with other individuals in their group (because selfishness can be selected for when competition occurs primarily within a group), but those selfish norms may be selected against when competition between groups arises. Cultural group selection is thought by many to explain several of the puzzling but apparently group-beneficial aspects of human cultures, from warfare (Turchin, 2015; Zefferman and Mathew, 2015) to religious institutions (Wilson, 2002; Norenzayan, 2013; Brooks et al., 2018) to music (Mehr et al., 2021; Savage et al., 2021; Moser et al., 2021). Of course, plenty of cultural evolution occurs without strong inter-group competition. Nevertheless, understanding the mechanisms that facilitate cultural group selection can help us to understand when it should be accounted for in our models.

Mechanisms that maintain between-group differences

A number of cultural and psychological mechanisms can contribute to the maintenance of fairly well-defined boundaries between groups, which can then provide variation for selection to act upon. These mechanisms include:

- *Conformist social learning.* Individuals exhibit tendencies to adopt behaviors, norms, and expectations that are common in their community.

- *Coordination benefits*. Individuals receive benefits from coordinating on the same behaviors, norms, and expectations as others in their community.
- *Punishment of deviant behaviors*. Individuals often punish norm violations with costly consequences, including economic or reproductive sanctions, ostracism, or even death.
- *Prestige-biased learning*. When individuals preferentially learn from prestigious or successful individuals in their community, it can facilitate correlations between what is learned by individuals throughout the community.
- *Symbolic markers*. Individuals may, whether consciously or not, adopt symbolic markers—including fashion or style choices as well as linguistic or semiotic nuances—that help delineate who they interact with and who they learn from.
- *Institutional complexity*. Cultural institutions can involve a dizzying array of norms that require hard-won tacit knowledge to navigate. This can make those institutions opaque and confusing to outsiders, and thereby limit cultural borrowing between groups.

In the multi-group model explored in this chapter, we focused on coordination benefits and prestige- or success-biased transmission. It is worth reiterating that the real world is always more complex and less easily demarcated than implied by models implementing these mechanisms, which serve as archetypal cases to which we can then add nuance.

Mechanisms by which cultural group selection can operate once between-group differences are established

Once differences exist between the norms of competing cultural populations, group selection can act on those norms. Cultural traits are subject to a variety of selection mechanisms above and beyond those that genetic traits are subject to. These include:

- *Natural selection*. Groups whose norms lead to faster reproduction may "swamp" competing cultural groups. Alternatively, groups may compete via warfare or conquest. Groups whose norms allow them to be better at war will tend to win wars and thereby spread their norms.
- *Selective imitation of successful groups*. Individuals are likely to adopt norms associated with members of highly successful groups. This may be particularly powerful if the new norm is adopted by influential group members (due to prestige bias).
- *Selective migration between groups*. Individuals in dysfunctional groups may leave for better groups. This may then weaken the former groups or encourage leaders to adopt more successful foreign norms, strengthened by the two other processes listed above.

In the multi-group model explored in this chapter, we focused on the selective imitation of successful groups. However, all of these mechanisms will eventually reduce inter-group variation, which will in turn reduce the strength of group selection. Thus, there is likely a balance to be struck between forces that differentiate groups and forces that make them more similar. All of these ideas can be explored with formal models (and many of them already have been). For more details on cultural group selection, see Henrich (2004), Smaldino (2014), and Richerson et al. (2016).

Table 7.3. Payoff matrix for a complementary coordination game, indicating payoffs to row player. Players do better by adopting complementary roles, but they may split the benefit of coordination unevenly.

	Norm 1	**Norm 2**
Norm 1	1	$1+\gamma G$
Norm 2	$1+(1-\gamma)G$	1

7.4 Division of Labor

The sort of coordination we've discussed so far is sometimes called **correlative coordination**, and it occurs when partners benefit by doing the same thing. Driving on the same side of the road, using the same words, and so forth. But working together effectively doesn't always mean doing the same thing. **Complementary coordination** involves sorting into, well, complementary roles.[12] I'll cook if you wash the dishes. Or perhaps you cook and I'll wash the dishes. Either is fine. But if we both cook and neither washes the dishes, we end up with overfull stomachs and a messy kitchen. It's best if we divide up the roles to get the job done right.

Division of labor involves sorting into specialized roles for tasks of complementary coordination. The result can be equally beneficial to all participants, where everyone does their part and receives an equal share of the benefits. For example, imagine that you and your roommate are having dinner together at home. If one of you wants to cook dinner and the other wants to do the dishes, that's perfect. If both of you want to cook dinner and neither of you wants to do the dishes, that's bad, since you end up with a dirty kitchen. If both of you want to do the dishes and neither of you wants to cook, that's also bad (unless you just get an exogenous pizza). Of course, you could simply take turns regarding who cooks and who does the dishes, so that neither gets stuck with their less-preferred job too often.

However, division of labor can also involve situations where participants become locked into inequitable equilibria in which some individuals benefit at the expense of others. This can happen if one person takes on a role that is more firmly associated with a particular job. What happens if both you and your roommate love to cook and hate to do the dishes, but your roommate adamantly refuses to ever do the dishes? You're certainly not going to do two jobs, so you leave the cooking to them and begrudgingly clean up their mess. They're a good cook, and obviously *someone* has to cook, but you can't help thinking that you have the raw end of this deal. This is a microcosm of something that happens often. Certain roles can become associated with specific classes of people, such as when people are sorted by gender, race, ethnicity, or caste. And if one role is associated with higher payoffs, inequalities can emerge and become canalized. In this section, we will briefly consider the emergence of division of labor and social classes in a complementary coordination game.[13]

We can model division of labor by making only minor changes to the coordination models that we've been using in this chapter. Consider the following payoff matrix for a complementary coordination game shown in Table 7.3. If both players take on the same role, they get a baseline payoff of 1. The parameter G can be seen as the surplus benefit created by division of labor. If $G > 0$, players do better by adopting complementary roles than

[12]This is sometimes called anti-coordination.

[13]For an extensive treatment of these and related games, with special attention given to gender inequality, see O'Connor (2019b).

they would if they both adopted the same role. The parameter γ is the proportion of this surplus that is allocated to the individual taking on the "norm 1" role, with the remainder going to the "norm 2" player. If $\gamma = 0.5$, the players divide the surplus fairly, otherwise one of the players takes a greater share.

If the division of the surplus is fair, then players will be equally motivated to take on whatever role complements their partner. If pairings occur at random, as we have often assumed, stable assortment into complementary roles can be difficult. In such a case, the best strategy may be to randomly select one of the two possible roles with equal probability. However, imagine instead that individuals are sorted into *types*, and that individuals often (though not necessarily always) interact with others of a different type than their own. This situation is pretty easily mapped onto domestic partnerships, but could also represent employment relationships, artistic or scientific collaborations, and others you might imagine. It is easy to imagine how different norms could become associated with each type, because once a relationship is established, neither type is incentivized to unilaterally switch norms. Associating specific types with specific roles does not necessarily imply any inequality, only that there is differentiation of normative roles that allows individuals to maximize their payoffs.

The situation becomes less rosy if $\gamma \neq 0.5$. For simplicity, we'll focus on the case where $\gamma > 0.5$, so that the player who takes on the "norm 1" role has an unfair advantage over the player taking on the "norm 2" role. As long as $(1-\gamma)G > 0$, playing norm 2 in a complementary role will still be more attractive (in a sense) than switching to norm 1. Either arrangement of complementary coordination is a Nash equilibrium of the game. However, if the population is divided into two types, the type stuck playing "norm 2" becomes effectively second class.[14]

When will this happen? Are there some scenarios that favor the emergence of social classes and other scenarios that do not? We can address these questions with only a minor modification of the two-group coordination model we studied in Section 7.3.[15] Suppose that each group represents one of two social types. In our previous model, individuals always interacted with members of the same type. Now we will assume that individuals sometimes interact with members of the *opposite* type. Formally, assume that an individual interacts with a member of the opposite type with probability d. If $d = 0$, individuals always interact with partners of their own type, as in the model we studied earlier. If $d = 1$, then individuals *always* interact with members of the opposite type, which should provide the optimal conditions for the division of labor.

Calculating expected payoffs for this model is slightly more complicated than with the previous model, because individuals can now interact with partners of either type. Let p_I and p_O be the frequency of norm 1 among individuals of the same and opposite type as the focal agent, respectively. The expected payoff to an individual of either type using norm 1 is

$$V_1 = (1-d)\left[p_I(1) + (1-p_I)(1+\gamma G)\right] + d\left[p_O(1) + (1-p_O)(1+\gamma G)\right]. \tag{7.17}$$

The expected payoff for an individual of either type using norm 2 is similarly

$$V_2 = (1-d)\left[(1-p_I)(1) + p_I(1+(1-\gamma)G)\right] + d\left[(1-p_O)(1) + p_O(1+(1-\gamma)G)\right]. \tag{7.18}$$

[14]Examining our assumptions, the game's payoff matrix seems to imply that an equitable scenario in which individuals all do equally poorly is inferior to an unequal scenario in which everyone receives a higher payoff but some individuals get much more than others. This preference ordering is not universally agreed upon.

[15]This analysis of the division of labor model is based on work by Henrich and Boyd (2008).

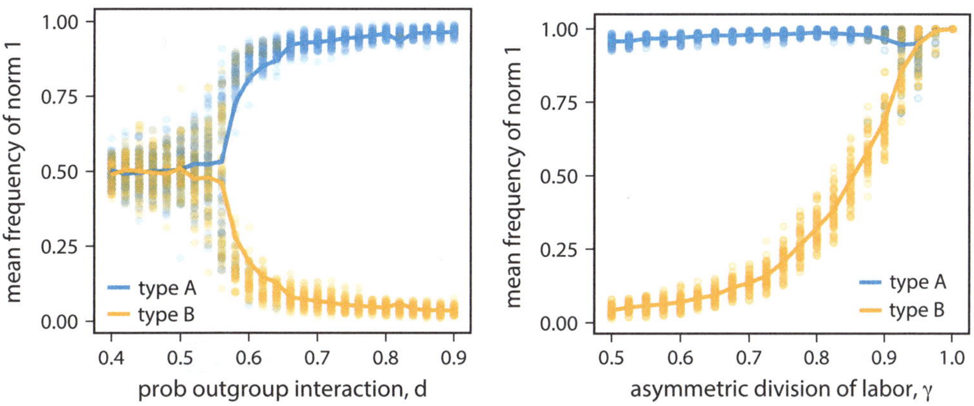

Figure 7.12 Division of labor. Left: equitable division of labor with $\gamma = 0.5$. The emergence of social classes occurs as d increases. Right: inequitable division of labor with $d = 0.8$. The subordinate social class increasingly takes on the role associated with the dominant class as the inequity of interactions increases. In all cases, $G = 3$, $m = 0.03$, $p_{A0} = 0.6$, $p_{B0} = 0.4$.

As with the groups in the previous model, we can label our types A and B. With these modifications in hand, we can proceed to code and analyze our agent-based model.

The NetLogo code for this model is **coordination_divisionoflabor.nlogo**. The code is extremely similar to the previous two-group model, so I will not review it here. I encourage you to play with this model and observe how several distinct behavioral regimes are possible. For now, let's go over two simple analyses. In both cases, I've initially made norm 1 slightly more common among type A individuals and norm 2 slightly more common among type B individuals.

First, we can examine the influence of outgroup interaction, d, on division of labor. If individuals mostly interact with members of their own type, they need to be generalists, because they have no way of knowing which norm their partner will use. As we increase the proportion of interactions that are with members of the opposite type, we see the emergence of social classes: types that consistently take on one role. It now pays to specialize. If one norm is overrepresented among members of the opposite type, the optimal strategy is to adopt the opposite norm, and the evolutionary dynamic automatically produces this outcome (Figure 7.12, left).

Second, let's observe what happens when the division of labor becomes increasingly inequitable. At least some division of labor is always present as long as $\gamma < 1$. Type A individuals continue to benefit from using norm 1, and they evolve to using norm 1 exclusively just as they did when the division of labor was equitable. However, as γ increases from 0.5, type B individuals do *not* conform to their prescribed roles as consistently as they did under equitable division of labor. Instead, they increasingly perform the norm 1 behavior associated with type A individuals, with a probability that increases as the underlying inequity of the division of labor grows (Figure 7.12, right). This is because they still occasionally interact with partners of their own type, and the occasional benefit of anti-coordination in these instances is enough to maintain norm 1 at relatively high levels among type B individuals. You might find this reminiscent of cultures in which behaviors associated with the opposite gender are both more common and more socially acceptable when adopted by the gender that has historically adopted the subordinate role in social relationships. The model provides a possible explanation for this pattern. When the division of labor is sufficiently unfair, type B individuals no longer benefit from anti-coordination, and completely adopt norm 1 (forgoing the surplus from division of labor) rather than be exploited.

7.5 Reflections

The models in this chapter rely on a number of important assumptions about norms and cultural transmission. All of the models assume that individuals are willing to consider an individual's success as a reason to copy their norms. This seems like a reasonable assumption, though success-biased copying can backfire if success is hard to ascertain or if the relationship between strategy and success is not consistent between individuals. The two-group model of group-beneficial norms further assumed that effectively all payoff-generating interactions occur within one's own group, even when observations can be made of out-group individuals. The implication is that individuals are willing to consider another person's success as a reason to copy them, even if that person is successful in a different social context. Based on the literature on social influence and prestige bias (Henrich and Gil-White, 2001; Kendal et al., 2018), this also seems like a reasonable assumption. However, there are certainly cases where it won't hold. If individuals from outside one's community are seen as foreign or "other," their behaviors may be dismissed out of hand, even if adopting them could provide benefits. There is a growing literature that describes how coalitional thinking influences social learning (Kish Bar-On and Lamm, 2023; Smaldino, 2022). When modeling scenarios where group identity works in this way, its influence on learning strategies must be taken into account.

The models in this chapter also relied on the assumption that norms and their associated payoffs are both easily observable. This may not always be the case. As many who have entered a new cultural environment know, the right way to behave isn't always obvious.

Sometimes observable norms can serve to signal a larger suite of norms that are themselves harder to observe. If you go to academic conferences and are male, you are reasonably likely to wear a tie if you are an economist, but extremely unlikely to do so if you are an anthropologist. The tie, or lack thereof, doesn't serve any function directly, but it signals to others the sort of person you are likely to be (among tie wearers, ties can also serve to signal status or trendiness). Likewise, norms often develop among communities or groups of friends that enable rapid communication between their members but are baffling to outsiders. The simple models we explored in this chapter do not capture these sorts of dynamics well.

It seems worth highlighting that the coordination benefits derived from norms need not reflect a net positive for the group. As a salient example for most in the developed world, imagine a social networking site that allows individuals to coordinate on a bunch of cool stuff like sharing family news and playing games, but also fosters surveillance capitalism and provides infrastructure on which misinformation and hate can spread. If there are costs to not using the site when most people do, it may be difficult for competing norms—like not using the site or even switching to a competing site—to spread. Some recent work has speculated that identity may serve as a powerful force to both preserve and disrupt such norms (Efferson et al., 2020).

Our culture is built around norms. This is possible because humans possess a psychology that has evolved to pay attention to normative behaviors from an early age (Chudek and Henrich, 2011; Legare and Nielsen, 2015). Understanding the emergence of norms and how they shape and are shaped by the tangled web of complex human social relations will require consideration of many factors beyond the sort of simple coordination discussed in this chapter. Nevertheless, there is a lot of power in thinking about norms as behaviors that solve coordination problems. If we could not coordinate, you could not be reading this, and the institutions that facilitate the learning you are doing would not exist.

7.6 Going Deeper

It is often hard to discover the norms of behavior someone will use until you are already engaged in interaction with them. If a person could signal information about their likely normative portfolio prior to an interaction, those signals would enable more effective assortment on norms and more efficient coordination. An early treatment of this idea was provided by the economist Michael Spence (1973) in the context of how employers evaluate potential employees. McElreath et al. (2003) used a variant of the models studied in this chapter to show that if individuals prefer to interact with partners who share arbitrary markers, an association between particular markers and particular norms could culturally evolve. Signaling also can facilitate the persistence of multiple norms in a population, with the norm–marker association strongest near the boundaries between populations who hold differing norms. Work by Bunce (2021) has extended these ideas to consider how power imbalance between majority and minority groups affects the spread of norms.

Cultural norms may vary between populations, but they also vary even within cultural groups. Assortment on norms can be desirable—surely you want friends, business associates, and romantic partners that have similar goals and worldviews—but avoiding individuals who don't share all your norms isn't always feasible. Within a society, we sometimes have to interact with those who don't share our norms, and we may want to avoid signals that highlight our differences. My colleagues and I have considered two types of signals—those sent overtly and covertly—and shown that covert signals should be dominant in cases where assortment is imperfect and uncertainty about shared norms between strangers is both common and consequential (Smaldino et al., 2018c; Smaldino and Turner, 2022). Related work by Hoffman et al. (2018) has explored conditions for intentionally noisy signals in which the act of obscuring is itself ascribed value.

Most communication requires some coordination, including the need for linguistic signals to represent similar meanings to both senders and receivers. Although language is in reality much more than a set of signaling conventions—its most important feature perhaps being the flexibility to convey almost any idea (Fitch, 2012)—its conventional aspect is nevertheless important to understand. Models have shown how repeated coordination games can give rise not only to shared signals but also shared categories (Contreras Kallens et al., 2018), as with conventions for color names (Puglisi et al., 2008) or safe versus dangerous foods (Cangelosi and Parisi, 1998).

There are many ways to study the division of labor and the emergence of social classes beyond the simple scenario considered in this chapter. A classic example is the so-called **battle of the sexes** game (Luce and Raiffa, 1957), in which both players benefit more from coordination than non-coordination, but each player benefits more from a different behavior. In the original formulation, the players are imagined to be a romantic couple. One prefers going to the ballet, the other to a boxing match, but they both prefer to be together rather than apart.[16] Another widely studied game is the **Nash demand game**, in which players "bargain" over particular ways to divide a resource. If the players' proposed division exceeds the total resource, neither player gets anything. A discrete three-option version of

[16] Another formulation characterizes the choice as one between two concerts featuring the music of either Bach or Stravinsky. While this formulation is perhaps less heteronormative, it strikes me as similarly problematic in terms of social class markers.

Table 7.4. Payoff matrix for a Nash demand game. Players can request either a high (H), medium (M), or low (L) share of the resource. Payoffs are to row player.

	H	M	L
H	0	0	7
M	0	5	5
L	3	3	3

the model, in which players can request either a high, medium, or low share of the resource, might look like Table 7.4. All divisions that fully exploit the resource are Nash equilibria, but those divisions can be either fair (M, M) or unfair (H, L). Several researchers have used models like this to understand the conditions in which unfair divisions become stabilized, often viewed in light of the emergence of social classes (Young, 1993; Axtell et al., 2001; O'Connor, 2019b). This game also demonstrates that games need not be limited to two options, but in fact can have an arbitrarily large number of them.

7.7 Exploration

1. **When in Rome.** Adapt the first model of symmetric coordination so that, instead of success-biased social learning, an agent observes several other agents and adopts whichever norm is more common among those observed. Does this assumption of conformity change any of your conclusions about either which norm will spread or how quickly it will do so? Compare how quickly norms spread as a function of the number of agents being observed at a time, and compare this to probabilistic copying using a sigmoid function.
2. **Do as I do, not as I say.** Update the first model of symmetric coordination so that agents can't accidentally copy the norm of an agent who has already had the opportunity to copy someone else that round. You may need to give agents a new variable so that they observe the norm that generated the payoff, not the norm another agent has only just copied but never used.
3. **Better and faster.** Consider the symmetric coordination model, starting with the initial frequency of norm 1 at $p_0 = 0.5$. Using batch runs, document the time it takes the model to reach fixation as a function of the benefit to coordinating, δ. Plot and explain your results.
4. **So many choices.** Extend the symmetric coordination model to allow for arbitrarily many norms, so that there are $k \geq 2$ norms. Again assume that coordinating on the same norm yields a payoff of $1 + \delta$, non-coordination yields a payoff of 1, and all norms are initially equally likely in the population.
   (a) Code this model and share its code.
   (b) How does the model behave differently as k increases? Does the population always converge to one norm? Does it take longer when k increases? Support your answers with data and/or plots.
   (c) What would happen if interactions were not random, but were instead spatially structured so that agents interacted with near neighbors? Speculate on this, or code it to find out.

5. **Structured coordination.** Update the one-community asymmetric coordination model to include within-group structure. In the model presented in the chapter, agents play many games at random, so we could calculate their expected payoff based on the frequencies of the two norms in the population. Alter the model so that agents instead play the coordination game only with their closest four neighbors, as in the prisoner's dilemma model in chapter 6. In other words, agent payoffs are derived from considering their own norm and the norms of their nearest neighbors. Similarly, you will update the `evolve` procedure so that, like the models in chapter 6, evolution works by deterministically copying the best neighboring strategy. Finally, add a switch for a Boolean variable that allows you to toggle between well-mixed and spatially structured populations. Do the results differ between these conditions? In particular, consider that in the well-mixed model, norm 1 could increase in frequency if and only if its initial frequency was at or above p^*. If the results are different, explain why. Support your answer with plots.

6. **Mathhtam.** Sometimes it's nice to prove things.

   (a) For the symmetric coordination model, show analytically that whichever norm is more common always has an advantage in expected payoff in a large population.
   (b) For the asymmetric coordination model, show analytically that norm 1 and norm 2 are ESSes against the other norm. Include full expected payoff equations for each norm, assuming a large, well-mixed population.

7. **Fair ain't fair.** Consider the division of labor model. Let $G = 7$, $\gamma = 0.7$, and $d = 0.9$. In other words, there is a large benefit for dividing into complementary roles, but the division is unequal, and individuals interact with members of the opposite type 90% of the time. Initialize your simulations with 60% of group A and 40% of group B using norm 1, so that group A is likely to become the preferred user of norm 1. The variable of interest here is m, the probability of outgroup observation.

   (a) Run batch simulations varying m between 0.0 and 0.4 in increments of 0.02. Plot the frequency of norm 1 in each group as a function of m. Describe these patterns verbally.
   (b) Explain why you observe the patterns you observe. For moderately large m, why is the prevalence of norm 1 in group B so high?

The Scientific Process

8

> Science is a social activity just like being a policeman, a factory worker or a politician.
>
> —Richard Lewontin

Science! We consume it, we are inspired by it, we do it, we love it. Science, we are taught, is the endeavor to probe the world carefully, methodically, and rigorously to determine what is in it, how it works, and, more generally, what is *true*. This endeavor has been tremendously successful in many regards—it's given us useful understanding of a huge range of phenomena from planetary motion to electricity, from biological development to cognitive perception. This record of success has led to claims like "the good thing about science is that it's true whether or not you believe it." This is an oft-repeated line usually attributed to the astrophysicist and TV presenter Neil deGrasse Tyson. The quote has become a rallying cry for supporters and defenders of science. And of course, science *should* be defended. Science not only increases our knowledge of the world but also serves as a bulwark against superstition and charlatanry. However, there is a counterpoint to Tyson's claim. Plenty of scientific results are wrong.

Published findings in many fields—from psychology and economics to neuroscience and oncology—have failed to replicate. The fact that a result was obtained using scientific methods is therefore not a guarantee that the claim it substantiates is correct. This is a problem, because our understanding of the world relies on facts. Charles Darwin (1871) understood the perniciousness of false facts, writing in *The Descent of Man*, "False facts are highly injurious to the progress of science, for they often endure long; but false views, if supported by some evidence, do little harm, for every one takes a salutary pleasure in proving their falseness; and when this is done, one path towards error is closed and the road to truth is often at the same time opened." What he is saying in his overwrought Victorian prose is this: we shouldn't worry too much about theories—which explain and organize the facts of the world—being wrong, because academics are competitive and love to take each other down a peg by demonstrating logical inconsistencies in one another's theories. Since logic is a common language in science, competition for theoretical explanations should remain relatively healthy, and theoretical explanations will improve over time via the marketplace of ideas. However, a coherent explanation must rely on a firm foundation of facts, so *facts* are what we should worry about. If our facts are false, we may end up wasting quite a lot of time arguing about how best to explain "facts" that aren't even true.[1]

Science involves both theory building and fact finding. This book is primarily about theory building. However, the pitch I am making in this chapter is that we can use one of the tools of theory building—modeling!—to help us understand the factors that determine the extent to which scientific findings are trustworthy. If we can address those factors, we can improve fact finding. How we conceptualize the scientific enterprise shapes how we conduct research as well as how we strive to improve scientific practices. And as always, the nature of those conceptualizations becomes clearer when we can translate them into formal models. In this chapter, I'll present several models of science, gradually adding complexity to build an increasingly rich picture of science as a process.

8.1 Science as Hypothesis Testing

Early in our schooling, many of us are taught a simple and somewhat naïve model of "the scientific method" as an activity primarily concerned with **hypothesis testing** (Figure 8.1). The scientist comes up with a hypothesis about some natural system. She cannot directly extract the essential epistemic value of the hypothesis (that is, whether it is true or false) from the environment. Instead, she investigates the hypothesis by experimentation or other empirical means, the results of which will either support the hypothesis (a positive result) or fail to support it (a negative result).

Here's the key thing: the alignment between the results and the epistemic state of the hypothesis is necessarily imprecise. For any investigation, there is some risk of a false positive, $\alpha = \Pr(+|F)$, and some risk of a false negative, $\beta = \Pr(-|T)$. Note that the expression $\Pr(A|B)$ is the probability of event or condition A being true under the assumption that event B has occurred, and is usually read as "the probability of A conditional upon B" or "the probability of A, given B." Thus, $\Pr(+|F)$ should be interpreted as "the probability of a positive result conditional on a false hypothesis" and $\Pr(-|T)$ should be interpreted as "the probability of a negative result conditional on a true hypothesis." The outcomes associated with these probabilities are sometimes called Type 1 and Type 2 errors, respectively.[2] It seems clear that scientists should strive to reduce their error rates. Nevertheless, the inevitable presence of some error means that there is uncertainty regarding the correspondence between a scientific result and the hypothesis it appears to provide evidence for or against. Assuming we could estimate the error rates, how confident should our scientist be in her results?

---

[1] The nature of "truth" is one of those topics that can easily keep the philosophically minded up at night. Personally, I have complicated views on the subject of truth and the nature of facts, which stem in part from the observation that any facts must rely on particular decompositions of the world (see chapter 1). For convenience, I've adopted a more staunchly realist perspective throughout this chapter.

[2] In reality, many scientific results carry explicit uncertainty, and scientists can interpret results more probabilistically than implied by these simple dichotomous characterizations of hypothesis states and experimental outcomes. I use these dichotomies for ease of explanation. I am in agreement with Healy (2017) that an excessive appeal to nuance can hinder the development of theoretical explanations. Readers interested in quantitatively characterizing nuance of this sort are directed to Gelman and Carlin (2014), and are more generally invited to better acquaint themselves with literature in the philosophy of science.

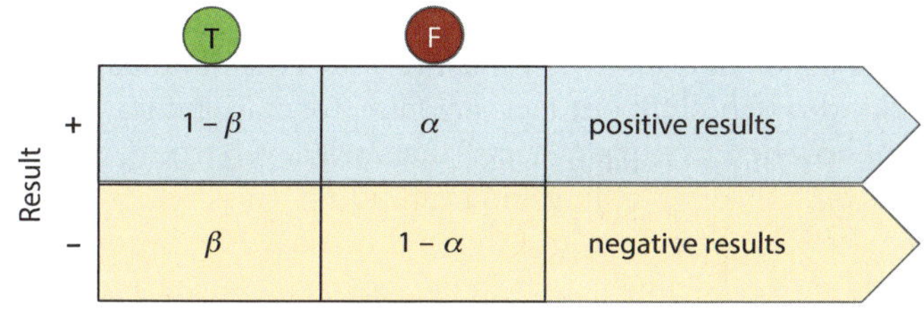

Figure 8.1 A first model of science. Hypotheses are investigated and results, with characteristic error rates, are recorded. The real epistemic state of each hypothesis, true or false (T or F), is unknowable except through this sort of investigation.

Consider the following scenario. The eminent scientist Dr. Pants investigates one of her many hypotheses. Using her well-established method, she estimates that the probability of a false positive, in which the analysis will yield a positive result when the hypothesis is false, is 5%. That is, $\Pr(+|F) = 0.05$. If the hypothesis is true, the probability that the analysis will correctly yield a positive result is 50%. That is, $\Pr(+|T) = 0.5$. The probability of correctly identifying a true result is sometimes called the analytical **power** of a method. The analysis is conducted, and the result is positive! Hooray! Now here's the crucial question: What is the probability that Dr. Pants's hypothesis is actually correct?

A commonly given answer is 95%. Surely, if the probability of a false positive is 5%, then a positive result will be correct 95% of the time, right? Wrong. Or at least it's wrong most of the time. The truth is that I haven't been playing fair. I have asked you a trick question. Indeed, I have not actually given you enough information to answer the question. Understanding both why the question is a trick and what information would be required to legitimately answer it requires the employment of **Bayes' theorem**. This simple theorem is so important in scientific reasoning that it warrants its own section.

8.2 Bayes' Theorem

When we test a scientific hypothesis, we'd like to know the probability that our hypothesis is true, conditional on the results of our experiment or test. This is quite similar to questions medical researchers face when they administer tests for diseases or other serious conditions: they want to know the probability that a patient has a disease given that their test has yielded a positive result. Since medical testing is a fairly intuitive milieu for many readers, we will use it as our narrative example.

Before we proceed, some terminology is in order (see Table 8.1). We will use as our example a fictional blood test for the dreaded Disease X, which afflicts a small portion of the population. The prevalence of the disease—that is, the probability that a randomly selected person from the population has the disease—is $\Pr(D)$, which we can read as "the probability of the disease," or sometimes "the base rate of the disease." The probability that a randomly selected person from the population tests positive for Disease X is $\Pr(R)$, which we can read as "the probability of a positive result." Note that this latter statistic is not independent of the prevalence of the disease—the more people have the disease, the more common positive tests will be. The probability of the simultaneous co-occurrence of Disease X and a positive test—that is, that someone both has Disease X and tests positive for it—is called the **joint probability**. Because co-occurrence simply requires both the disease and the result to be present, the order in which we write them does not matter: $\Pr(D, R) = \Pr(R, D)$. Order *does*

Table 8.1. Notation for pure, joint, and conditional probabilities

Notation	Definition
$\Pr(D)$	The probability that a randomly selected person has Disease X.
$\Pr(R)$	The probability that a randomly selected person receives a positive test result.
$\Pr(R, D) = \Pr(D, R)$	The joint probability that a person both has Disease X and receives a positive test result.
$\Pr(R\|D)$	The conditional probability a person receives a positive test result, given that they have Disease X.
$\Pr(D\|R)$	The conditional probability a person has Disease X, given that they receive a positive test result.

matter, however, when it comes to **conditional probabilities**. The probability of receiving a positive result conditional on having the disease, $\Pr(R|D)$, is not the same as the probability of having the disease conditional on a positive test result, $\Pr(D|R)$. The former is the probability that someone with the disease will test positive. The latter is the probability that someone who receives a positive test result really has the disease.

The joint probability of two events is equivalent to the conditional probability of the first event given the second event, multiplied by the prior probability of the second event. In our medical example, we can write this as:

$$\Pr(R, D) = \Pr(R|D) \times \Pr(D) \tag{8.1}$$

This means that the probability that some individual has both a positive test result and Disease X is equal to the probability of having a positive test result conditional upon actually having Disease X (that is, how good is the test at correctly identifying the presence of the disease), times the prior probability of having Disease X (that is, how prevalent is this disease in the population). If the test almost always comes up positive for people who have the disease (say, $\Pr(R|D) = 0.99$), but the disease is very rare in the population (say, $\Pr(D) = 0.0001$), then, if tests are administered at random, it will be quite rare to find a person who both has the disease and tests positive ($\Pr(R, D) = 0.000099$).

The joint probability of having Disease X and testing positive can also be written

$$\Pr(D, R) = \Pr(D|R) \times \Pr(R). \tag{8.2}$$

This is the conditional probability that person who tests positive has the disease, multiplied by the baseline probability of a positive test result. Recall that for joint probabilities, order doesn't matter. That is,

$$\Pr(D, R) = \Pr(R, D). \tag{8.3}$$

By substitution, this means that

$$\Pr(D|R) \times \Pr(R) = \Pr(R|D) \times \Pr(D). \tag{8.4}$$

Let us take the perspective of a medical practitioner treating a patient who has tested positive for Disease X. Our patient, now extremely worried, asks us how likely it is that she actually has this dreaded disease, given her positive test result. In other words, she wants to know $\Pr(D|R)$. We can answer this by taking Equation 8.4 and dividing both sides by $\Pr(R)$.

$$\boxed{\Pr(D|R) = \frac{\Pr(R|D) \times \Pr(D)}{\Pr(R)}} \tag{8.5}$$

The general form of this equation, where R and D can be exchanged for any two events, is **Bayes' theorem**, named for the eighteenth-century philosopher, mathematician, and Presbyterian minister Thomas Bayes, who first derived it.[3] Bayes' theorem allows us to update our estimate of an event's probability in light of new evidence. In this example, the base prevalence of the disease, $\Pr(D)$, is called the **prior** probability of the disease—our estimate prior to evidence. Upon seeing the positive result, we calculate the **posterior** probability, $\Pr(D|R)$. The application of Bayes' theorem in this way is not always intuitive. To help build up some more intuition, let's work through an example of a real-world scenario where this kind of calculation matters.

8.2.1 Testing for Chromosomal Disorders

MaterniT21 is a noninvasive blood test administered to pregnant women to analyze fetal DNA for signs of chromosomal disorders (it can also be used to assess the sex of the fetus). In 2014, controversy arose. The makers of MaterniT21 had claimed that the screening had a 99% detection rate for trisomies—chromosomal disorders involving an extra copy of one chromosome, including Down syndrome (trisomy 21) and Edwards syndrome (trisomy 18). Based on this claim of accuracy, some women who tested positive for trisomy 18 chose to terminate their pregnancies after the MaterniT21 screening, while others received further testing. An investigation found that in over a third of these latter cases, the additional testing failed to find evidence of trisomy 18. A headline in the *Boston Globe* declared the shocking statistic: "Studies show a positive test is only accurate 64% of the time" (Daley, 2016). That is, only 64% of women who had tested positive were later confirmed to be carrying infants with trisomies. What's going on? Were the makers of MaterniT21 lying?

What's going on is the nonequivalence of conditional probabilities. Specifically, the probability of a positive result given a true presence of the disorder, $\Pr(+|T)$, is not equivalent to the probability of having the disorder conditional on a positive test result, $\Pr(T|+)$. Among women with fetal trisomy 18, the test *does* almost always yield a positive result, $\Pr(+|T) = 0.99$. The rate of false positives for the test—the probability of a positive result given the absence of the condition—is also quite low, $\Pr(+|F) = 0.00022$. The reason there were so many false positives—the reason a staggering 36% of positive results were false alarms—is that the condition itself is very rare, occurring in about one out of every 2,500 pregnancies ($\Pr(T) = 0.0004$). With these statistics in hand, we can use Bayes' theorem to derive the probability of having the disorder conditional on a positive result:

$$\Pr(T|+) = \frac{\Pr(+|T)\Pr(T)}{\Pr(+)} \tag{8.6}$$

Wait a minute. We have values for the terms in the numerator, but what about the denominator, $\Pr(+)$? This is the overall probability of observing a positive result. There are actually

[3] The theorem was independently derived around the same time by the French mathematician Pierre-Simon Laplace, who was apparently unaware of Bayes' work.

two ways to get a positive result: a true positive (when the disorder is present), and a false positive (when the disorder is not present). The probability of a true positive is the numerator of Equation 8.6. We can calculate the probability that a result is a false positive by first recognizing that a patient either has the condition or she doesn't, so the probability of *not* having a condition is one minus the probability of having the condition, $\Pr(F) = 1 - \Pr(T)$. By substitution, we get the following:

$$\Pr(T|+) = \frac{\Pr(+|T)\Pr(T)}{\Pr(+|T)\Pr(T) + \Pr(+|F)(1 - \Pr(T))} \tag{8.7}$$

We know the values for each of these terms, so we can calculate the posterior probability of trisomy 18 conditional on a positive test result:

$$\Pr(T|+) = \frac{(0.99)(0.0004)}{(0.99)(0.0004) + (0.00022)(0.9996)} = 0.643 \tag{8.8}$$

This confirms that there is a greater than one-in-three chance that a positive result indicates nothing but error, even though the claim of 99% accuracy was valid. Even with a highly accurate test and a seemingly low risk of false positives, *actual* false positives may be common if the base rate prevalence of the condition is low. There is an important lesson in all this concerning scientific hypothesis testing, as we will discuss shortly. But first, a brief aside about p-values.

8.2.2 A Problem with p-Values

For those familiar with the use of p-values in null hypothesis significance testing, the medical example just given illustrates why a low p-value does not necessarily indicate strong support for a particular hypothesis. This is because the p-value tells us the probability of the data conditional on a null hypothesis,[4] but it does not tell us what we really want to know: the probability of the null hypothesis conditional on the data. That is,

$$\underbrace{\Pr(\textbf{data}|\textit{null})}_{p\text{-value}} \neq \underbrace{\Pr(\textit{null}|\textbf{data})}_{\text{what we usually want to know}} \tag{8.9}$$

A low p-value tells us that the data we obtained are unlikely, given the null hypothesis. But if the null hypothesis itself is well supported by other evidence (and thus has a high prior probability of being true), or if the data themselves are unusual and unlikely to be obtained again in a similar study, rejecting the null hypothesis may not be advisable.[5] This logic is central to arguments favoring Bayesian approaches to statistical analysis, which explicitly weigh the probabilities of different models of the data-generating process against one another.

[4] Technically, it tells us the probability of obtaining data *at least* as extreme as what was actually obtained, but close enough.

[5] This argument was central to critiques of a 2011 study that appeared in a mainstream psychology journal purporting to show evidence for psychic forecasting of the future; see Wagenmakers et al. (2011).

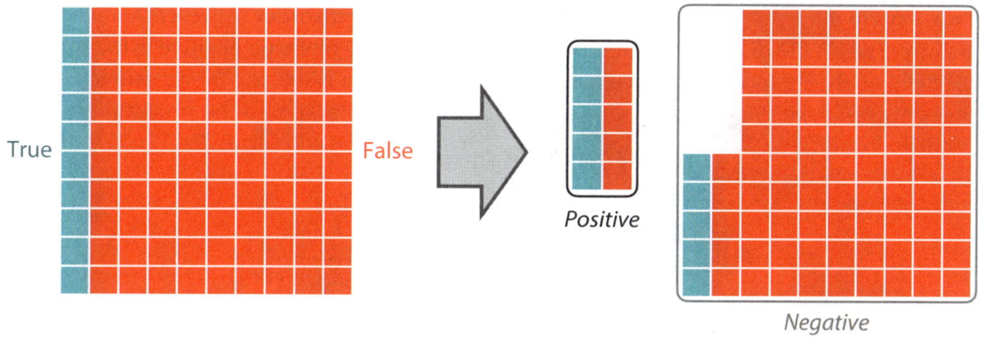

Figure 8.2 The importance of base rate in hypothesis testing. Left: 100 hypotheses are tested, of which 10 are true (the base rate is $b = 0.1$). Right: 50% of the true hypotheses and 5% of the false hypotheses yield positive results, producing a posterior probability of $\Pr(T|+) \approx 0.5$.

8.3 Science as Bayesian Inference

Let's return to our pop quiz from Section 8.1. You should now be able to see that the question about Dr. Pants's hypothesis could not be answered because we did not know the prior probability of a hypothesis being true. In the science of science, this probability is sometimes referred to as the **base rate**, $\Pr(T)$. In practice, the base rate is usually difficult to estimate. Its value depends on the field of inquiry, the types of hypotheses being considered, and the quality and imagination of the researchers. In theory, we can use modeling to explore how different assumptions about the base rate should influence our conclusions about hypothesis testing. While we may not be able to exactly calculate the probability that a real hypothesis is true, we can gain intuitions about the factors that make results more or less reliable.

Imagine that instead of testing only one hypothesis, Dr. Pants tests one hundred hypotheses. Unbeknownst to her, ten of these are true and ninety are false. To make this example more concrete, suppose that Dr. Pants runs a behavioral genetics lab and is looking for single nucleotide polymorphisms (SNPs) that correlate with a heritable behavioral disorder. She tests 100 SNPs, of which ten are actually involved in processes that reliably increase or decrease the probability of the disorder onset. Each SNP being tested represents a hypothesis. The base rate is therefore $\Pr(T) = b = 0.1$. Every association tested, every interaction considered, every statistical test run is a hypothesis that may or may not be supported. As mentioned, Dr. Pants tests her hypotheses using her well-calibrated methods, which are known to have a false positive rate of $\Pr(+|F) = \alpha = 0.05$ and an analytical power of $\Pr(+|T) = 1 - \beta = 0.5$.

What is the probability that a positive result actually reflects a true hypothesis? In this case, it's closer to 50% than 95% (see Figure 8.2).

$$\Pr(T|+) = \frac{(0.5)(0.1)}{(0.5)(0.1) + (0.05)(0.9)} \approx 0.526 \tag{8.10}$$

The lower the base rate, the lower this posterior probability gets. This means that when many false hypotheses are tested, false positives are likely to be rampant, even with seemingly low error rates. Worse yet, in reality we can never know for certain the epistemic states of our hypotheses, nor can we easily estimate the base rate. Our results are all we have.

It should now be clear that a model of science benefits from including a process of **hypothesis selection** in addition to the experimental investigation of that hypothesis. Such

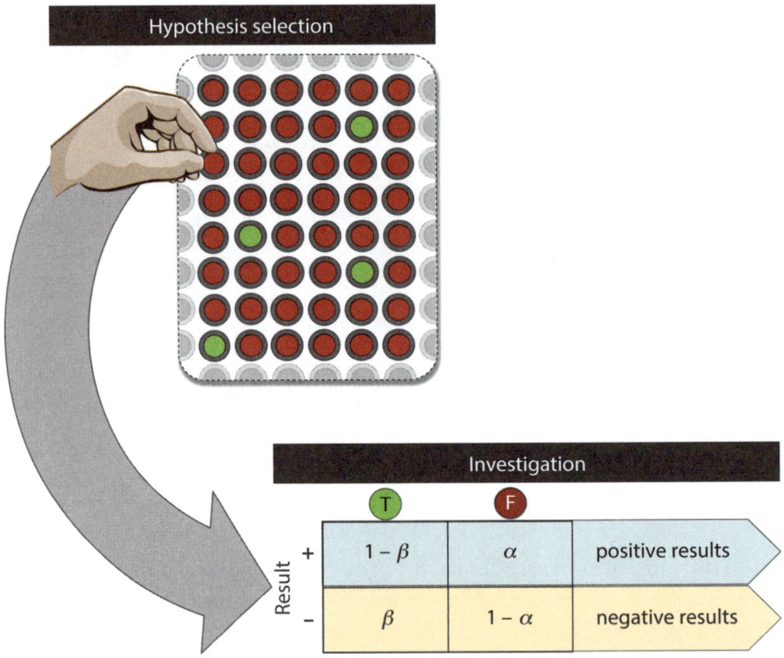

Figure 8.3 Science as Bayesian inference. Investigation is preceded by hypothesis selection, illustrating the critical role of base rate. The inner circles indicate the real epistemic value of each hypothesis. Red indicates false, green indicates true. The gray outer circle represents the fact that these epistemic values are unknown before investigation.

a model is illustrated in Figure 8.3.[6] We can represent this model as an equation for the posterior probability that a positive result indicates a true hypothesis, using the notation introduced so far:[7]

$$\Pr(T|+) = \frac{(1-\beta)b}{(1-\beta)b + \alpha(1-b)} \tag{8.11}$$

If the base rate b is low, even a moderate false positive rate (such as 5%) will lead to a low posterior probability and a large number of false positives. The lesson here is that, in the quest to avoid false positives, carefully considered hypothesis generation is just as important as methodological care. It is my opinion that learning to model processes like this is critical to forming intuitions about complex systems and therefore to constructing good hypotheses. Well done, you, for working to improve your modeling skills.

A possible concern about the model I have just presented is that it treats each hypothesis in isolation. It ignores the social and public aspects of science. Scientists don't just produce results, they also try to publish them, and some results are easier to publish than others. Once published, results can then be replicated, and with new information from replication efforts comes the opportunity for new estimates of the underlying hypotheses' epistemic states. In other words, science is a population process.

[6]The model portrays hypothesis selection as preceding investigation. In reality, new hypotheses can arise at various stages in the scientific process. For simplicity I have organized the model of science in this stylized way.

[7]This Bayesian model of science was originally introduced by Ioannidis (2005).

8.4 Science as a Population Process

Hypotheses are not tested in isolation. Evidence for or against a hypothesis accumulates in a body of literature. The promise of meta-analysis is to use the literature to more carefully weigh all the evidence. However, there are many factors that influence the power of meta-analysis to give us an accurate assessment of hypotheses. I want to highlight two of these factors: replication and publication bias. The first of these amplifies the power of scientific inquiry to reveal truth, while the second factor can vastly diminish that power.

8.4.1 Replication Reduces Uncertainty

Over time, a research community builds up a body of evidence for a hypothesis. Some studies may support it, while others may not. If we consider the totality of evidence, we should be able to gain increasing levels of certainty about our hypothesis. Indeed, Bayes' theorem allows us to use new evidence to update the probabilistic estimate of some event's occurrence. Bayes' theorem gains significant power from the fact that it can be applied *iteratively*, so that the posterior estimate can become the new prior. This process of algorithmic continual learning is known as **Bayesian updating**.

Let's explore how replication can increase our confidence regarding the true epistemic state of a hypothesis, even when individual studies are error prone. If the base rate b is known, we can use that as our initial prior probability that our specific hypothesis is true. However, we don't actually need to assume that we have a good estimate of the prior. If our data accurately represents our system of interest, our posterior probability will eventually converge on an accurate estimate, even if we start with a biased prior probability.[8] When the information we receive from investigations is unbiased, our initial prior affects only the time it takes to obtain an accurate portrait of the hypothesis's likelihood. Scientists don't have unlimited time, so starting with an accurate estimate is often useful. Later, we will see that initial priors can matter quite a lot when our information is biased.

As before, we'll keep things simple and consider just two types of results: positive results (supporting the hypothesis) and negative results (failing to support the hypothesis). In the case of a positive result, the posterior is updated as follows:

$$\Pr(T|+) = \frac{\Pr(+|T)\Pr(T)}{\Pr(+|T)\Pr(T) + \Pr(+|F)\Pr(F)} = \frac{(1-\beta)\Pr(T)}{(1-\beta)\Pr(T) + \alpha(1-\Pr(T))} \tag{8.12}$$

In the case of a negative result, the posterior is updated thusly:

$$\Pr(T|-) = \frac{\Pr(-|T)\Pr(T)}{\Pr(-|T)\Pr(T) + \Pr(-|F)\Pr(F)} = \frac{\beta\Pr(T)}{\beta\Pr(T) + (1-\alpha)(1-\Pr(T))} \tag{8.13}$$

In subsequent rounds of updating, the most recent posterior probability becomes the new prior.

We can build a simple model to see the power of Bayesian updating in an idealized form of replication. In this model, we will consider a scientist testing a hypothesis that is known

[8] It is also worth noting that the order in which the evidence is received does not affect the final estimate, though it will affect intermediate estimates derived before the full set of evidence is observed.

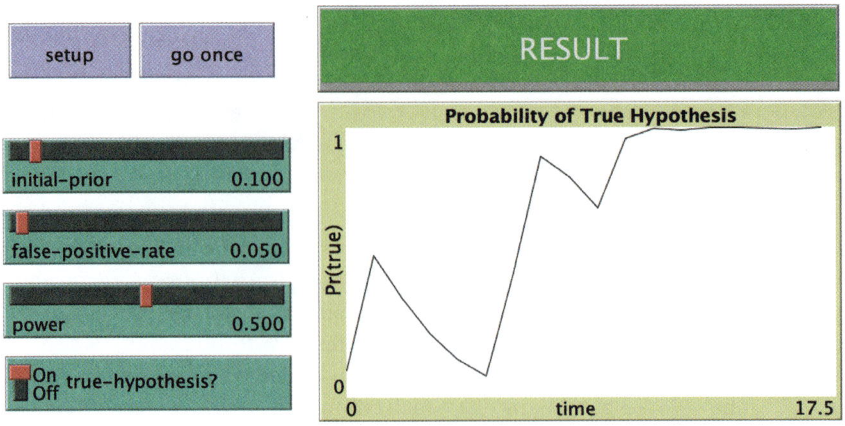

Figure 8.4 NetLogo interface for a simple model of science as Bayesian updating. Here the estimate eventually converges to represent the actual epistemic value of the hypothesis (true).

to us, the modelers, to be true or false. The scientist starts with an initial prior, which is her confidence that the hypothesis is true. Through iterated tests of the hypothesis, this estimate gets updated, hopefully toward ever-increasing certainty.[9]

The NetLogo code for this model is **science_bayesianupdating.nlogo**. There are no agents in this model, so NetLogo has no special advantage as a coding language here. However, for consistency, we'll stick with it. We begin with the global parameters that are declared in the Interface tab (Figure 8.4). We must specify the initial prior (`initial-prior`). This may or may not be equal to the estimated base rate of true hypothesis in the community—it doesn't matter for this simulation as the model takes the perspective of a researcher updating their estimate for a single hypothesis that is categorically either true or false. The point is to show how Bayesian updating converges on an accurate estimate. The Boolean variable `true-hypothesis?` indicates whether our hypothesis is actually true or false. Finally, we specify the character of our methods, indicating their power (`power`) and false positive rate (`false-positive-rate`). Note that I have purposefully not made the `go` button a "forever" button, so that the posterior probability must be estimated one iteration at a time.

The purpose of this code is to document the change in the estimated probability that a hypothesis is true or false. To do this, we will introduce a global variable, `prob-true`, that records this estimate.

NetLogo code 8.1

```
globals [ prob-true ]
```

The model is initialized by setting `prob-true` equal to our initial prior probability, `initial-prior`.

[9]In addition to the point estimates used in this chapter, Bayesian updating can also be used to estimate continuous probability distributions, and thereby capture some of the uncertainty around those estimates. For a thorough introduction to this approach, see McElreath (2020).

NetLogo code 8.2

```
to setup
  clear-all
  set prob-true initial-prior
  reset-ticks
end
```

On each tick of the model, the hypothesis is tested using the specified methods, with their characteristic power ($1-\beta$) and false positive rate (α). If the hypothesis is true, the method will yield a positive result with probability $1-\beta$ and a negative result otherwise. If the hypothesis is false, the method will yield a positive result with probability α and a negative result otherwise. The value of `prob-true` is then updated accordingly, using Equation 8.12 or 8.13. The colored patches in the visualization reflect whether the result is positive (green) or negative (red).

NetLogo code 8.3

```
to go
  ifelse true-hypothesis? [ ;;true hypothesis
    ifelse random-float 1 < power [ ;;true positive
      ask patches [set pcolor green]
      set prob-true (power * prob-true) / ((power * prob-true) +
      (false-positive-rate * (1 - prob-true)))
    ]
    [ ;;false negative
      ask patches [set pcolor red]
      set prob-true ((1 - power) * prob-true) / (((1 - power)
      * prob-true) + ((1 - false-positive-rate) * (1 - prob-true)))
    ]
  ]
  [ ;;false hypothesis
    ifelse random-float 1 < false-positive-rate [ ;;false positive
      ask patches [set pcolor green]
      set prob-true (power * prob-true) / ((power * prob-true) +
      (false-positive-rate * (1 - prob-true)))
    ]
    [ ;;true negative
      ask patches [set pcolor red]
      set prob-true ((1 - power) * prob-true) / (((1 - power)
      * prob-true) + ((1 - false-positive-rate) * (1 - prob-true)))
    ]
  ]
  tick
end
```

When we run the model iteratively and plot the value of `prob-true` over time, we witness Bayesian updating.[10] Although the posterior probability may rise and fall early on in the process, it will always converge, with certainty, on the true epistemic state of the hypothesis. Play around with this model. You will see that with a low base rate, false hypotheses will be

[10]This chapter presents a simple form of Bayesian updating, in which probabilities are represented as point estimates. In more complex approaches, probabilities are often represented as distributions, which allows all possible values to be represented.

identified fairly quickly, but true hypotheses require more tests to achieve confidence (that is, high probability) in their truth. And the larger the rates of Type 1 and Type 2 errors (α and β, respectively), the longer convergence to near certainty will take. In all cases, replication reduces uncertainty by allowing us to more accurately estimate the truth or falseness of a hypothesis.

8.4.2 Publication Bias Increases Uncertainty

Modeling cumulative scientific knowledge acquisition as a Bayesian updating process illustrates the power of replication. Our model also assumed a system in which all results are made available, or, failing that, in which results are omitted from the record without bias. However, all results are *not* made available with equal probability. Some results are more likely to be omitted (i.e., remain unpublished) than others. It is widely acknowledged that scientific publishers, particularly fancy or "high impact" journals, often exhibit a bias in favor of positive results (those supporting a novel hypothesis) and against null or negative results (those failing to support the novel hypothesis, or, perhaps, those merely confirming established knowledge). If that is the case, we will end up with a biased sample in the published literature for our meta-analysis. Our Bayesian estimation will therefore receive an overabundance of positive claims and a dearth of negative claims.

This seems like a problem. How serious of a problem is it? We can assess this question with some small modifications to our model. First, let's add a term representing publication bias, ρ. Assume that positive results are always published. When $\rho = 0$, all negative results are published as well. Otherwise, negative results will fail to be published with a probability equal to ρ. Equivalently, the probability that a negative result is published is $1 - \rho$. If a result is not published, it cannot contribute to updating the posterior probability that the hypothesis is true. In reality, publication bias can result from either failure to submit on the part of authors or failure to accept on the part of editors. Our model makes no distinction between these cases and merely assumes that certain results are more likely to be published than others on the basis of whether or not they support the hypothesis being tested.[11]

To assess the severity of publication bias, we can consider how often we assign a high probability of truth to a false hypothesis (the reverse scenario, assigning a low probability of truth to a true hypothesis, is far less likely—proving this is left as an exercise). Humans must constantly make decisions under uncertainty. Rarely can we be 100% sure that our decision is the correct one. Instead, we usually take some level of certainty as "good enough" and proceed *as if* the decision was correct. For scientific results, some level of certainty may be enough to **canonize** a hypothesis as fact. If our Bayesian estimation reveals that a hypothesis has a 99% probability of being true, it may be advisable to proceed as if it is indeed true, for example, in the design of subsequent empirical and theoretical research. The assumption of "close enough to true" is important if that assumption informs policy, education, or otherwise socially relevant decisions. We will further extend our Bayesian updating model by adding one additional parameter, τ, which is the threshold for canonization. If we represent the posterior probability of a hypothesis being true as p, then our model will have

[11]The analysis in this section is based on Nissen et al. (2016), which considers the publication history of a single hypothesis. Readers interested in a comparatively more sophisticated model are directed to McElreath and Smaldino (2015), which considers the distribution of all possible publication histories and the associated probability that a hypothesis with a particular history is true.

the following stopping rule: when $p \geq \tau$, the hypothesis is canonized as true, and when $p \leq 1 - \tau$, the model is canonized as false.

The NetLogo code for this model is **science_pub-bias.nlogo**. We add two global parameters to the Interface tab: the extent of publication bias, ρ (`pub-bias`), and the canonization threshold, τ (`canonization-threshold`). For our outcome measure, we will add a global parameter `canonized-true`, initialized in the Code tab. This variable will be assigned a value of zero unless the hypothesis is canonized as true, when it will be assigned a value of one (it will remain equal to zero if the hypothesis is canonized as false). You might wonder why this isn't a Boolean variable, since it can take only two possible values. A real-numbered value makes it easier to assess the overall proportion of simulation runs that ended up with a hypothesis canonized as true: we simply take the average value of `canonized-true` across runs.

The `setup` procedure works exactly as in the previous model, with the probability that the hypothesis is true initialized to `initial-prior`. The dynamics work slightly differently, as we must factor in publication bias and canonization. The `go` procedure first checks the posterior probability of the hypothesis being true. If it exceeds the threshold for canonization, the value of `canonized-true` is updated and the simulation ends. Otherwise, a new result is considered and used for Bayesian updating. To keep the code clean, this step is accomplished by calling the procedure `update-hypothesis`.

NetLogo code 8.4

```
to go
  if prob-true >= canonization-threshold [ ;;canonized TRUE!
    set canonized-true 1
    stop
  ]
  if prob-true <= (1 - canonization-threshold) [ ;;canonized FALSE!
    set canonized-true 0
    stop
  ]
  update-hypothesis
  tick
end
```

Updating the probability that the hypothesis is true works the same as in the simpler updating model, as long as the result is positive. If the result is negative, it is only published with probability $1 - \rho$. If publication bias prevents a negative result from being published, the estimate of the hypothesis's epistemic value is not updated.

NetLogo code 8.5

```
to update-hypothesis
  ifelse true-hypothesis? [ ;;true hypothesis
    ifelse random-float 1 < power [ ;;true positive -- PUBLISH
      ask patches [set pcolor green]
      set prob-true (power * prob-true) / ((power * prob-true) +
      (false-positive-rate * (1 - prob-true)))
    ]
    [ ;;false negative -- PUB BIAS
      ask patches [set pcolor red]
      if (random-float 1 < (1 - pub-bias)) [
        set prob-true ((1 - power) * prob-true) / (((1 - power)
```

```
        * prob-true) + ((1 - false-positive-rate) * (1 - prob-true)))
      ]
    ]
  ]
  [ ;;false hypothesis

    ifelse random-float 1 < false-positive-rate [
      ;;false positive -- PUBLISH
      ask patches [set pcolor green]
      set prob-true (power * prob-true) / ((power * prob-true) +
      (false-positive-rate * (1 - prob-true)))
    ]
    [ ;;true negative -- PUB BIAS
      ask patches [set pcolor red]
      if (random-float 1 < (1 - pub-bias))[
        set prob-true ((1 - power) * prob-true) / (((1 - power)
        * prob-true) + ((1 - false-positive-rate) * (1 - prob-true)))
      ]
    ]
  ]
end
```

The model takes the perspective of a researcher who naïvely assumes that the publication record can be trusted and accurately reflects the evidence for and against a hypothesis. As such, the researcher calculates the posterior probability that the hypothesis is true conditional on that (biased) publication record. We can use simulation to see how often a false hypothesis will be mistakenly canonized as true as a function of publication bias and error rates. Figure 8.5 shows the results of these simulations,[12] plotting the proportion of runs in which a false hypothesis was mistakenly canonized as true (instead of false) as a function of publication bias (ρ). The plots are presented in a way that allows us to simultaneously examine the influence of several other model parameters: the false positive rate (α), the power $(1-\beta)$, and the canonization threshold (τ).

Several things become clear from this analysis. First, increasing publication bias increases the probability that a false fact is canonized as true. Second, publication bias becomes more of a problem with more error-prone methods—when power is lower and the false positive rate is higher. And third, the less certainty we demand of our facts—that is, the lower the canonization threshold τ is—the more likely it is that false facts will be canonized as true. When we fail to publish negative outcomes, our ability to assess the truth of hypotheses is not only hindered, but can be subverted to the extent that we gain certainty over the truth of false claims. If we only see supportive (or only negative) claims, we are likely to form incorrect conclusions.

8.5 Science as a Cultural Process

Scientists are human, and humans are cultural beings. Scientific communities have norms of operation just as all cultural communities do. Some of those norms are methodological: the practices, data collection techniques, statistical tools, etc. that scientists use. These norms

[12]Derivation of a closed-form mathematical solution is also possible, requiring understanding of Markov chains and binomial distributions (Nissen et al., 2016).

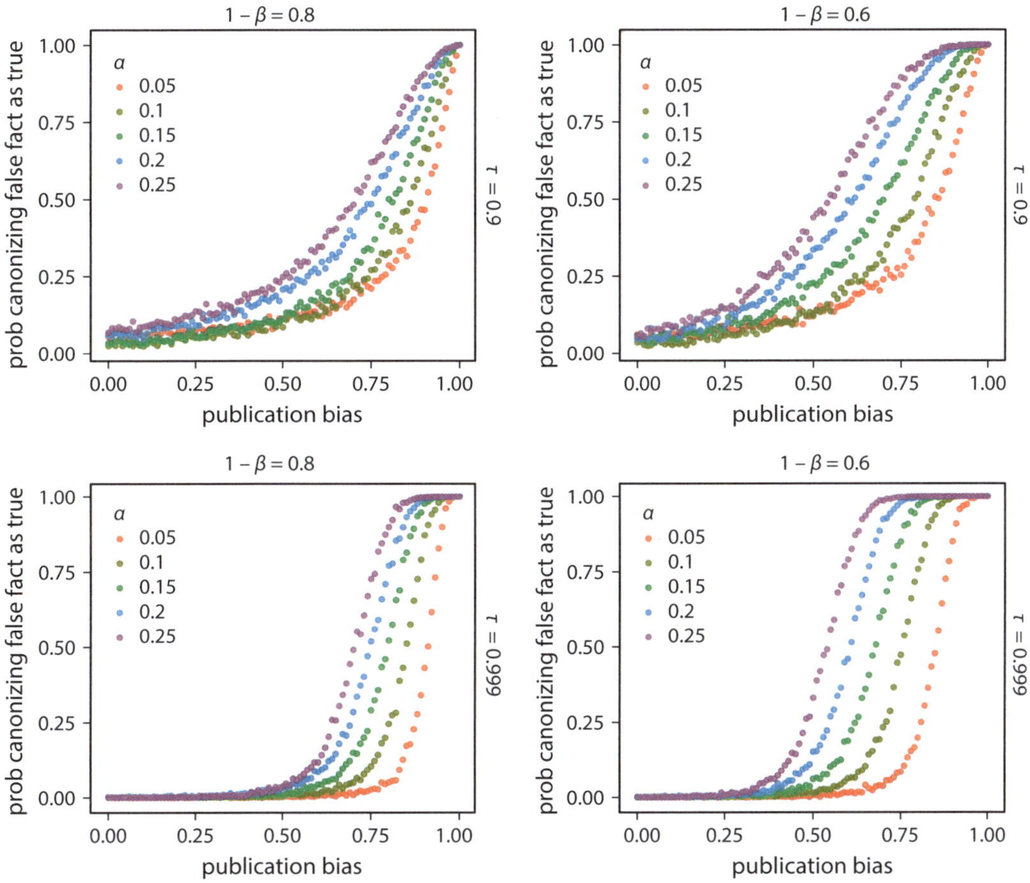

Figure 8.5 Publication bias and the canonization of false facts. The probability that a false hypothesis is incorrectly canonized as true, as derived from 1000 simulations of the Bayesian updating model for each data point. We see that things improve when there is greater methodological power ($1-\beta$), lower false positive rate (α), and more rigorous canonization threshold (larger τ), but in all cases sufficient publication bias (ρ) promotes the canonization of false claims as true. In all cases, `initial-prior` $= 0.5$.

are shaped by cultural forces. In this light, we can interpret an observation that is otherwise puzzling. Many of the problems that increase the likelihood of false discoveries have been known to the scientific community for decades or longer.

Here are some examples. False claims are more likely to appear in the literature when:

- Negative results aren't published, distorting the publication record by eliminating disconfirmatory evidence.
- Statistical techniques are misunderstood, leading to false positives and ambiguous results.
- Studies are underpowered, because small sample sizes lead to false positives and ambiguous results.
- Surprising results are the easiest to publish, because such results have a low prior probability of being true.

Although the factors in this list may be new to some readers, scientists have, in general, been aware of them for some time. Nevertheless, they are still prevalent in the way science is conducted. Why? Why isn't science better? You would think that a community dedicated to accurately describing the universe would have rapid and efficient self-correction. Yet that isn't always the case. Scientific practices are performed by scientists, who are humans that respond to incentives. Understanding how scientific practice—and not just scientific knowledge—changes over time requires a new model that includes the scientists themselves in the model dynamics.

All models start with key assumptions about the world in order to propose how those assumptions lead to observable patterns. We will model not only the investigation of scientific hypotheses, but also how the results of those investigations influence the careers of scientists. I want to include in our new model a core but often overlooked assumption: *there is a bottleneck*. That is, not everyone who wants to be a scientist gets to be one, at least in the case of coveted academic jobs at research universities. There are more applicants than jobs. Further, selection through this bottleneck is nonrandom. Certain traits are selected for and others selected against, such that individuals who get jobs and promotions in academia come to have some things in common. In the current landscape of academia, it is often the case that statistics like number of publications, impact of journals published in, and grant dollars awarded are used as proxy measures in the calculus of decision-making for hiring and promotion. While disciplines and departments vary in their reliance on these metrics, surveys indicate that academics tend to value these things and expect their colleagues to value them as well (Niles et al., 2020).

It may not be obvious that preferentially rewarding things like productivity and impact factor[13] is bad. It seems like we should *want* scientists to be productive and we should *want* their work to have a wide impact. Don't we want our scientists to be awesome? The difficulty is that awesomeness is in reality quite complicated and multidimensional. The importance of research may not be manifest for quite some time, and a lack of productivity can just as easily reflect careful study of a difficult problem as it can a lack of drive. This difficulty becomes a serious problem when awesomeness is assessed with crude, quantitative metrics like paper count, journal impact factor, and h-index.[14] It has been widely noted by savvy social scientists that, in the words of Donald Campbell (1976, p. 49), "The more any quantitative social indicator is used for social decision-making, the more subject it will be to corruption pressures and the more apt it will be to distort and corrupt the social processes it is intended to monitor." When incentives to publish drive scientists, scientists may adapt their behavior to maximize publication at the cost of research quality.

Perhaps you agree that this sounds bad. If scientists are trying to game the system and are strategically altering their methods to get ahead, it's clear that problems will arise. Surely that's the issue. If we can convince individual scientists to simply ignore those temptations and focus on just doing their best science, we shouldn't have a problem, right? Right? Maybe not. Let's take a look at an evolutionary model of science to see whether or not this sort of optimism is justified. This model uses the same sort of cultural-evolutionary logic we saw in the previous two chapters, albeit with a slightly more complicated dynamic.

---

[13]The impact factor of a journal is a number that reflects the extent to which papers published in that journal tend to be cited elsewhere. It is a low-precision but widely used proxy of prestige.

[14]The h-index of a scientist is the largest number h satisfying the claim that the scientist has h papers each with at least h citations.

8.5.1 An Evolutionary Model of Science

Science, like many cultural phenomena, can evolve through a Darwinian process. Several influential philosophers of science have discussed how scientific theories evolve by variation and selective retention (Campbell, 1965; Popper, 1979; Hull, 1988). But scientific *methods* can also evolve. As we have noted in previous chapters, Darwinian evolution requires three conditions to occur. First, there must be variation. Second, that variation must have consequences for survival or reproduction. And third, that variation must be heritable. In other words, traits that have positive consequences for survival or reproduction relative to other traits will tend to spread. Do scientific methods fit these criteria?

Research practices and methods certainly vary. Variation in methods leads to differences in the sorts of results that are produced by different researchers and, consequently, the publications that arise from those researchers. These publications have consequences in determining who is successful in terms of getting hired and promoted, securing grants, attracting graduate students and postdocs, and placing those trainees in positions heading their own research groups. In other words, variation in methods has fitness consequences. Finally, variation in methods is at least partly heritable, in the sense that trainees acquire research habits and analytical techniques from mentors and peers. Researchers also acquire research practices from successful role models in their fields, even if they do not personally know them. Therefore, when researchers are rewarded primarily for publishing, habits that promote publication are likely to be passed on.

Let's examine a model based on the idea that a scientific community is a system that rewards productivity, and consider the sorts of methods—and ramifications thereof—that evolve as a consequence.[15] This model is a bit more complicated than those presented previously throughout this book, and yet it is still quite a simplified view of the practice and culture of science.

The agents in this model are scientific laboratories, or *labs* for short: collections of scientists who work together on research projects. Each lab i has a characteristic method that is defined by it *power*, $W_i = \Pr(+|T)$, reflecting the method's ability to obtain true positives. Each lab also has a characteristic *effort* level, e_i. Together, power and effort determine the lab's characteristic false positive rate, $\alpha_i = \Pr(+|F)$. This is an operationalization of the idea that increasing power can also increase false positives unless effort is exerted. This may seem counterintuitive, but consider that it is easy to have perfect power if *every* result is declared positive—you won't miss any true hypotheses! Identifying those true positives while correctly eliminating the false hypotheses requires additional work. Indeed, this is essentially the logic of signal detection theory.[16] To formalize this relationship, we can specify a mathematical relationship between W_i, e_i, and α_i that has the following properties: as power goes up, so does the false positive rate, but increasing effort will decrease the false positive rate for a given power. A function that fits the bill[17] is this:

$$\alpha_i = \frac{W_i}{1 + (1 - W_i)e_i} \tag{8.14}$$

[15]This is a simplified version of a model presented in Smaldino and McElreath (2016).

[16]See Lynn and Barrett (2014) for a brief introduction to signal detection theory.

[17]In general, a familiarity with functional forms will often come in handy as you develop your own models.

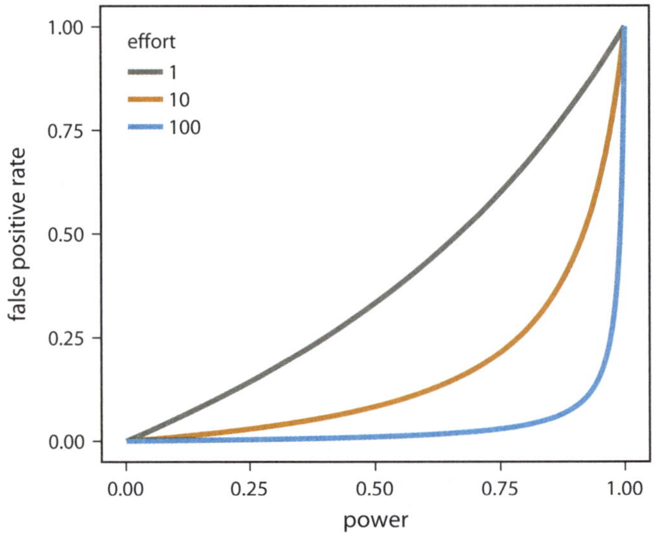

Figure 8.6 The relationship between power and the rate of false positives, moderated by effort (per Equation 8.14).

This function is plotted in Figure 8.6. Initially, we can assume that labs exert high levels of effort and hence have relatively low false positive rates. Common standards in some disciplines aim for power to be at least 0.8 and for the rate of false positives to be no more than 0.05. For these values to co-occur, effort must be set to 75 (you should confirm this).

Our population of N labs will evolve (in the cultural sense) by using their methods to produce a body of published research, which will then be used to determine which labs will transmit their methods to new researchers. Each time step of the model has two distinct stages: *Science* and *Evolution* (Figure 8.7). In the Science stage, labs conduct research and attempt to publish their results. In the Evolution stage, old labs retire and new labs arise, seeded with the academic "progeny" of successful labs. These two stages repeat until the model reaches an equilibrium distribution of methods in the population. Let's look at each of these in more detail.

8.5.1.1 Science

In the Science stage, each research lab has the opportunity to investigate a hypothesis and to publish the results of that investigation. Labs that exert more effort are less likely per unit time to tackle new hypotheses, under the assumption that more rigorous work takes longer to complete. To capture this, we need a function that is maximal when effort is minimal and decreases with effort. A simple linear function would probably suffice, but we can also use a function that captures a somewhat more realistic relationship between effort and productivity. While labs exerting more effort should have a lower probability of starting a new study than labs exerting less effort, this probability shouldn't get *too* low. I've used the following function:

$$\Pr(\text{new study}) = 1 - \eta \log_{10} e_i \tag{8.15}$$

Here, η represents the influence on effort on a lab's productivity. A default value of $\eta = 0.2$ yields a probability of 1 when effort is minimal ($e_i = 1$) and a probability of 0.625 for labs

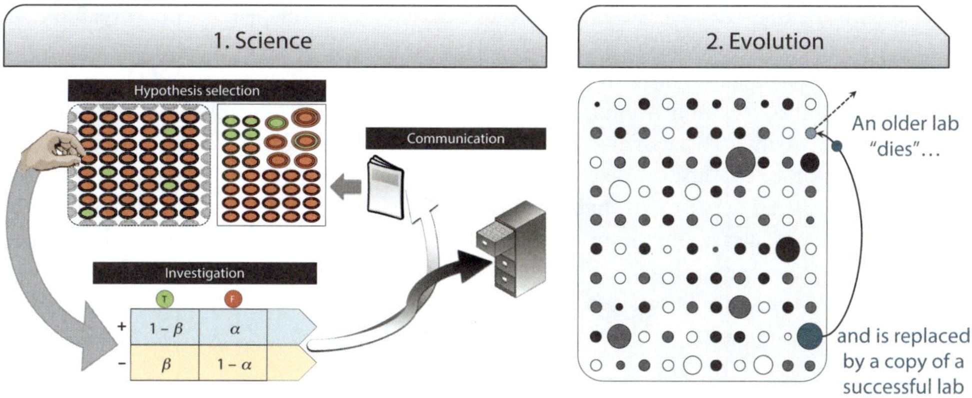

Figure 8.7 An evolutionary model of science. Dynamics occur in two stages: Science and Evolution. In the Evolution stage, labs compete for research positions. A lab's methods are represented by its shading, and its prestige is represented by its size. When a new position opens, it is more likely to be filled by someone using the methods of more prestigious labs.

using our initially effortful method ($e_i = 75$). I encourage you to plot this function in order to see how it differs from a simple line, and also to vary the function to see its effect on the model outcomes. A linear function with a shallow slope and the same maximum and minimum values will perform similarly. A lab that investigates a new hypothesis selects a true hypothesis with probability b, and a false hypothesis otherwise. A true hypothesis yields a positive result with probability equal to the lab's characteristic power, W_i, while a false hypothesis yields a positive result with a probability equal to the lab's characteristic false positive rate, α_i.

In order to translate publications into evolutionary fitness, each publication is associated with a payoff. We can model publication bias in two ways: differential probability of publication and differential payoffs for positive and negative results. For simplicity, let's assume that positive results (which confirm the hypothesis being tested) are always published, while negative results are published with a probability $p \leq 1$. When published, positive results yield a payoff of 1, while negative results yield a payoff of $v \leq 1$. The idea is that negative results are less likely to be published, and when they are published, researchers receive a smaller increase in credit or prestige than if they had published a positive result. Labs keep track of their total payoffs over time.

8.5.1.2 Evolution

In the Evolution stage, an older lab is chosen to "die," representing retirement or otherwise ceasing to actively produce research. Next, another lab is chosen to have its methods reproduced in a new lab, such that labs that have produced more positive results (and hence have higher cumulative payoffs) are more likely to be chosen. This represents the advantage that productive labs have in training or otherwise influencing grad students, postdocs, and other junior researchers.

Algorithmically, there are many possible ways to implement this idea. I've used a computationally simple mechanism for a population of 100 agents. First, ten labs are chosen at random, and the oldest of these dies. Next, another ten labs are chosen at random, and the one with the highest payoff is chosen to reproduce.[18] An "offspring" lab is created that inherits the characteristic methods (power and effort) from its "parent" lab. When you build evolutionary models with continuous traits, it's important to allow for some random variation, or mutation, in the inheritance of traits, so that values not present in the original population can evolve. This can also make the model more realistic, as it represents imperfect transmission of methods. To separate the evolutionary dynamics of power and effort, we can assign each a separate probability of "mutating": μ_W and μ_e, respectively. When a parameter mutates, a random quantity is added to its value, drawn from a normal distribution with a mean of zero and a standard deviation of σ_W or σ_e. This formalization allows us to turn evolution on each of our two traits "on" and "off" independently by setting either of the mutation rates to zero.

8.5.2 Coding the Model

The NetLogo code for this model is **science_NSOBS.nlogo**. The model involves a fairly large number of global parameters, many of which I have instantiated as sliders so that their values may be meaningfully explored. Due to this large number, I have used NetLogo's notes feature to mark which variable symbols described above correspond to the global parameters used in the model code (Figure 8.8). Of course, we don't *have* to make sliders for all of the global parameters if we don't anticipate changing their values. Along these lines, I have declared two additional global parameters in the Code tab: the influence of effort on productivity, η (`effort-influence`), and the number of agents sampled for death and reproduction (`reproduction-sample-size`). Values for these variables can later be assigned in the `setup` procedure. Even when a number is intended to be fixed in a model, it is still usually advisable to code it as a variable, which will make the code easier to interpret and will also make it easier to implement future extensions or robustness tests.

NetLogo code 8.6

```
globals[
  effort-influence
  reproduction-samplesize
]
```

You'll notice from an inspection of Figure 8.8 that, although this is an agent-based model, I have not included any visual representation of the individual agents. I could have done so, perhaps with color representing agents' effort or power values. However, I have chosen not to, in part to highlight that visualization is a choice about model analysis, not about model design or dynamics. You are invited to add this sort of visualization as an exercise.

The agents in this model are labs, and each lab must keep track of several variables. These include the lab's characteristic power and effort, as well as the false positive rate that will be derived from the previous two parameters. Each lab must also keep track of its total payoff

[18]It can be shown that a more conceptually elegant but computationally costly algorithm, in which each agent is assigned a probability of reproducing based on its relative payoff, produces qualitatively similar results. See Smaldino et al. (2019c). This sort of robustness test is important for making sure that your results aren't an artifact of some arbitrary modeling choice. As you gain experience with modeling, you may gain an intuition for likely outcomes, which is a benefit of learning a wide range of modeling approaches.

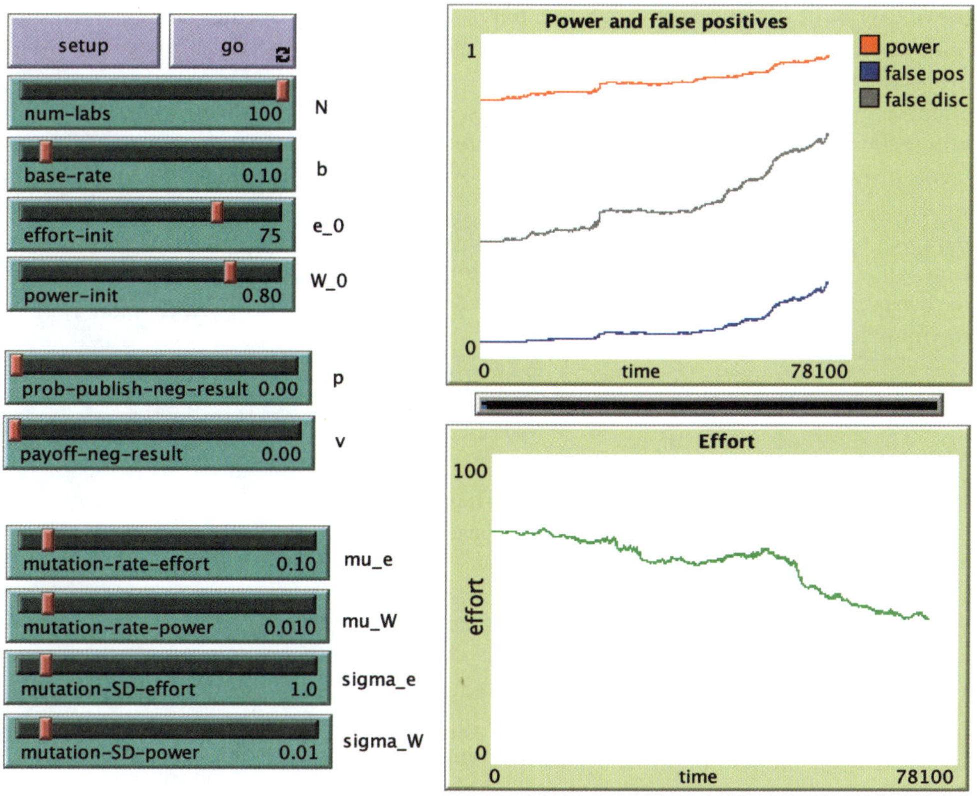

Figure 8.8 NetLogo Interface for the evolutionary model of science.

from publications, as well as its age, which determines the probability that it "dies" on a given time step. We therefore declare a number of agent-level variables.

NetLogo code 8.7

```
turtles-own[
  power
  effort
  false-pos-rate
  pub-payoff
  age
]
```

Initialization of this model is fairly straightforward; we need to create the agents and initialize some parameter values. In order to keep things streamlined, I have added a secondary procedure called `make-labs`, in which the individual labs are created. The `setup` calls this procedure to create labs and then assigns values to the fixed global.

NetLogo code 8.8

```
to setup
  clear-all
  make-labs
```

```
    set effort-influence 0.2
    set reproduction-samplesize 10
    reset-ticks
  end
```

The make-labs procedure creates each agent and assigns values to each of the agent-level parameters. Notice that the procedure uses Equation 8.14 to calculate each lab's false positive rate. In this way, all labs start out identical, perhaps with the high effort and high power we desire in a functioning scientific community.

NetLogo code 8.9

```
to make-labs
  create-turtles num-labs [
    set power power-init
    set effort effort-init
    set false-pos-rate (power / (1 + ((1 - power) * effort)))
    set pub-payoff 0
    set age 0
  ]
end
```

The model dynamics proceed as follows. First, the Science stage occurs, in which labs produce and publish research. Second, the Evolution stage occurs, in which older labs are replaced by the progeny of productive labs. And finally, the ages of all labs are incremented. The two main stages of the model dynamics are coded in separate procedures that are called from the go procedure.

NetLogo code 8.10

```
to go
  science
  evolution
  ask turtles [set age age + 1]
  tick
end
```

In the science procedure, each agent tackles a new hypothesis with a probability given by Equation 8.15. Sometimes, for bespoke functions like this one, it can be useful to create a separate procedure for its calculation. This increases the readability of the code and makes it easier to swap out other mathematical functions. The new-hypothesis procedure returns the probability of tackling a new hypothesis as a function of the effort of the agent that calls it.

NetLogo code 8.11

```
to-report new-hypothesis
  report 1 - (effort-influence * (log effort 10))
end
```

If an agent embarks on a new investigation, a hypothesis is selected that is either true or false according to the parameter `base-rate`. The epistemic nature of that hypothesis dictates the parameters that determine whether the investigation yields positive or negative results (power vs. false positive rate). Positive results always lead to publication, which increases the lab's payoff by 1. Negative results lead to publication with probability `prob-publish-neg-results`, which increases the lab's payoff by `payoff-neg-result`. These dynamics are all captured by the `science` procedure below.

NetLogo code 8.12

```
to science
  ask turtles[
    if random-float 1 < new-hypothesis [ ;;tackle a new study?
      let actual-truth? false
      if base-rate < random-float 1
        [ set actual-truth? true ]
      ifelse actual-truth?
      [ ;;publish pos result or maybe publish negative result
        ifelse (random-float 1 < power) ;;true positive
        [ set pub-payoff pub-payoff + 1 ]
        [
          if (random-float 1 < prob-publish-neg-result)
          ;;false negative
        [ set pub-payoff (pub-payoff + payoff-neg-result) ]
      ]
    ]
    [ ;;publish pos result or maybe publish negative result
      ifelse (random-float 1 < false-pos-rate)
      ;;false positive
      [ set pub-payoff pub-payoff + 1 ]
      [
        if (random-float 1 < prob-publish-neg-result)
        ;;false negative
      [ set pub-payoff (pub-payoff + payoff-neg-result) ]
    ]
  ]
]
```

After each lab has had the opportunity to publish new research, we move on to the Evolution stage of the model dynamics. First, a sample of `reproduction-samplesize` agents is selected at random, and from these the oldest lab is selected to die. Next, another sample of the same size is generated from the remaining agents, and from these the lab with the highest `pub-payoff` is chosen to reproduce. This lab creates an exact copy of itself using the `hatch` procedure. We then set the newly created lab's age and payoff to zero.

NetLogo code 8.13

```
to evolution
  ;;death
  let death-labs n-of reproduction-samplesize turtles
  let oldest-lab max-one-of death-labs [age]
  ask oldest-lab [die] ;;kill off the oldest lab.
```

```
    ;;birth
    let birth-labs n-of reproduction-samplesize turtles
    let fanciest-lab max-one-of birth-labs [pub-payoff]
    ;;the oldest in the set
    ask fanciest-lab [
      hatch 1 [ ;;create a copy that inherits its attributes
        set pub-payoff 0
        set age 0
        mutate
        set false-pos-rate (power / (1 + ((1 - power) * effort)))
      ]
    ]
  end
```

Notice that each newly created lab calls the procedure `mutate`, which adds random variation to the methods the lab has inherited from its "parent" based on the model's mutation parameters. Both power and effort are mutated separately. Mutation involves the addition of random noise, drawn from a normal distribution with a mean of zero, meaning that mutation is just as likely to increase as it is to decrease the value of these parameters.

One thing to keep in mind when mutating a continuous variable is whether the values to which the variable can be assigned are **bounded**—that is, constrained to remain within some range. Here both power and effort are bounded—power is bounded in [0, 1] and effort is bounded in [0, 100]. We therefore add code to keep the variables within their allowed ranges. One should be careful about this sort of **truncation**. If truncation occurs very often, it can produce trait distributions in which values near the boundaries are overrepresented.

NetLogo code 8.14

```
to mutate
  ;;mutate power
  if random-float 1 < mutation-rate-power [
    set power power + (random-normal 0 mutation-SD-power)
    if power > 1 [set power 1]
    if power < 0 [set power 0]
  ]
  ;;mutate effort
  if random-float 1 < mutation-rate-effort [
    set effort effort + (random-normal 0 mutation-SD-effort)
    if effort > 100 [set effort 100]
    if effort < 1 [set effort 1]
  ]
end
```

That's it for the dynamics of the model. With this code, we can examine the evolution of scientific methods under incentives to publish. We can track the changes to power and effort associated with various labs. However, another thing we might be interested in, besides the quality of our scientists, is the quality of the published literature. The false positive rate is the probability that a lab produces a positive result conditional on a false hypothesis. We might also want to know the **false discovery rate** (*FDR*), which is the proportion of *published* findings that are incorrect. Notice that the false discovery rate reflects properties of both the individual scientists (the extent to which tested hypotheses tend to be correct) and the institutional infrastructure around publication (the extent to which negative results are published).

In this model, an incorrect finding means there is a mismatch between the experimental result and the true epistemic value of a hypothesis. False discoveries can reflect false positives, but also false negatives (a negative result for a true hypothesis) as long as the probability of publishing negative results is nonzero. The false discovery rate for the publications produced in a given time step is the number of incorrect results published divided by the total number of results published. This will be a function of the base rate, b, the probability of publishing negative results, p, the mean power, W, and the mean false positive rate, α.

$$FDR = \frac{bp(1-W) + (1-b)\alpha}{b\left[W + (1-W)p\right] + (1-b)\left[\alpha + (1-\alpha)p\right]} \tag{8.16}$$

We can implement this as a reporter called `false-discovery-rate` that returns this value.

NetLogo code 8.15

```
to-report false-discovery-rate
  let pow (mean [power] of turtles) ;;mean power
  let fpr (mean [false-pos-rate] of turtles) ;;mean fpr
  let total-pubs (base-rate * (pow + (1 - pow) *
             prob-publish-neg-result) + (1 - base-rate) *
            (fpr + (1 - fpr) *  prob-publish-neg-result))
  let false-pubs (base-rate * (1 - pow) * prob-publish-neg-result) +
             ((1 - base-rate) * fpr)
  report false-pubs / total-pubs
end
```

8.5.3 Analyzing the Model

With the model coded, we can explore its dynamics by considering how power, effort, and the overall false discovery rate evolve. Recall the purpose of the model: to investigate how scientific methods and the literature that results from those methods evolve in response to incentives to publish. We have multiple parameters coevolving, which can get complicated. In such cases, it is often useful to start by holding all but one parameter constant so we can see how it evolves on its own. Let's start by holding effort constant at its initially high value by setting $\mu_e = 0$ and allowing power to evolve. Try it. You should observe that power evolves to its maximum value and the false discovery rate skyrockets (Figure 8.9, top). Everything is deemed "true," and we have no information about anything.

This scenario is pretty unrealistic. Scientists actually have reasonably good ways of assessing the power of their research methods, and it's unlikely that anyone would ever allow this situation to arise. However, *effort* is notoriously difficult to assess. It's not often easy to tell how rigorously a lab is conducting its research, especially if their methods are somewhat obscured in their written reports. If we hold power constant and allow effort to evolve, we find that effort reliably evolves to its minimum value, and once again the false discovery rate balloons (Figure 8.9, bottom).

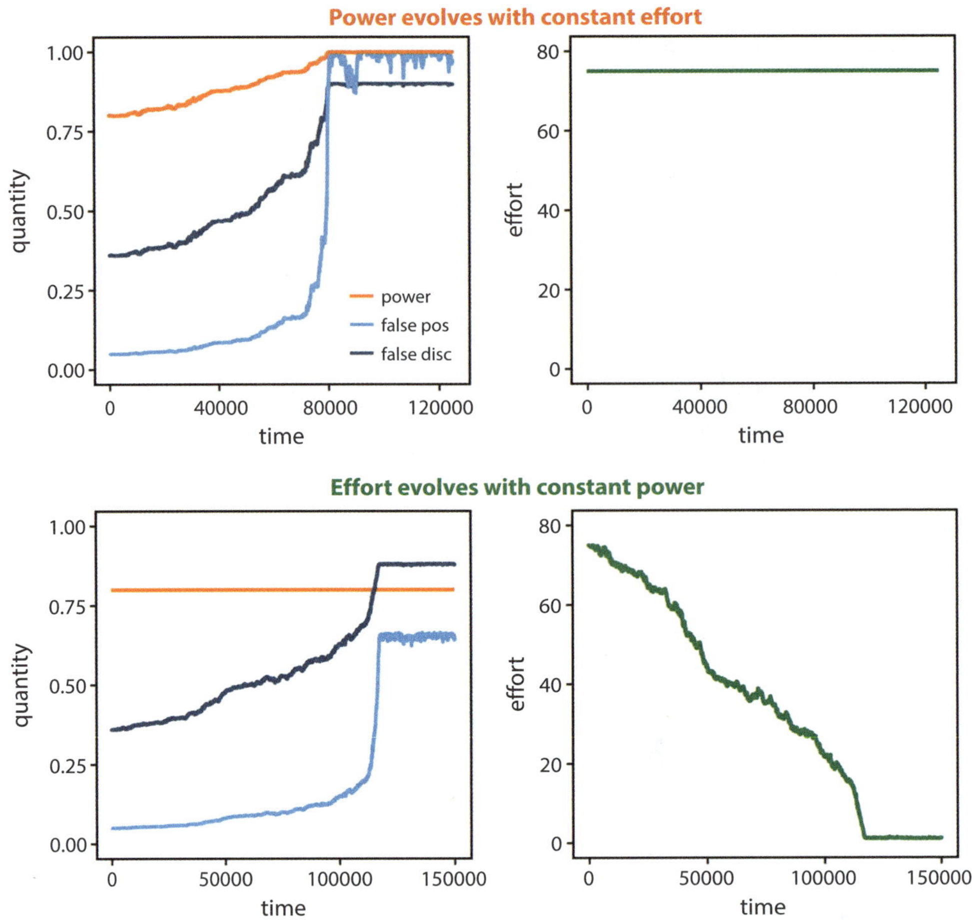

Figure 8.9 Results from our evolutionary model of science. Top: when effort is held constant but power evolves, all results are declared positive. Bottom: When power is held constant but effort evolves, all researches exhibit minimal effort and false discoveries soar. Except for where mutation rates for power or effort were zero, all parameter values are those shown in Figure 8.8.

It's important to emphasize that the agents are not strategically altering their behavior to maximize their payoffs. Each lab merely embraces the methods they were taught. It is the institutional selection for publication that drives the dynamic. My collaborators and I have referred to this dynamical process as *the natural selection of bad science* to highlight that this is a population process rather than the result of individuals striving for success at all costs.

Many other analyses and extensions are possible for this model.[19] Some of these are suggested as exercises for the reader. For the present, let's pause to consider the results we've produced so far. What do these results mean? If our model of science is at least moderately realistic, and incentives for publishing do drive selection on research methods, then we should see evidence for impediments to the improvement of scientific methods on the

[19] Several have been explored by myself and others (Smaldino and McElreath, 2016; Smaldino et al., 2019c; Barnett et al., 2018; O'Connor, 2019a; Stewart and Plotkin, 2021).

time scale of generations. If, on the other hand, incentives are rewarding methodological rigor, we should see a steady increase in the quality of methods for scientific inquiry.

In 1967, Paul Meehl cautioned about the widespread misuse of p-values, warning that scientists were wrongly interpreting their meaning and consequently generating lots of false positives (Meehl, 1967). Nearly a half-century later, in 2016, the American Statistical Association published their "Statement on p-values," once again cautioning about their misuse, and warning that scientists were wrongly interpreting their meaning and consequently generating lots of false positives. They bemoaned, "Let us be clear. Nothing in the ASA statement is new. Statisticians and others have been sounding the alarm about these matters for decades, to little avail" (Wasserstein and Lazar, 2016, p.130).

In 1962, Jacob Cohen published a meta-analysis of abnormal and social psychology experiments, noting the frustratingly low statistical power of most published research. He cautioned that many studies were not sufficiently powered to adequately provide confirming or disconfirming evidence, leading to an excess of spurious results. In the late 1980s, two studies provided new meta-analyses investigating whether there had been any improvement to the average statistical power of psychological research (Sedlmeier and Gigerenzer, 1989; Rossi, 1990). They found no improvement. More recent studies have similarly found no improvement to the average statistical power in the social and behavioral sciences, with an average power to detect small effects of only about 0.24 (Smaldino and McElreath, 2016; Szucs and Ioannidis, 2017). Since very many effects in these fields are expected to be small, this result is concerning.

Incentive structures that push scientists to boost quantitative metrics like publication counts and impact factors can lead to the degradation of methods. This dynamic requires no fraud or ill intent on the part of individuals, only that successful individuals transmit their methods. From this, we might conclude that changing individual behavior—each of us improving our own methods—is insufficient to improve scientific methods; improvement requires institutional change. Specifically, it requires that the selection bottlenecks of hiring and promotion, as long as they occur, are able to filter researchers using assessments of quality that maintain high methodological integrity. Unfortunately, institutional change is usually difficult. For the most part, institutions are not *meant* to change easily. They provide a stable framework that structures social interactions and exchanges, and they ensure some consistency to the operation of a society in the absence of enforcement by specific individuals (North, 1990). This is often a good thing, but it also means that we run into trouble when our institutions are unhealthy. Changing the institutional incentives for publishing in academic science may require patience.

8.6 Why the Most Newsworthy Science Might Be the Least Trustworthy

A very simple model[20] can help us think about the importance of selection in other aspects of science—and in other aspects of life more generally. It often seems like the most newsworthy results are the least likely to be true. The splashy paper getting all the news coverage that shows that potatoes cure Lyme disease, or that reports the discovery of a new bacterium that makes its genetic material out of tungsten, seems likely to be overturned sooner or later. Meanwhile, the most rigorous, careful studies rarely seem to end up on the front page of

[20]The material in this section is based on joint work with Richard McElreath. For more detailed discussion of how the selection-distortion effect influences statistical analysis, see McElreath (2020).

the newspaper's science section[21] or to go viral on social media. This perception is probably accurate, as multiple studies have now shown that measures of attention correlate negatively with the reliability of the research results. For example, Brembs (2018) found that journal impact factor correlated negatively with multiple indexes of quality, while Serra-Garcia and Gneezy (2021) found that among papers whose results were eventually replicated, those that failed to replicate were cited at substantially higher rates than those that were successfully replicated. Why should this be the case? Is it necessarily true that newsworthy results reflect science that is inherently less trustworthy?

Suppose that the editor at the prestigious *Journal of Nature and Science* (JONAS) only publishes papers that are deemed to be both rigorous (trustworthy) and important (newsworthy). There is no reason to suspect a priori that there is any connection between the importance of a result and the rigor with which it was investigated. However, if newsworthiness is sufficiently important, the editor may relax their standards just a smidge for those findings deemed most important. This is sometimes referred to as "big if true"—that is, an editor may publish a result that would be transformative if it were true even if the evidence isn't airtight, gambling on the potential rewards for publishing a landmark paper. In contrast, a study that is slightly less newsworthy (but still newsworthy enough to warrant publication in JONAS) must be absolutely rock solid to be published. In this case, *among studies that are published*, there will be a negative relationship between newsworthiness and trustworthiness, even if no such relationship exists in the population of studies overall.

To see how this works, consider a population of 500 studies, each with newsworthiness (N) and trustworthiness (T) scores independently drawn from a normal distribution with a mean of zero and a standard deviation of one (Figure 8.10, left). There is, understandably, no correlation between N and T in the general population. The studies are submitted to JONAS, and the journal selects for publication the top 10% of studies based on their combined (summed) scores for both N and T. Among the published studies, there is now a strong negative correlation between newsworthiness and trustworthiness (Figure 8.10, right).

This is an example of a general phenomenon in which selection on multiple properties induces a correlation in the post-selection population that wasn't there in the pre-selection population. This phenomenon goes by a few different names: it is sometimes called **Berkson's paradox**, **collider bias**, or the **selection-distortion effect**. It can pop up anywhere samples are selected on multiple additive criteria, from restaurants and dating to college admissions and the characteristics of professional athletes. The next time you hear that there is a correlation (or lack thereof) between two variables in a sample, consider whether the sample was itself selected based on those variables.

8.7 Reflections

Science is a direct confrontation with the unknown, a gauntlet thrown down to nature proclaiming, "We will *know* you!" It is an attempt to both impose structure on nature and to describe the structure that is already present. To test an idea is an exciting moment, when the possibility exists that we have discovered something true. To be validated is a triumph, to be proven wrong a disappointing return to the status quo. However, our validations may be illusory, and our disappointments fleeting.

[21] For younger readers, a newspaper is an archaic form of information transmission in which important news is printed in bundles of paper and sold by local merchants.

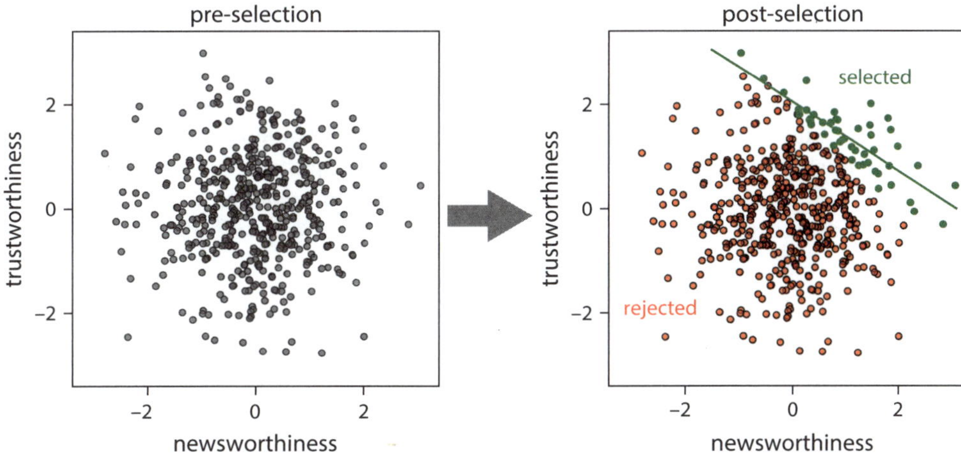

Figure 8.10 Why the most newsworthy science might be the least trustworthy. 500 studies are ranked by their combined newsworthiness and trustworthiness, and the top 10% are selected for publication. Although there is no correlation before selection, there is a strong negative correlation between newsworthiness and trustworthiness after selection (in this example, $r = -0.84$). The R code to reproduce this figure is in the code repository.

To see scientific investigation as a process of Bayesian inference is to acknowledge the uncertainty inherent in testing ideas in an open system, and that to be a scientist is to live with uncertainty. It is to acknowledge that knowledge generation is cumulative, and that failure is an integral part of the process.

Of course, science is so much more than epistemological inference. For starters, hypotheses must come from somewhere. In fact, they usually come from models, whether formal models like those we have studied in this book or mental models that embrace more of the vagueness and ambiguity ever present in how we usually see the world. Indeed, it is our models—how we divide the world into parts and relationships between those parts—that determine what we observe and what we hypothesize. And it is the same with how we conceptualize the act of science itself. How we model science, whether formally or just mentally, will determine how we approach the business of *doing* science. There are many ways to look at the business of doing science. Science is hypothesis testing, yes, but it is also model building, it is theory construction, it is measurement. It is a social enterprise that is shaped by social networks and institutional structures. There is therefore an ongoing need for more ecological, dynamic models of science in order to examine the consequences of how we structure this important activity. In the meantime, I encourage you to strive to be deliberate in how you characterize what science *is*, because that characterization will shape the science you *do*.

The type of dynamics we have explored in this chapter also apply to other domains beyond academic science. One striking example involves the news media. In 1988, Edward Herman and Noam Chomsky published their influential book *Manufacturing Consent: The Political Economy of the Mass Media*. In this book, they discussed how corporate and political interests shape the landscape of news that gets reported as well as the individuals that do the reporting. This is accomplished by the establishment of multiple filters of education, opportunity, and hiring that weed out views and personalities that don't reflect the interests of those in power. In an interview with Chomsky, the British journalist Andrew Marr pushed back at this thesis, upset at the perceived implication that he was self-censoring what he reported.[22] Chomsky quipped, "I'm sure you believe everything you're saying. But if you believed something different, you wouldn't be sitting where you're sitting." Chomsky himself recognized the similarity between cultural selection in the media and in academia. In a 2019 interview with the journalist Matt Taibbi, Chomsky revealed that he was initially interested in the role of selection in shaping academia, but found it too challenging to study. He instead applied the idea to the media, for which there was richer data available at the time (Taibbi, 2019).[23] This dynamic of selection may be quite general. Cultural and institutional forces shape the character of who becomes influential in many domains, not only by shaping those people directly, but more perniciously by applying filters that determine the sort of people who become influential.

8.8 Going Deeper

Using the methods of science, including formal modeling, to study science has become a field unto itself, known as **metascience** or the science of science. Researchers have been increasingly aware of problems in academic science, and have often turned to models to help them understand where those problems come from and how they might be remedied. Many of the questions being asked extend those that have been asked by philosophers and sociologists of science for decades, but the methods, available data, and questions being asked have advanced substantively.

The evolutionary model of science presented in this chapter considered the consequences of selection for publications. A complementary approach by Higginson and Munafò (2016) used rational actor models to show that utility-maximizing actors were likely to increase their false positives to maximize productivity. Of course, selection for publications is not the only force shaping the cultural evolution of science, and methodological rigor is not the only behavior that cultural selection acts upon. Holman and Bruner (2017) showed how industry funding can shape results to align with corporate interests. More direct extensions have also been fruitful. For example, Stewart and Plotkin (2021) extended the model to permit investment in theory, and showed that the ability to translate effort into improved hypothesis selection could offset the evolution of low effort. O'Connor (2019a) used a variant of this model to consider how different methods might correspond to differences in the distributions of risk and reward, and examined the conditions that might select for more risky or more conservative methodologies.

A Bayesian approach is also more generally useful in models where agents learn over the course of numerous events. Some cognitive scientists have argued that Bayesian updating is either a valid description of how learning works in humans and other animals or is at least a good-enough approximation for testing related theories of behavior (Tauber et al., 2017). In this vein it has been used to model a wide range of social phenomena, including the emergence of social categories (Falandays and Smaldino, 2022), the inference of pragmatic meaning in conversation (Goodman and Frank, 2016), and the evolution of social learning strategies (Perreault et al., 2012).

[22] The full interview is available here: https://www.youtube.com/watch?v=GjENnyQupow

[23] The interview is available here: https://taibbi.substack.com/p/preface-an-interview-with-noam-chomsky-the-fairway

Science is a particularly interesting social phenomenon to study, not least because anyone studying it is also engaging in science and so is automatically a subject of their own investigation.[24] Other than the fact that science is an activity that scientists are by definition involved with, and therefore have an intrinsic motivation to better understand, it is also an activity characterized by relatively clear norms, institutions, and goals. This makes it somewhat easier to model than many other social phenomena. Existing models of science make it clear, however, that there are many interesting problems to be considered beyond what we have considered in this chapter. Among other topics, researchers have considered how scientists select research problems (Bergstrom et al., 2016), how scientific norms involving coordination emerge (Akerlof and Michaillat, 2018; Smaldino and O'Connor, 2020; Weatherall and O'Connor, 2021), and how scientific credit assignment and norms of priority (in which the first to make a discovery gets the credit) incentivize or distort methodological practices (Zollman, 2018; Tiokhin et al., 2021). There's still a lot to learn about how we learn about the world.

8.9 Exploration

1. **She blinded me with models of science.** Consider the models of science in this chapter, which focused on hypothesis testing. How well does this view reflect your own experience with science? Do you find any aspects of these models problematic? How do you think a model could be better set up, not only to more accurately reflect reality, but also to answer important questions about the practice of science? How would you decompose the system of science and scientists to better answer your question(s)?

2. **Fancy pants science.** Professor Pantaloons tests a hypothesis with her Fancy Science Machine. The machine is extremely fancy. It yields positive results 99% of the time when hypotheses are true, and only 1% of the time when hypotheses are false.

   (a) The first hypothesis Professor Pantaloons tests returns a positive result! She is elated. She asks you to calculate the probability that her hypothesis is correct. What should you tell her?

   (b) The professor uses her machine to conduct a mass test on one million individual hypotheses. Each hypothesis is whether a particular genetic polymorphism is significantly associated with the dreaded behavioral disorder of compulsive nose picking. She has good reason to believe that most of the polymorphisms are not associated with the disorder, but that 100 of them are. How many of her one million tests should she expect to come out positive? Of these, what proportion will be false positives?

3. **Rinse and repeat.** You are testing a hypothesis. Your method has a power of $\Pr(+|T) = 0.7$ and a false positive rate of $\Pr(+|F) = 0.02$. Before you begin your studies, you believe there is a 20% probability that the hypothesis is correct. All computations in this problem should ideally be performed by hand.

   (a) You perform your study and get a positive result! Assuming that you are a rational Bayesian agent, what probability do you now assign to the truth of the hypothesis?

[24] Whoa.

(b) You perform a total of ten tests of this hypothesis, each using the exact same method. Your results are $\{+, -, +, +, -, -, +, +, +, +\}$. Plot the estimated probability that the hypothesis is correct over time, as each result comes in. Explain your technique. After all ten trials, what is the final probability that the hypothesis is correct?

(c) Oh no, your lab assistant messed up your files, and your results are now out of order! You gather your results in the following order: $\{-, -, +, +, +, +, +, -, +, +\}$. Starting with your initial prior estimate of 20%, repeat the previous analysis. Plot the estimated probability that the hypothesis is correct over time as each result comes in. After all ten trials, what is the final probability that the hypothesis is correct? Does the order of the results matter to the final estimate?

4. **Impossible truths.** Consider the model in Section 8.4, on the canonization of false facts. To assess the negative consequences of publication bias, we used a model to investigate how often we should expect to incorrectly assign a high probability of truth (canonization) to a false hypothesis. We did not consider the reverse scenario, the dismissal of true facts (canonizing them as false). Explain why this is not a concern under the assumptions of the model.

5. **I want to believe.** Consider the model in Section 8.4, on the canonization of false facts. Analyze the influence of the initial prior probability that a hypothesis is true on the canonization of false facts. Plot the probability of canonization of a false fact as true as a function of the initial prior (test between 0.05 and 0.95 in increments of at least 0.05) for several values of publication bias ($\rho = \{0.025, 0.05, 0.2, 0.4\}$). Perform batch runs, averaging across 20 runs per parameter combination. Use $\alpha = 0.05$, $1 - \beta = 0.8$, $\tau = 0.99$. How does the initial prior change the outcome?

6. **The plot thickens.** Plotting functional relationships is often valuable for gaining intuition as to the nature of those relationships. Consider the evolutionary model of science in Section 8.5.

(a) Plot Equation 8.14 to show the relationship between false positive rate and power for different values of effort, $e = \{1, 10, 75\}$. Describe the relationship between these three variables. How well does this function capture the desired relationship between power, false positives, and effort?

(b) Plot Equation 8.15 to show the relationship between the probability of conducting a new study and the effort exerted for different values of influence of effort on productivity, $\eta = \{0.1, 0.2, 0.3, 0.5\}$. Describe the relationship between these three variables. How well does this function capture the desired relationship between effort and the probability of conducting new research?

(c) Run the evolutionary model with different values for the influence of effort on productivity, $\eta = \{0, 0.4\}$. In one case, there is no effect of effort on productivity, in the other case there is a strong effect. Hold power constant and allow effort to evolve. How do the results differ between cases, if they differ at all?

7. **Publish or perish.** Consider the evolutionary model of science in Section 8.5, and focus on the evolution of effort and false discoveries. Under the default parameters, effort evolved to its minimum value and false discoveries increased. These parameters assumed that publication bias was very strong, such that only positive results would be published. Run a batch of simulations in which the probability of publishing negative results and the associated payoff for those publications are varied. Try

$p = \{0.1, 0.5, 0.9, 1\}$ and $v = \{0.1, 0.5, 0.9, 1\}$. Consider that effort tends to evolve more slowly than the false discovery rate. Why is that? Run sample simulations to estimate the time needed for your simulations. How robust is the natural selection of bad science to publication bias?

8. **Now you try.** Consider the evolutionary model of science in Section 8.5 with its default parameter values, focusing on the evolution of effort. In this model, all studies are investigations of novel hypotheses. However, many scientists and philosophers of science have discussed the importance of replication to ensure robust science. Could replication inhibit the natural selection of bad science?
   (a) Add replication to the model. There are many possible ways to do this. Describe how you will implement it. What sorts of new analyses does this allow you to perform?
   (b) Implement replication in the model code as you described in part (a). Perform some preliminary analyses. What do you find?
   (c) Labs are rewarded for publications in the form of payoffs that translate into reproductive fitness (at least as far as their methods go). What if they were punished by receiving a negative payoff whenever another lab failed to replicate their previously published results? Would a sufficiently severe punishment (say, -100) inhibit the natural selection of bad science? If you are able, implement this extension in the code and report your findings.

Networks

9

> Our lives are not our own. We are bound to others, past and present, and by each crime and every kindness, we birth our future.
>
> —David Mitchell, *Cloud Atlas* (2004)

Almost everything interesting about human behavior stems from each individual's entanglement with the lives of others. We are connected to other individuals through myriad relationships. Those connections have connections, which have connections of their own, and so on. In our treatment of social behavior, we ignore the vast web of relationships, and the details of its structure, at our peril. We have already explored the importance of social structure in our lattice models of opinion dynamics and cooperation, and in the group-structured models of norm evolution. We saw that random interactions often yielded very different population dynamics than did highly structured interactions. In this chapter, we will take a deeper dive into how networks of social relationships may be organized.

This chapter represents a very quick overview of **network theory**, which is a rich and rapidly developing field. Many technical and scholarly books have been written focusing solely on the formal study of networks. The interested reader should pursue these.[1] Much of the success of network science stems from the utility and versatility of the network as a metaphor. A "network" originally referred to what we now just call a net, like that thing used by fishermen (Figure 9.1). The intricate lattice of knots and connections between those knots provides fuel for countless analogies in which individuals, objects, and even ideas are similarly connected.

An early precursor to network science is the branch of mathematics known as **graph theory**, which goes back at least as far as the 1700s and the work of Leonhard Euler (See Box 9.2: The Seven Bridges of Königsberg). Modern network science draws from a wide range of mathematical contributions from Euler in the 1700s to Erdös in the 1950s, as well as from sociological contributions from the latter half of the twentieth century (e.g., Granovetter, 1973; Coleman, 1988; Burt, 1992). However, network science really came into its own

[1] Examples include Easley and Kleinberg (2010), Barabási (2016), Newman (2018), and Menczer et al. (2020).

Figure 9.1 A network in the original sense of the word.

in the late 1990s when computational models of network formation were introduced that showed how simple algorithms could generate networks with structural elements much like those seen in the real world. In this chapter, I'll introduce the basic building blocks of network models, discuss important network topologies and metrics, and explore simulations of social interactions on networks.

9.1 Network Building Blocks

The basic building blocks of networks are **nodes** and **edges**. A network requires a set of nodes (also called vertices), which can represent anything that can be connected (physically or metaphorically) to other things—from individual people to firms, nations, neurons, or genes. An edge (also called a link or tie) defines a connection between two nodes. The exact nature of the connection is not specified a priori, and it is exactly this flexibility that has given network science its power and widespread applicability. When nodes represent individual people, edges usually represent some type of social relationship, which can include friendship, kinship, business relationships, or simply physical proximity.

The simplest type of network is defined by a single set of nodes and a set of undirected, unweighted edges that represent connections between nodes. In such networks, the existence of an edge between two nodes is binary: an edge either does or does not exist. Other, more complex, types of networks are possible, including those with directed edges (so that A's relationship to B can be different from B's relationship to A), weighted edges (so that not all relationships carry equal influence), multipartite networks (in which different types of nodes exist), and multilayer networks (in which multiple sets of edges coexist, with each defining a particular relationship type). These more complex network types will be briefly

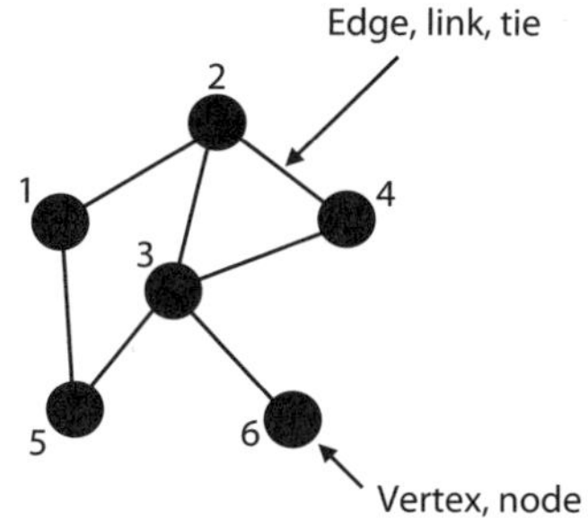

Adjacency matrix:

$$A = \begin{pmatrix} 0 & 1 & 0 & 0 & 1 & 0 \\ 1 & 0 & 1 & 1 & 0 & 0 \\ 0 & 1 & 0 & 1 & 1 & 1 \\ 0 & 1 & 1 & 0 & 0 & 0 \\ 1 & 0 & 1 & 0 & 0 & 0 \\ 0 & 0 & 1 & 0 & 0 & 0 \end{pmatrix}$$

Figure 9.2 Left: the building blocks of networks: edges and nodes (and their aliases). Right: the network's mathematical representation as an adjacency matrix. Rows and columns represent pairs of nodes. A 1 or 0 represents the presence or absence of an edge, respectively.

discussed toward the end of this chapter. For the majority of the chapter, we will limit our discussion to single-layer, unipartite, unweighted, undirected networks. This sort of network can be represented mathematically as a square matrix called an **adjacency matrix**, in which a 1 or 0 represents the presence or absence of an edge between each pair of nodes, respectively[2] (Figure 9.2). In an undirected network, adjacency matrices are always symmetrical. That is, $a_{ij} = a_{ji}$ for all pairs of nodes (i, j), where a_{ij} represents the edge connecting node i to node j. The formalism of the adjacency matrix affords many different ways of quantifying and comparing the structural characteristics of networks. One of the things that makes network science so useful is the presence of meaningful quantitative metrics for characterizing network structure. We will consider several of these throughout this chapter.

The simplest and most straightforward way to characterize a network's structure is to describe the distribution of its nodes' **degree**. The degree of a node is simply the number of edges that node shares with other nodes. Formally, the degree of node i is

$$k_i = \sum_{j \neq i}^{n} a_{ij}, \tag{9.1}$$

where n is the number of nodes in the network, and $a_{ij} = 1$ if there is an edge between nodes i and j, and zero otherwise. Degree is a node-level characteristic. However, several network-level measures can be derived from knowing the degree of each node in the network. The total number of edges in the network, m, is the sum of all the nodes' degrees, divided by two to avoid double counting:

$$m = \frac{1}{2} \sum_{i=1}^{n} k_i \tag{9.2}$$

The **average degree**, $\langle k \rangle$, is just the total number of edges divided by the number of nodes, double counted this time because each edge involves two nodes. This can be viewed as the

[2]For sparse networks in which most pairs of nodes do not share an edge, a more computationally efficient representation is an **adjacency list**, which eliminates the need to account for all the zeros in the adjacency matrix by simply keeping track of the existing edges. The distinction is not particularly important for the discussion in this chapter, but understanding how networks are represented is useful if you continue to work with them.

expected number of social connections a random node has. It has an upper bound of $n-1$.

$$\langle k \rangle = \frac{1}{n} \sum_{i=1}^{n} k_i = \frac{2m}{n} \tag{9.3}$$

Average degree is often a useful metric with which to characterize a network. However, one limitation of comparing networks based on differences in their average degree is the fact that networks may vary greatly in their overall size, and thus in the total number of edges that could possibly exist—that is, on the upper bound for degree. We can overcome this limitation by calculating the **network density**, ρ, which is the proportion of all dyads (pairs of nodes) for which an edge exists. Density reflects the extent to which possible connections are actual connections.

$$\rho = \frac{m}{\binom{n}{2}} = \frac{2m}{n(n-1)} = \frac{\langle k \rangle}{n-1} \tag{9.4}$$

The expression $\binom{n}{2} = \frac{n(n-1)}{2}$ is the number of ways that n things can be arranged two at a time, which here reflects the number of pairs of nodes (see Box 9.1: Permutations, Combinations, and the Binomial Distribution). A network in which every node is connected with every other node is called "fully connected," and has a density of $\rho = 1$. Networks with low values of ρ are often called "sparse."

In addition to scalar measures like average degree and density, a network's structure is often characterized by its **degree distribution**. The average degree may not be a meaningful metric if degree varies widely, across orders of magnitude. Network scientists often look at the degree distribution, which can be represented by a histogram or density plot reflecting the relative distribution of different degrees. For some network structures, the degree distribution can be approximated by well-known mathematical functions, as we shall discover.

BOX 9.1: Permutations, Combinations, and the Binomial Distribution

Modeling complex processes sometimes reduces to simply understanding how often certain outcomes are expected to occur or knowing all the ways in which some set of outcomes can be arranged. We saw this in chapter 2 in our discussion of random walks and the central limit theorem. When considering network structure, it is often useful to know how many different ways a particular set of nodes and edges can be arranged, as well as the statistical properties of those arrangements. The concepts reviewed in this box are foundational in probability theory, and are therefore valuable to review. We will now do this in the most serious of ways.

Suppose we are putting together a seating chart for a monster wedding, and are considering who will sit with the bride and groom (Frankenstein's monster and bride, obviously). The wedding party involves wolfpeople, vampires, zombies, swamp creatures, ghosts, and sasquatches (Figure 9.3). If we have room for four guests at the table, how many different seating arrangements are possible?

The first step is to ask whether order matters. Do the bride and groom care about the order in which the others are seated at their table, and who sits next to whom? If order matters, then we are dealing with a **permutation**. If order doesn't matter (the bride and groom only care about who is or isn't at the table), we are dealing with a **combination**. Let's take each of these in turn.

Figure 9.3 Arranging seating for guests at the Frankenstein wedding.

Permutations

We might first suppose that items are drawn *with replacement*. This means that each item can be chosen from the full set each time. Suppose there are n different types of monsters, and we need to choose r of them. In our bride example, $n = 6$ and $r = 4$. If each type of monster can be chosen over and over, then we have r instances where we must make a choice, with n options each time. There are n choices for the first seat, n choices for the second seat, and so on. Multiplying these all together, there are n^r total possibilities. Frankenstein's bride can choose from among a whopping $6^4 = 1{,}296$ possible table arrangements.

Now suppose that instead of types of monsters, we are talking about individual monsters. So, once the vampire has been chosen for the first seat, she cannot sit at the second seat; a different monster must be chosen instead. This is permutation *without replacement*. Here, there are again n choices for the first seat, but only $n-1$ choices for the second seat, $n-2$ choices for the third seat, and so on. If $r = n$, the number of possibilities is simply

$$n \times (n-1) \times (n-2) \times \cdots \times 1 = n!$$

(The last term, $n!$, is read as "n factorial"). However, if $r < n$, we have to stop seating guests once we've reached the rth seat. To account for this, we can divide the factorial by all the numbers we *don't* want to multiply by, so they will cancel out in the fraction. So if $n = 6$ and $r = 4$, we want

$$\frac{6 \times 5 \times 4 \times 3 \times 2 \times 1}{2 \times 1} = 6 \times 5 \times 4 \times 3 = 360$$

You can see that there are substantially fewer choices to consider when choices are made without replacement. The number of options for permutation without replacement is more generally expressed as

$$\frac{n!}{(n-r)!}$$

Combinations

We often want to know the number of ways some process can unfold. For example, how many ways can a set of k edges be assigned to n nodes in a network? Or, to return to our monster wedding, what monsters might be seated at Frankenstein's bride's table? Let's again assume that we are dealing with a situation without replacement, but that the order in which the items are arranged no longer matters. Only one of each monster can sit at the table, and the bride and groom only care about who is there, not about the order in which they are seated (more germanely, a node may share only one edge with any other node, but there is no ordering to its edges in an unweighted network). In this case, the seating arrangement {vampire, zombie, ghost, sasquatch} is equivalent to {ghost, zombie, sasquatch, vampire}. How many ways can this one set of four be rearranged? We already know the answer: 4!. More generally, the number of ways one can arrange r items is $r!$. Therefore, if we want to know the number of ways there are to choose r items from n possibilities when order doesn't matter, we simply divide the number of ways it could be done when order *does* matter—the permutation without replacement—and divide by the number of ways there are to arrange any particular set of choices, $r!$. This calculation comes up in probabilistic modeling so often that it has its own special notation. It is often pronounced as "n choose r," and is sometimes called the **binomial coefficient** due to its importance in the binomial distribution.

$$\binom{n}{r} = \frac{n!}{r!(n-r)!}$$

The binomial distribution

The binomial distribution, which we encountered in chapter 2, is hugely useful in statistics and probability theory. Relevant to this chapter, it describes the degree distribution of a random network, so it seems worth understanding in greater detail. If you understood the bit above about permutations and combinations, you're already most of the way there.

Imagine some process with only two possible outcomes—call them successes and failures. Now imagine a series of independent trials of this process, where the outcome does not depend on previous trials. On any given trial, success occurs with probability p, and failure occurs with probability $1 - p$. If we run the process n times, how many successes should we expect? Relatedly, what is the probability of exactly k successes? If we can answer this question, we can determine the expected distribution of outcomes.

This question is exactly equivalent to asking about the degree distribution in a random network. Each of n pairs of nodes is independently considered, and an edge is generated between them with probability p. It will be easier to understand how this works if we consider all of the edges from a single focal node. The node could have a degree of any value between zero and $n - 1$. Imagine every other node to which our focal node can connect laid out in a long list of length $n - 1$. The probability of generating an edge with every one of the first k nodes in the list is the joint probability of forming an edge with the first node, forming an edge with the second node, and so on up to the kth node, which is equal to p^k. The probability that the focal node then *fails* to form an edge with all of the remaining $n - 1 - k$ nodes is $(1 - p)^{n-1-k}$. The joint probability of this particular outcome is the product of the two probabilities: $p^k(1 - p)^{n-k}$. This probability describes a particular permutation for the focal node with a degree of k. However, when considering the node's degree, we don't actually care *which* pairs of nodes are connected, we just want to know the probability that the node is connected to exactly k other nodes. And because the formation of each edge is independent of all the other edges, the probability of every other arrangement of the focal node's connections involving exactly k edges is also $p^k(1 - p)^{n-k}$. So, in order to get the probability that any one of these combinations occur, we need to multiply the probability that one of them occurs by the total number of possibilities involving degree k. This is a combination without replacement. So, in a random network, the probability that any given node has degree k in a random network is given by the binomial distribution:

$$\Pr(k; n, p) = \binom{n-1}{r} p^k (1 - p)^{n-1-k}$$

9.2 Network Architectures: Order and Chaos

A foundational insight of modern network theory is that regularities exist among connected systems across many domains and scales. For example, networks of protein interactions, power grids, websites, and scientific collaborations all exhibit heavy-tailed degree distributions, with many low-degree nodes and a small number of high-degree nodes (Newman, 2003). If we can computationally generate artificial networks with similar properties, we can explore the consequences of those properties in greater depth. Models of network architectures allow us to build dynamic models of structured interactions on those networks—networks whose properties are known and that allow meaningful analogies to be readily drawn to real-world scenarios.

NetLogo has a very nice feature for modeling networks: a primitive class of agents called **links**. A link is a connection (or edge) between two turtles, which can serve as nodes in a network. A network can then be constructed algorithmically using turtles and links. We will go over several network formation algorithms in this chapter. We'll start out with two rather simple structures that were among some of the first network structures to be studied (due in part to their simplicity): regular lattices and random networks. We'll then move on to more complex structures designed to mimic real-world networks. We'll focus at first on the algorithms for constructing these networks, but we'll also discuss modeling dynamic social interactions between nodes.

9.2.1 Regular Lattices

A lattice is a symmetrical structure in which all nodes have the same degree and can be drawn as a regular tiling. These are sometimes called "regular networks" because there is a strict regularity to their structure. The constraint that each node must have exactly the same degree yields a patterning that looks like a lattice or crystalline structure. Other constraints beside degree are necessary to define any particular lattice network. For example, it is possible to construct both a square lattice and a ring lattice with degree 4 but with otherwise different topologies. In a network of finite size, the constraint of fixed degree may not hold at the ends for certain lattice types, so for the purpose of modeling, researchers often assume either a lattice of infinite size or one with periodic (e.g., toroidal) boundaries. The most common lattice structures used in models of social dynamics are square lattices and ring lattices.

9.2.1.1 *Square lattices*

Many models of structured social interaction use a square lattice. We've already encountered square lattices several times in this book. Square lattices are common in the modeling literature because they are computationally easy to code, mathematically tractable for at least some cases,[3] and visually intuitive and appealing. Ring lattices are popular for similar reasons. Thus, when some minimal spatial or network structure is needed, square or ring lattices are often used. The downside to such network models is that few real-life networks exhibit such regular structure. Thus, lattices are often used as a first pass to understand the importance of structure for some dynamic process, but are rarely the final word on the subject.

Constructing a square lattice in NetLogo is trivially easy, because the default grid of patches is already laid out in this way. The NetLogo primitive `neighbors4` returns the four nearest patches to the focal patch, making modeling network dynamics possible without using the `link` primitive. A square lattice is also easy to code in other programming languages because it is easily represented as a square matrix.

9.2.1.2 *Ring lattices*

A ring lattice is a structure in which nodes can be arranged on a circle, with each connected to the nearest k neighbors, where k is an even number (because each node has an equal number of neighbors to either side). A ring lattice is one of the simplest ways to model structured interactions, and is often used to test the robustness of a previously well-mixed model to stricter constraints of social structure. A square lattice involves two spatial dimensions, while a ring lattice only requires one. And because a ring is a closed loop, there are never boundary effects. To create a ring lattice, we can imagine our agents laid out consecutively along a line, with neighbors connected to neighbors. If we let the first and last agents on our line be connected to each other, we've got ourselves a ring. Ring lattices become only

---

[3]There is a technique called the pair approximation by which one can estimate outcomes of at least some dynamic processes on square lattices; see Matsuda et al. (1992).

slightly more complex when we increase the degree, such that agents are connected not only to their immediate neighbors, but to their neighbors' neighbors, and so on.

To begin to get comfortable with coding network models, let's create a simple ring lattice in NetLogo.[4] The code for this model is **RingNetwork.nlogo**. This code doesn't involve any social interactions; there are no dynamics and hence no `go` function. Instead, the goal is just to demonstrate how one can create a ring lattice. Later on, we will learn to use this sort of network as the social structure for a dynamic model.

The structure of the network is fully determined by the number of nodes (`num-nodes`) and the degree (`degree`). The `setup` function calls two subordinate functions: `make-turtles`, which creates the nodes and arranges them in a circle, and `wire`, which creates the edges. The circular layout is for visualization purposes only, of course, as it is possible to study a ring lattice without ever visualizing it.

The `make-turtles` procedure employs two NetLogo primitives, `layout-circle`, which automatically organizes the turtles in a circle of a given radius (in this case, `max-pxcor-1`, which causes the circle to nearly fill the grid), and `sort`, which sorts the set of turtles according to their `who` number—the index that reflects the order in which they were created. This latter computation is useful for a number of network models, because it allows us to use consecutive `who` numbers to connect neighbors without repetition.

NetLogo code 9.1

```
to make-turtles
  create-turtles num-nodes [ set color orange ]
  layout-circle (sort turtles) max-pxcor - 1
end
```

To create the edges between nodes, I've created a procedure called `make-edge`. This procedure takes two arguments—in this case two turtles called `node1` and `node2`—and creates an edge between them using the NetLogo primitive `create-link-with`. I've made the links black for visualization purposes.

NetLogo code 9.2

```
to make-edge [node1 node2]
  ask node1 [ create-link-with node2 [set color black] ]
end
```

The `make-edge` procedure is called from a procedure called `wire`, which iterates over all the nodes. The following is really the meat of the algorithm for constructing a ring lattice network with degree k. Each node forms an edge with the $k/2$ nodes whose `who` numbers are higher than its own, using modular arithmetic to start again from zero once the maximum `who` number has been reached. Note that each node only creates half of its total edges, going along one "direction" on the ring. The other half of its connections will be created by the nodes in the other "direction," which connect to it. Note also that I've included a constraint so that the code will not attempt to produce a ring lattice with a degree of more than $n - 1$,

[4]The NetLogo Models Library, which is included with NetLogo, contains versions of many of the network models described in this chapter. For instructional purposes, I have provided my own versions of these.

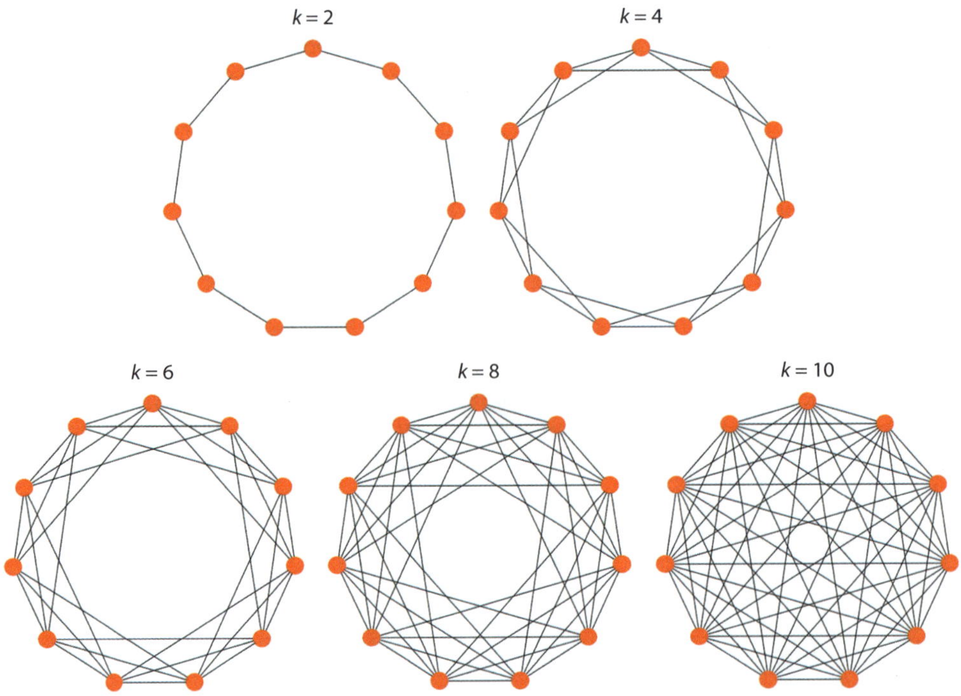

Figure 9.4 Ring lattices with $n = 11$ nodes and degree k.

because that would be impossible. This code produces lovely ring lattices of arbitrary size and degree (Figure 9.4).

NetLogo code 9.3

```
to wire
  if degree >= num-nodes
  [set degree num-nodes - 1]
  let n 0
  while [n < count turtles]
  [
    let k 1
    while [k <= degree / 2]
    [
      make-edge turtle n turtle ((n + k) mod count turtles)
      set k k + 1
    ]
    set n n + 1
  ]
end
```

9.2.2 Random Networks

Lattices are regularly structured such that each node has exactly the same number of neighbors. This eliminates any effects of heterogeneity in terms of how connected individuals are. But there is often variation in the number of connections individuals have, and this sort of heterogeneity often matters. The regular structure of lattices also means that information flow across the network requires traversing a large number of connections, because all connections are local, so to speak. It will therefore be convenient to have a network structure that provides some heterogeneity and non-local connectedness while still allowing us to maintain some control over the density (and therefore the average degree) of the network.

A simple solution is to begin with a set of nodes and then form edges between them at random. This sort of network was made popular by the mathematical analysis of Paul Erdös[5] and Alfréd Rényi (1959), and for this reason is often called an Erdös-Rényi random network or sometimes just an E-R network. These networks often serve as a minimal model for structured populations when the homogeneous degree and regular structure of lattices are undesirable.

There are actually two variations of the E-R network. The network formation algorithm first introduced by Erdös and Rényi is now usually referred to as the $G(n, M)$ algorithm (the "G" stands for "graph"), which works as follows:

$G(n, M)$ algorithm for an Erdös-Rényi random network

1. Create n nodes.
2. Choose a pair of nodes i and j at random, and create an edge between them if one does not exist already.
3. Repeat step 2 until the network has M edges.

This method keeps the total number of edges constant but is somewhat computationally costly, at least for relatively dense networks, as it requires continuously sampling pairs of nodes that may already be connected. The problem of repeated sampling is avoided by the use of a slightly different algorithm for creating random networks, which considers each pair of nodes exactly once. The $G(n, p)$ algorithm was introduced a few years later but has since become the de facto industry standard. When people talk about generating random networks these days, they are almost always talking about using this second algorithm:

$G(n, p)$ algorithm for an Erdös-Rényi random network

1. Create n nodes.
2. For each pair of nodes i and j, create an edge between them with probability p.

This sort of network is very easy to implement computationally, and also has some nice mathematical properties. In particular, although individual nodes will vary in their degree, the expected average degree of a random network is exactly equal to:

$$\langle k \rangle = p(n - 1). \tag{9.5}$$

[5]Paul Erdös is a fascinating character in the history of mathematics, who published over 1,500 mathematical papers in his career with over 500 collaborators. As such, he is often considered a key node in the network of mathematicians, who keep track of their "Erdös number" to indicate their minimum distance from Erdös in the network of published collaborations. See Hoffman (1998) for more details. As of the time of this writing, your humble author has an Erdös number of 3.

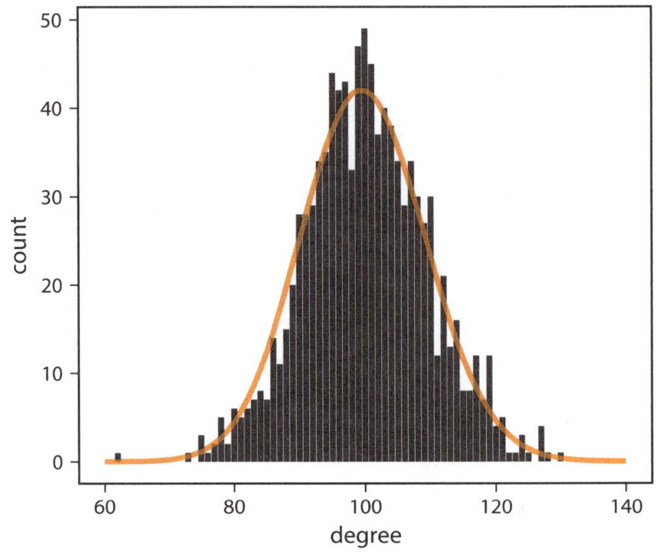

Figure 9.5 Histogram showing distribution of node degrees in a 1000-node random network with $p = 0.1$. The orange line shows the predicted degree distribution.

This is because every node considers each of the $n - 1$ other nodes and forms an edge with each with probability p. Moreover, because the formation of each edge occurs with a fixed probability, the degrees in the network will be approximately binomially distributed (see Box 9.1).

The NetLogo code for a random network is **RandomNetwork.nlogo**. We begin in the same way as with the ring lattice, by creating a population of nodes. The difference is in how we wire the nodes together. The global parameter `wiring-prob` is the probability of connecting each pair of nodes, p. All links are first erased in order to create a new network. Each node then considers every other node with a who number greater than its own, and creates an edge with that node with probability `wiring-prob`. By having each node consider only those other nodes with larger indices, we ensure that each pair of nodes is considered only once.

NetLogo code 9.4

```
to wire
  ask links [ die ]
  ask turtles [
    ask turtles with [ who > [ who ] of myself ] [
      if random-float 1.0 < wiring-prob [
        create-link-with myself
      ]
    ]
  ]
end
```

Figure 9.5 shows the degree distribution for a 1000-node network with $p = 0.1$, illustrating that degree is approximately binomially distributed.

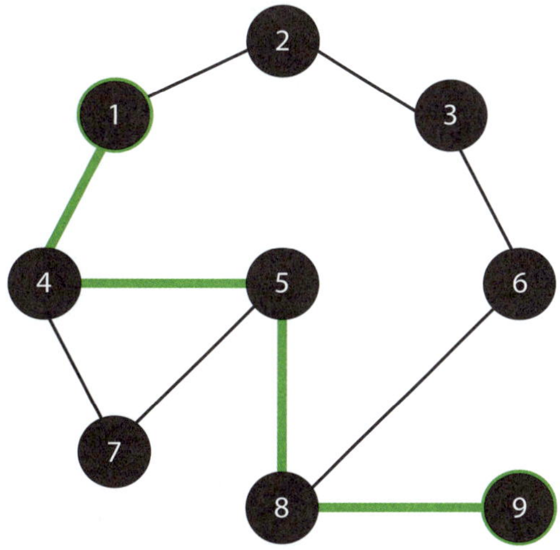

Figure 9.6 What is the shortest path between nodes 1 and 9?

9.3 Small-World Networks: Short Paths and Strong Clustering

Most real-world systems—be they social, biological, or physical—are not well characterized by either lattices or random networks. Lattices are too ordered, and random networks are too, well, random. However, each has certain properties that many real-world systems *do* exhibit. Perhaps a happy medium can be found. In this section, we'll see that we can hit just such a sweet spot by combining elements of the lattices and random networks. Before we do so, however, we need two additional measures for characterizing network structure: path length and clustering.

9.3.1 Path Length

As soon as we think of networks as connections between social actors, we can imagine them as conduits for communication between those actors. Imagine that each individual node can communicate information only with its neighbors—that is, with those nodes with which it shares an edge. The same logic applies if we think about contagion along a network. For information to travel from one node to another node that is not its direct neighbor, the information must go through intermediate nodes along existing edges.[6] Any such route between two nodes in a network is called a **path**, wherein each section of the path is between two nodes that share an edge.[7]

The **length** of a path between nodes i and j is equal to the total number of nodes in the path, including j but excluding i; this is equivalent to the number of edges in the path. The **distance** between nodes i and j, ℓ_{ij}, is the length of the shortest path connecting the two nodes (there need not be a unique shortest path). Consider the network in Figure 9.6. There are several paths connecting nodes 1 and 9, including $\{1, 2, 3, 6, 8, 9\}$ and $\{1, 4, 7, 5, 8, 9\}$. However, the shortest path is $\{1, 4, 5, 8, 9\}$ and has a length of 4. The **average path length**, L, of a network is the mean distance over each pair of nodes in the network. To calculate this,

[6] The field of percolation theory deals explicitly with understanding when information will flow across a network; for an example, see Sander et al. (2002).

[7] A network in which a path exists between every pair of nodes is called a *connected* network. This is distinct from a *fully connected* network, in which every node shares an edge with every other node.

we can add up the path lengths for each pair of nodes and then divide by the total number of pairs:

$$L = \frac{1}{2\binom{n}{2}} \sum_{i \neq j} \ell_{ij} = \frac{1}{n(n-1)} \sum_{i \neq j} \ell_{ij} \tag{9.6}$$

This is also the expected distance between a randomly chosen pair of nodes.

Many real-world networks have short path lengths, like random networks. Consider the well-known "six degrees of separation" experiments of the 1960s (Travers and Milgram, 1969). Letters were given to participants in Nebraska, who were asked to get the letter to a man working as a stockbroker in Boston. Participants were given minimal information about the man: his name, address, occupation and place of employment, military service dates, and his wife's maiden name and hometown. Of course, today this would be quite enough information to allow most of us to contact the man directly—we could find him with a search engine and message him via email or social media. However, communication with strangers was a bit more difficult in those olden days before personal computers and the Internet. Most importantly, participants in the study were asked to only send the letters to someone they knew personally (and were on a "first name" basis with), and were advised to select someone who might have a shot at knowing the target person. Even without the aid of Google searches, LinkedIn profiles, or email, letters that reached the target involved, on average, just a hair under six mailings. More recent studies of online social networks show that global connectivity has, if anything, increased (Bakhshandeh et al., 2011).

A random network also tends to have a relatively small average path length. This is due to the presence of random edges that connect far-flung parts of the network. In contrast, a regular lattice tends to have a relatively large average path length, because all connections are local.

BOX 9.2: The Seven Bridges of Königsberg

In the early eighteenth century, a devilish question began to spread among the intelligentsia of Königsberg, in Prussia (now Kaliningrad, Russia). The Pregel River, which ran through the city and had created two small islands, had seven bridges that facilitated crossings throughout the city. The question was this: Was it possible to map out a walk through the city that would involve crossing each bridge exactly once?

In 1736, the mathematician Leonhard Euler proved that the answer was a definitive *no*. Euler realized that he could abstract away all of the messy details of the city's topography and model it as a graph: a set of nodes, each representing a sector of the city, and edges, each representing a bridge connecting two sectors (Figure 9.7; note that this graph differs from the networks we explore in this chapter in that a pair of nodes can be connected by multiple edges). The shapes of the sectors and the exact locations of the bridges are irrelevant. The problem then becomes one of finding a path that crosses each edge only once.

The phrasing of the problem requires that each node is entered and exited via a unique edge. Therefore, except for the nodes where the journey begins and ends, each node must connect to an even number of edges. The number of edges that connect a node to other nodes in a graph or network is called the node's degree. If the journey starts and ends on different nodes, then those two nodes must be the only nodes with an odd degree. If the journey begins and ends on the same node, then all nodes must

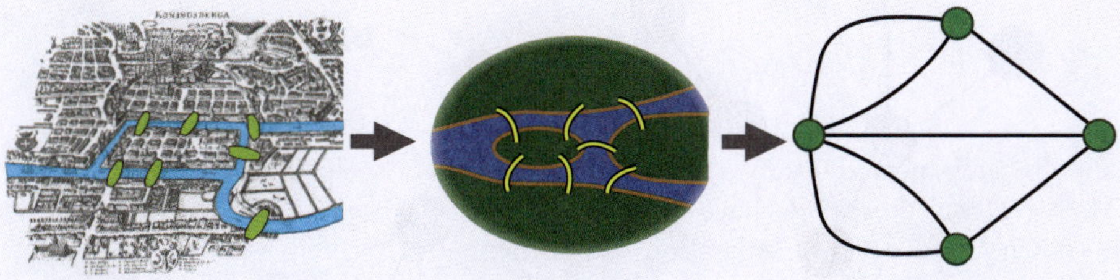

Figure 9.7 Euler's simplification of the seven bridges of Königsberg, which launched graph theory.

have an even degree. A glance at the graph in Figure 9.7 shows that *all* the nodes have an odd degree. Constructing a path that crosses each edge only once is therefore impossible.

9.3.2 Clustering

Your social network is defined not only by the people to whom you are directly connected. It's also defined by how those people are connected to each other. You may have a close-knit group of friends in which most of your friends are also friends with each other. However, you might also sometimes hang out with friends from work or school, who are unconnected to your old hometown friends. Interconnectedness in social networks can create synergies by virtue of shared knowledge, or it can create problems, as when diseases spread rapidly in tight-knit communities. This idea is captured by the concept of **clustering**.

The subnetwork involving all of a particular node's connections is usually called that node's **ego network**. In a network representing friendship ties, your ego network is maximally clustered if all your friends are also friends with each other. If none of your friends know each other, your ego network is minimally clustered. We can formally define a measure of local clustering for a given node i by considering all possible pairs of i's neighbors and reporting the proportion of those pairs that share an edge. The **local clustering coefficient** for node i is

$$C_i = \frac{\text{number of pairs of neighbors of } i \text{ that are connected}}{\text{number of pairs of neighbors of } i} \tag{9.7}$$

For example, in Figure 9.8, nodes A and B have the same degree ($k_A = k_B = 3$), but node A has low local clustering ($C_A = 0/3 = 0$) while node B has high local clustering ($C_B = 3/3 = 1$). By convention, $C_i = 0$ if $k_i \in \{0, 1\}$. To get a network-level statistic, we can average over all the nodes in the network so that the clustering coefficient for the entire network is:

$$C = \frac{1}{N}\sum_{i=1}^{N} C_i \tag{9.8}$$

A set of three nodes that are all connected to one another is called a **closed triangle**. An alternative measure of clustering that is sometimes used is the proportion of triplets in the network that are closed triangles.

Most real-world social networks are highly clustered. People assort into cliques of people who know one another. A random network tends to have low clustering, due to the fact

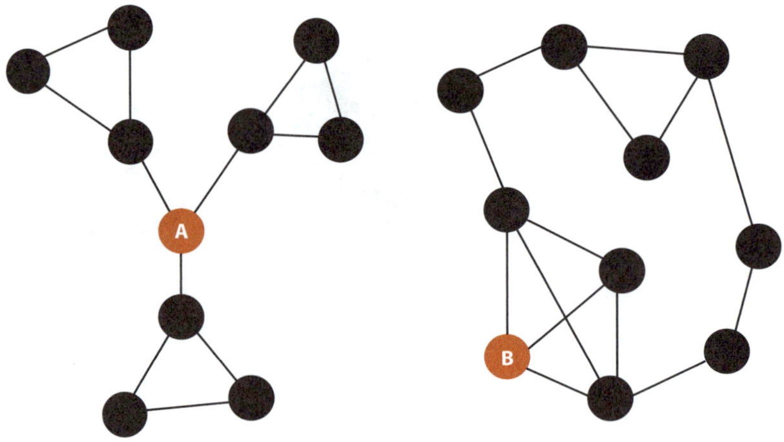

Figure 9.8 Node *A* has the same degree but lower local clustering than node *B*.

that there is no mechanism for correlating ties between multiple nodes. Lattice networks, in contrast, often exhibit relatively high clustering.

9.3.3 The Small-World Algorithm

To recap, many real-world networks are characterized by short path lengths, like random networks, and strong clustering, like lattices. Ideally, it would be nice to be able to generate a simple network structure that took the best of both worlds, so that our network models could be based on more realistic social structure. Networks that exhibit both short path lengths and strong clustering are often called **small-world networks**.[8] Luckily, these networks can be generated with a simple algorithm, based on the convenient fact that as one starts to randomly rewire edges on a highly clustered lattice, the average path length decreases faster than does the clustering. The algorithm, which was first introduced by Watts and Strogatz (1998), works as follows:

Small-world network algorithm

1. Create a ring lattice with N nodes and degree k.
2. Select a node and the edge that connects it to its nearest neighbor in a clockwise sense. With probability p, destroy this edge and form a new edge between the node and another randomly chosen node chosen in the network, with duplicate edges forbidden; otherwise leave the edge in place. Repeat this process for each node in the network by moving clockwise around the ring, so that each node is considered in turn until one full lap around the ring is completed.
3. Next, consider the edges that connect nodes to their second-nearest neighbors clockwise. As before, randomly rewire each of these edges with probability p, and continue this process, circulating around the ring and proceeding outward to more distant neighbors after each lap, until each edge in the original lattice has been considered once. Repeat this process until each node's kth-nearest neighbor has been considered.

[8]Several metrics are used to characterize the extent to which a network is a "small world," all of which quantify the gap between the normalized path length and clustering coefficient (Neal, 2017). See Watts (1999) for a discussion of other network structures exhibiting small-world-like properties.

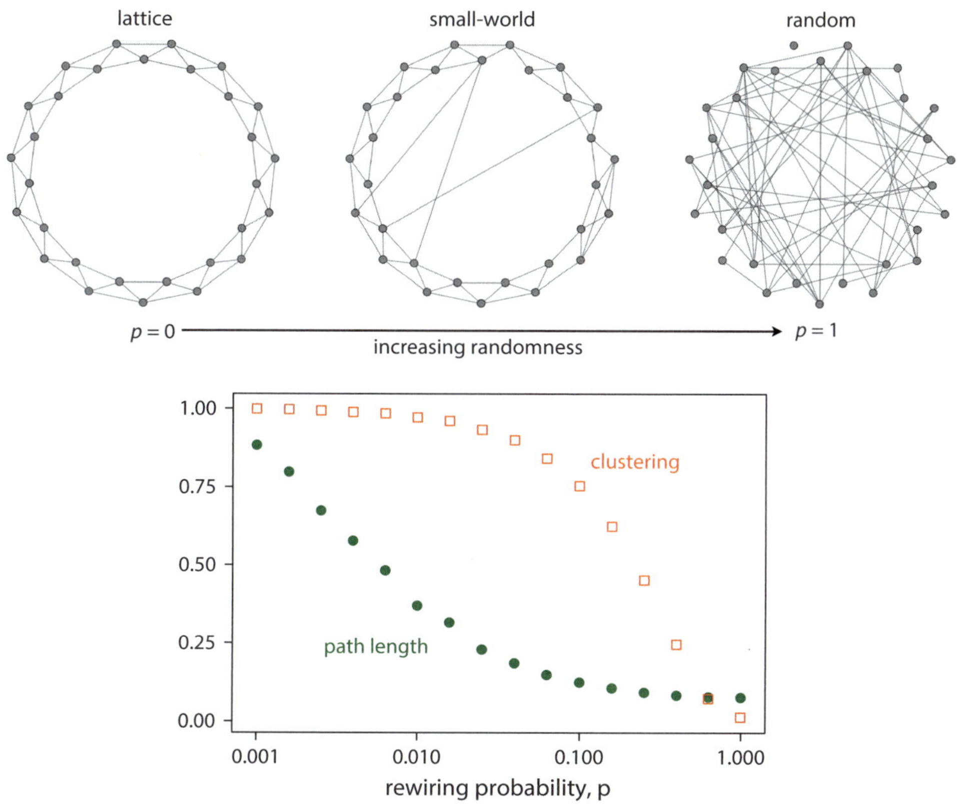

Figure 9.9 The small-world algorithm. Left: Small-world networks arise from intermediate levels of random rewiring applied to a ring lattice. Right: Clustering (C) and average path length (L) normalized to their values for a ring lattice as a function of the rewiring probability, p. Note that both clustering and path length decrease with p, but path length decreases more rapidly, leading to the small-world property for intermediate values of p. These data are averages from 20 runs for each value of p, using $N = 500$ and $k = 4$.

In other words, each edge is randomly rewired with probability p. Obviously, $p = 0$ leaves the initial ring lattice intact, and $p = 1$ yields a fully random network. The interesting bits occur for intermediate values of p. As p is increased from zero, the average path length of the network decreases more rapidly than average clustering, yielding networks with high clustering and short path lengths, as illustrated in Figure 9.9. Note that the x-axis is on a logarithmic scale, indicating that a very small amount of rewiring goes a long way.

9.3.4 Coding a Small-World Network

The NetLogo code to create a small-world network is **SmallWorldNetwork.nlogo**. We will begin with the creation of a ring lattice of degree $k = 4$ (this value is somewhat arbitrary, but is common for models of small-world networks). We already know how to do this. From there we simply rewire some of the links to create a small-world network. If you look over the code, you will see that it is very similar to the previously seen **RingNetwork.nlogo**.

In the next section, we are going to use our small-world network to model social dynamics between connected nodes. In particular, we are going to implement a network model of contagion.[9] As usual, it will be helpful to visualize the model dynamics—in this case, to see the path of the contagion. As the number of nodes in a ring lattice grows, it can be harder to see the connections between neighbors when $k > 2$. An aesthetically appealing solution, at least in my opinion, is to stagger the agents in two concentric circles so that each edge is more clearly visible. This is a little bit trickier than just putting all the agents in a single circle, but I encourage you to always build the model you want, not just the model that's easier to code.

Just as before, we use a procedure called `make-turtles` to create the nodes and arrange them in a circle. Recall that the NetLogo command `sort turtles` returns a list of all turtles in order of their `who` number so that they will be arranged in order on the circle. The `who` number is an index reflecting the order in which each node was created. To make the lattice prettier, we will move every other node a little bit inward toward the center of the circle. To select only the even-numbered nodes, modular arithmetic comes to the rescue once more. We first ask each agent to turn toward the center of the circle. Then, if their `who` number is divisible by 2, they take a step toward the center, creating a nicely staggered ring lattice.

NetLogo code 9.5

```
to make-turtles
  create-turtles num-nodes [
    set color gray + 1.5
    set size 4
  ]
  layout-circle (sort turtles) max-pxcor - 8
  ask turtles
  [
    facexy 0 0
    if who mod 2 = 0 [fd 10]
  ]
end
```

Edges between pairs of nodes are initially created in exactly the same way as in the ring lattice example given earlier in this chapter, using the procedure `wire-lattice`. If we stopped here, we'd just have a pretty ring lattice. But we must forge ahead!

To turn our ring lattice into a small-world network (Figure 9.10), we need to rewire some of the edges. This is accomplished in the procedure `rewire-network`, which is called after the ring lattice is created. This procedure loops over each link and rewires it with probability `rewiring-probability`. Much like humans traveling via transporter in *Star Trek*, the link doesn't actually move, but is instead destroyed, with a new copy taking its place somewhere else. If a link is to be rewired, it asks one of the turtles it connects, `end1`, to create a new link with another turtle it is not already connected with, after which the original link dies.

[9] The principles at work here are general and can be extended to any of the other modeling frameworks we've explored in this book, as well as many others we have not explored. In other words, you can model all sorts of social dynamics on a network.

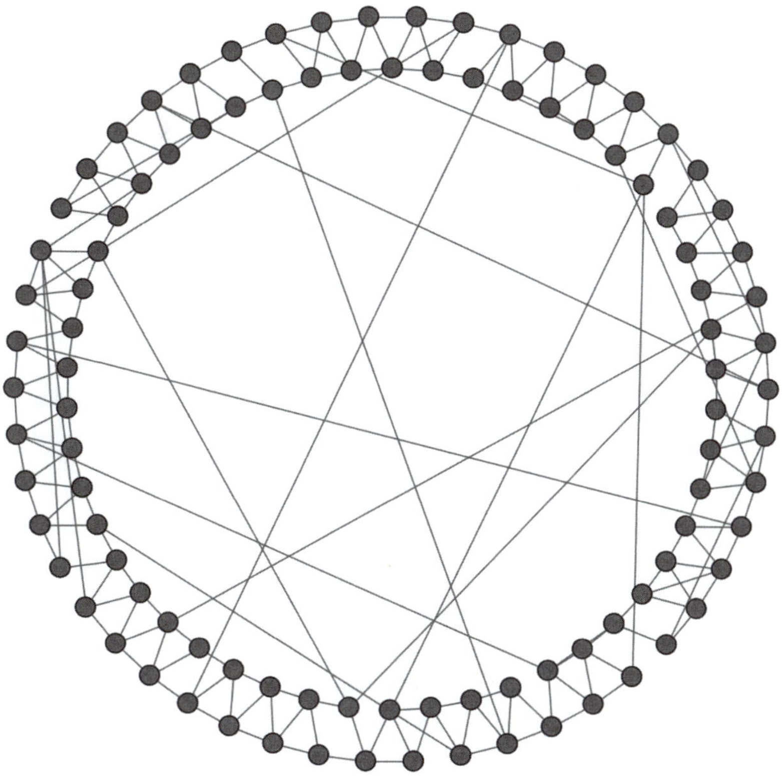

Figure 9.10 A 100-node small-world network based on a ring lattice of degree $k = 4$, with rewiring probability $p = 0.1$.

NetLogo code 9.6

```
to rewire-network
  ask links
  [
    if (random-float 1) < rewiring-probability
    [
      ask end1
      [
        create-link-with one-of other turtles with
          [not link-neighbor? myself ]
          [set color gray + 1.5]
      ]
      die
    ]
  ]
```

With intermediate values of p (`rewiring-probability`) we have a network with a low average path length and high average clustering, just like many real-world social networks. We can use this network to study models of social dynamics using a somewhat more realistic population structure than either a well-mixed population or a regular lattice. In the next section, we will return to our old friend the SI model, and introduce a new way to model the spread of socially transmitted information: complex contagion.

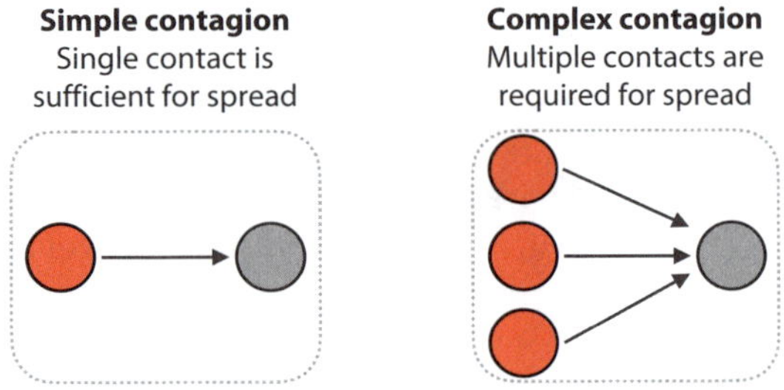

Figure 9.11 Simple vs. complex contagion.

9.4 Simple vs. Complex Contagion on Small-World Networks

Now that we have an algorithm for constructing more realistically structured networks, we can start thinking about how to simulate social behavior on those networks. We'll return to contagion and introduce a new distinction between simple and **complex contagion**. A simple contagion is of the type we explored in chapter 4. A single contact with an infected individual is sufficient to become infected, even if contact doesn't *always* lead to infection. The idea is drawn from disease transmission, though simple contagion can also apply to the spread of behaviors or information. If information is easily transmitted and is unlikely to be false or overly idiosyncratic, we can model its spread as a simple contagion. For other types of ideas or behaviors to spread, social reinforcement from multiple contacts might be required. One friend tells you that organic banana peel supplements are great for your health, but you are skeptical. Another friend starts taking the banana peel supplements, and you begin to warm to the idea. Five friends are all raving about banana peels, and you find yourself seeking out the banana peel section of your local health food store. A complex contagion is one that requires contact with more than one "infected" individual to spread (Figure 9.11). The term was first coined by Centola and Macy (2007) and builds on earlier work on threshold models of behavior by the sociologist Mark Granovetter (1978). Complex contagion is, however, more accurately described as a type of frequency-dependent social learning strategy (Boyd and Richerson, 1985), and is just one of many possible strategies for social learning (reviewed in Kendal et al., 2018).

Are the dynamics of a complex contagion different from the dynamics of a simple contagion? How do these dynamics respond to the network structure of the population? We will explore these questions with some agent-based modeling. Our previous investigations of dynamics in structured populations focused on regular lattices (mainly square lattices). We will begin our analysis here using a ring lattice network, with its high clustering and large path lengths. We will then gradually rewire the network using the small-world algorithm, yielding networks that are increasingly less clustered and have more long-range ties. Things are starting to get complicated, so we'll make the contagion models as simple as we can. We'll ignore recovery (or, equivalently, disadoption), and begin with a simple SI contagion model in which agents are nodes on a network and are either susceptible (S) or permanently infected (I). Contagion can spread only through social ties, so a susceptible agent must share edges with infected agents to become infected. We will define complex contagions so that they spread only when a susceptible agent shares an edge with two or more infected agents.

The model will work as follows. We will initialize a population of size N on a small-world network with degree $k = 4$ and rewiring probability p. At initialization, two randomly chosen agents who share an edge will be infected with the contagion; all other agents will be uninfected. It is important that the initially infected agents share an edge—otherwise, a complex contagion would be unlikely to spread at all due to the low probability of any other agent having two infected neighbors.

The dynamics proceed as follows. Each time step, each infected agent will consider each of its uninfected neighbors. In the case of a simple contagion, each uninfected neighbor becomes infected with probability τ. In the case of a complex contagion, infection works the same way, but with the additional provision that an uninfected neighbor can become infected only if at least two of its neighbors are infected (exploring variations on this rather strict assumption is left as an exercise to the reader). In the simplest case of $\tau = 1$ and a regular ring lattice, simple contagions will spread rapidly, while complex contagions will spread slightly more slowly due to the need for two neighbors to become infected before the contagion can spread further. The case of $\tau = 1$ is also the easiest to analyze, because the spread of the contagion becomes deterministic and so we can stop the model after one time step passes without any new infections. Once we code the model, we will focus our analysis on how the rewiring probability p influences the spread of simple and complex contagions.

9.4.1 Coding Contagion in a Small World

The NetLogo code for modeling contagion on a small-world network is **SmallWorldDiffusion.nlogo**. This model uses the small-world network model we developed in the previous section and adds contagion as described above. The basic idea behind the contagion models should be understandable based on the material we covered in chapter 4. Some of the model code is related to purely aesthetic choices regarding the model visualization. I will not review or justify those choices here, but I do recommend working through the code to understand how the visualization is produced. I will instead focus on code that generates the model's contagion dynamics.

First, let's consider the agents. Each agent will keep track of whether or not it is infected, so we assign each a Boolean variable `infected?`.

NetLogo code 9.7

```
turtles-own [ infected? ]
```

I have also added a global variable called `num-infected`, which is simply a count of the number of infected agents. While this variable isn't strictly necessary, because we can always count the number of infected agents directly, it does make it easier to report the infection rate during data collection.

The main addition to the `setup` procedure is a call to a procedure called `infect-two`. This, naturally, initializes the contagion by infecting two neighboring agents. We accomplish this by selecting one turtle at random, who then selects one its neighbors at random. Both agents are then infected, with the visualization and infection count updated as needed.

NetLogo code 9.8

```
to infect-two
  ask turtles [reset-node]
  ask links [set color gray + 1.5]
```

```
  ask one-of turtles
  [
    set infected? true
    set color yellow
    set size infected-size
    ask one-of link-neighbors [
          set infected? true
          set color yellow
          set size infected-size
    ]
  ]
  set num-infected 2
end
```

The model dynamics rely on two global parameters. The first of these is `prob-infection`, which is the transmissibility τ. We will simplify things further by restricting our analysis to the case of $\tau = 1$. Recall from chapter 4 that the speed at which the contagion spreads in an SI model depends on τ. Here we hold this constant in order to focus on the influence of network structure. The second parameter is the Boolean variable `complex-contagion?`, which indicates whether the contagion we are modeling is simple or complex.

The dynamics are straightforward and are contained entirely in the `go` procedure. We first set the model to stop running if all of the agents become infected so that we can study the time needed for the contagion to spread throughout the entire population (though, as we will see, this will not be a sufficient stopping condition for complex contagions). Otherwise, each infected agent considers each of its susceptible neighbors and causes them to become infected with a probability `prob-infection`. If the contagion is complex, we add the additional requirement that the susceptible neighbor must have at least one other infected neighbor. If the contagion is transmitted, the susceptible agent becomes infected and changes color and size (for visualization purposes). At the end of the time step, the simulation updates its count of infected agents.

NetLogo code 9.9

```
to go
  if all? turtles [infected?] [stop]
  ask turtles with [ infected? = true ]
  [
    ask link-neighbors with [not infected?]
      [
        if (random-float 1 <= prob-infection and
          ((not complex-contagion?) or (count link-neighbors with
          [infected? = true] > 1)))
        [
          set infected? true
          set color yellow
          set size infected-size
        ]
    ]
  ]
  set num-infected count turtles with [infected? = true]
  tick
```

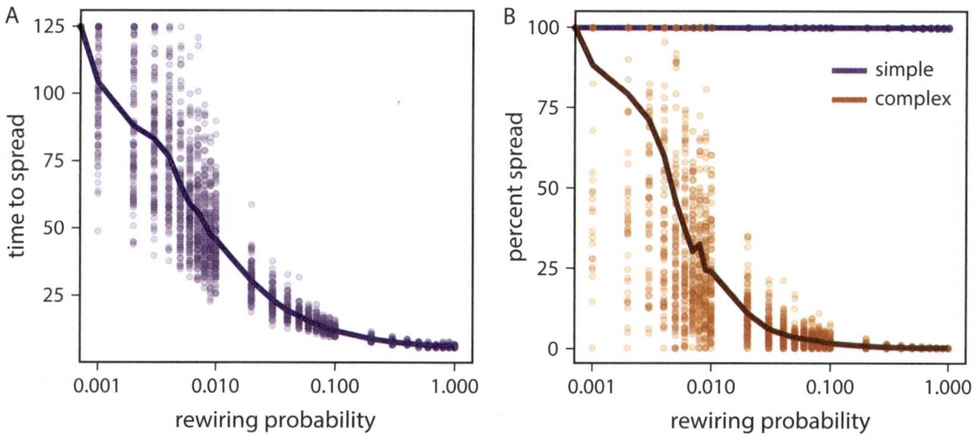

Figure 9.12 Simple and complex contagions on small-world networks. (A) Time for a simple contagion to spread throughout the entire population. (B) Percent of the population infected at equilibrium for both simple and complex contagions. All simulations used $N = 500$ and $\tau = 1$. Data is from 100 runs of each parameter combination. Circles are individual runs, dark lines are means. Note that the x-axis is on a logarithmic scale.

9.4.2 Analyzing the Model

We can now perform some interesting computational experiments. Let's start simple, with a simple contagion on a regular ring lattice ($p = 0$). You should observe that the contagion spreads along the ring in both directions, eventually infecting the entire network. Now re-run this simulation several more times, gradually increasing the rewiring probability p each time. Take notice of the time it takes for the contagion to fully saturate the network. You should notice that as more edges are rewired—and therefore as the average path length decreases—the time it takes for the contagion to spread throughout the population decreases substantially. This is because the contagion can easily hop across the network, multiplying the speed at which it can spread. This result should be intuitive: the more interconnected disparate parts of a population are, the faster a contagion can spread through that population. Figure 9.12A plots the time to spread as a function of the rewiring probability for a 500-node network. Notice that for very small rewiring probabilities, there is a lot of variation in the time to spread, which indicates a large dependence on the particular distribution of the small number of random ties and their location relative to the initial infections.

Let's now examine a complex contagion. If you run some simulations on the regular ring lattice ($p = 0$), you should observe that the time it takes for the contagion to fully saturate the network is slightly longer for the complex contagion than it was for the simple contagion. This is due to the fact that the contagion can spread to fewer neighbors of an infected node at any given time. A more interesting difference between simple and complex contagions becomes apparent when we begin to rewire the network using the small-world algorithm. Instead of speeding up the spread of the contagion, rewiring often causes the spread to stall completely. That is, the contagion regularly fails to spread throughout the entire network. Figure 9.12B plots the proportion of the population that becomes infected in each simulation, and shows that this failure to spread becomes more likely and more severe as the probability of rewiring increases.

What's happening? When the contagion requires multiple unique contacts to spread, the heterogeneity in the degree distribution creates occasional gaps in which neighboring communities are only weakly connected through a single edge. In other words, there are "structural gaps" in the network—communities without strong ties between them. This creates situations in which the contagion fails to spread. That is, networks that are more random and less strongly clustered often fail to facilitate the spread of social behaviors when those behaviors require reinforcement from multiple sources. This phenomenon has received some empirical support for the adoption of health-related behaviors (Centola, 2010). Once again, we see a lot of variation in the outcomes for small rewiring probabilities, which in this case reflects variation in the distance between the starting location of the contagion outbreak and the gaps created by the rewiring.

The small-world algorithm is not the only way to produce artificial networks with realistic properties. Furthermore, it has at least two limitations. First, it yields a network with a fairly narrow distribution of degrees. This makes sense, because it is bounded on the one end by a regular lattice, in which every node has the same degree, and on the other end by a random network, which exhibits a binomial distribution of degrees. In contrast, many real-world networks exhibit a **heavy-tailed** degree distribution, which means there are a very small number of high-degree nodes (sometimes called **hubs**), and many low-degree nodes. The second limitation of the small-world algorithm, and indeed of every algorithm we've discussed so far in this chapter, is that it requires us to specify the size of the network in advance and tells us nothing about the mechanism by which nodes became so arranged. The small-world algorithm is not a dynamic model of how networks form as new nodes enter the system. The algorithms we have explored so far therefore provide us with ways to produce networks with certain properties, but they are not models of how those networks form. In the next two sections, we will discuss models that are not limited in these ways.

9.5 Preferential Attachment

Many real-world networks exhibit heavy-tailed degree distributions. A commonly used analytical technique for approximating these heavy-tailed degree distributions is a **power law**. Mathematically, a power-law degree distribution is captured by the assumption that the probability $P(k)$ that a node in the network has degree k will follow a distribution of the form

$$P(k) \sim k^{-\gamma} \tag{9.9}$$

where $\gamma > 1$. Networks of this sort are often called **scale-free**, because subnetworks of the main network will continue to yield the same power-law degree distribution. In other words, there is no defining "scale" of the degree distribution.

Power-law relationships between two variables will appear as straight lines when plotted on a log-log scale. This is due to a convenient property of logarithms whereby multiplication becomes addition and exponentiation becomes multiplication. That is,

$$\log(ab) = \log(a) + \log(b) \tag{9.10}$$

and

$$\log(x^y) = y \log(x). \tag{9.11}$$

These are not model assertions, but rather immutable mathematical properties of logarithms (which, as you might expect, is one of the main reasons people use logarithms in the first place). Consider the following power-law relationship for the probability of finding nodes of degree k, where P is the probability distribution function and a is a constant:

$$P = ak^{-\gamma} \tag{9.12}$$

If we take the logarithm of both sides and apply our logarithm rules, we are forced to conclude that the relation between $\log(P)$ and $\log(k)$ is linear, with a slope equal to the exponent γ:

$$\begin{aligned} \log(P) &= \log\left(ak^{-\gamma}\right) \\ &= \log(a) - \gamma \log(k) \end{aligned} \tag{9.13}$$

While heavy-tailed degree distributions are common in social networks (and in many other kinds of networks), there is increasing evidence that most social networks are not really scale-free and are often better characterized by log-normal distributions than by power laws (Jones and Handcock, 2003; Broido and Clauset, 2019). Of course, neither distribution is in itself a generative model of network structure. What we would like is a model that includes processes by which new nodes join a network and form edges with existing nodes.

One of the simplest and most influential models with these properties is based on the process of **preferential attachment**. The idea is simple: new nodes preferentially attach to higher-degree nodes over lower-degree nodes. This idea has a long history in twentieth-century science (Yule, 1925; Simon, 1955). Notably, de Solla Price (1976) used a simple urn model[10] to explain heavy-tailed distributions of academic citations; that is, why most papers were cited only a few times but a few papers were cited thousands of times or more. In his model, there are many urns, each representing a paper. Initially, all urns have equal numbers of red beads (representing success) and black beads (representing failure). An urn is selected at random, and a bead is extracted. If the bead is black, it is replaced, and another urn is selected. If the bead is red, more red beads are added, and another bead is selected from the same urn. In this way, urns that are initially successful by chance become increasingly likely to experience longer and longer streaks.

Preferential attachment was first introduced to network science by Barabási and Albert (1999). They developed a dynamic model of network formation, which generates heavy-tailed degree distributions that are well approximated by power laws when the networks are sufficiently large. As with de Solla Price's model, the preferential attachment algorithm operates on a principle of "the rich get richer." When a new node enters the network, it forms a connection with an existing node. Nodes that currently have lots of connections are more likely to be the targets of new connections. This dynamic probably applies in a lot of contexts, whether we are talking about technological, biological, or social systems. Nodes with many connections are likely to be robust and may generate additional benefits (such as coordination benefits or economies of scale) based on their popularity. For example, a brand-new website is more likely to link to Google or Amazon than to *howtodrawvampirerabbits.net*. The algorithm works like this:

[10] Pólya urn models, named after the Hungarian mathematician George Pólya, are commonly used to model cumulative stochastic processes.

Preferential attachment algorithm

1. Start with m_0 nodes
2. Create a new node with m edges ($m \leq m_0$) to m extant nodes. For each new edge, choose a connecting node with a probability proportional to the current degree of those nodes, such that the most likely node with which to connect is the one that is already the most highly connected.
3. Repeat Step 2 until the desired number of nodes is reached.

This is a very simple algorithm! It's made even simpler if we restrict ourselves to the most commonly used case of $m_0 = 2$ and $m = 1$, so that one new node with one connection is added on each time step.

The NetLogo Models library includes code to generate networks using the preferential attachment algorithm, in the Networks folder, called **PreferentialAttachment.nlogo**. Here I'll walk you through the essential parts of the code, ignoring those aspects that deal with visualization.[11] First, we create a procedure called `make-node`, which takes as an argument an existing node (`old-node`) and creates a new node that will connect to that existing node. If the existing node doesn't exist (it is `nobody`), the procedure simply creates a new node without any connections.[12]

NetLogo code 9.10

```
to make-node [old-node]
  create-turtles 1
  [
    set color orange
    if old-node != nobody
      [
      create-link-with old-node [ set color grey ]
      ]
  ]
end
```

The model is initialized by using the `make-node` procedure to create two nodes connected to one another.

NetLogo code 9.11

```
to setup
  clear-all
  set-default-shape turtles "circle"
  make-node nobody          ;; first node, unattached
  make-node turtle 0        ;; second node, attached to 1st node
  reset-ticks
end
```

The model dynamics are very simple. Every time step, we create a new node and connect it to an existing node. We can use the `make-node` procedure to create a new node, we just

[11] Netlogo has a number of nice features for dynamic visualization of networks, which are illustrated in the full code for this model.

[12] The colors used in this code are arbitrary. I am partial to orange nodes with grey links.

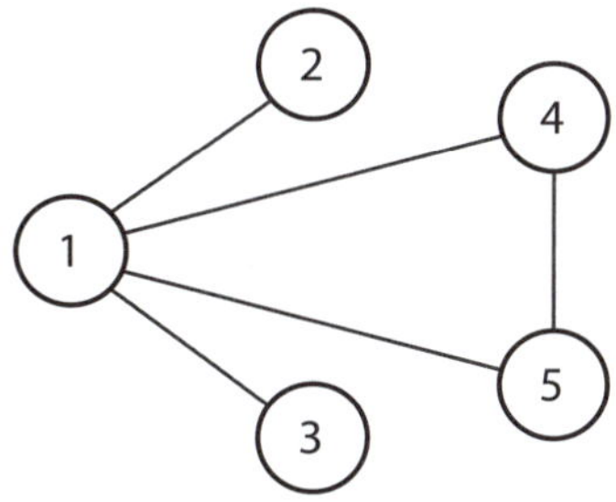

Figure 9.13 An example network.

need to know which of the existing nodes the new node is to be connected to. *That* node is chosen using a procedure called `find-partner` procedure.

NetLogo code 9.12

```
to go
  make-node find-partner
  tick
end
```

If the existing node was chosen purely at random, the resulting network would have a random structure, and would not exhibit a heavy-tailed degree distribution. However, that node is *not* chosen at random. In truth, the `find-partner` procedure is the heart of the preferential attachment algorithm. It considers all existing nodes in the network, and the probability that any given node will be chosen is exactly proportional to its degree. It might seem like doing this would be complicated. In fact, the procedure contains only a single line of code. Here it is:

NetLogo code 9.13

```
to-report find-partner
  report [one-of both-ends] of one-of links
end
```

The way it works is quite extraordinary. Let's unpack what this code is doing. Reading right to left can help. The procedure selects `one-of links`. That is, it chooses an edge of the network at random—any edge has equal probability of being chosen. The procedure then selects `one-of both-ends` of that edge. An unweighted, undirected edge is defined entirely by the two nodes it connects. We simply choose one at random, so that each end node has an equal probability of being chosen. So, to recap, we choose one edge at random, and then we choose one of that edge's two nodes at random. How can this process generate preferential attachment?

The algorithm works because the probability that a given node will be selected is exactly proportional to its degree. It may help to imagine a lottery in which a ticket is selected at random from out of a hat. Each edge is a ticket, on which is written the names of two nodes. So nodes vary in how many tickets they are associated with. As a simple example, consider the five-node network shown in Figure 9.13. Node 1 has degree of four ($k_1 = 4$), while the other nodes have lower degrees ($k_2 = k_3 = 1$, $k_4 = k_5 = 2$). With our `find-partner` algorithm in mind, let's consider the probability that each node is selected. There are five links

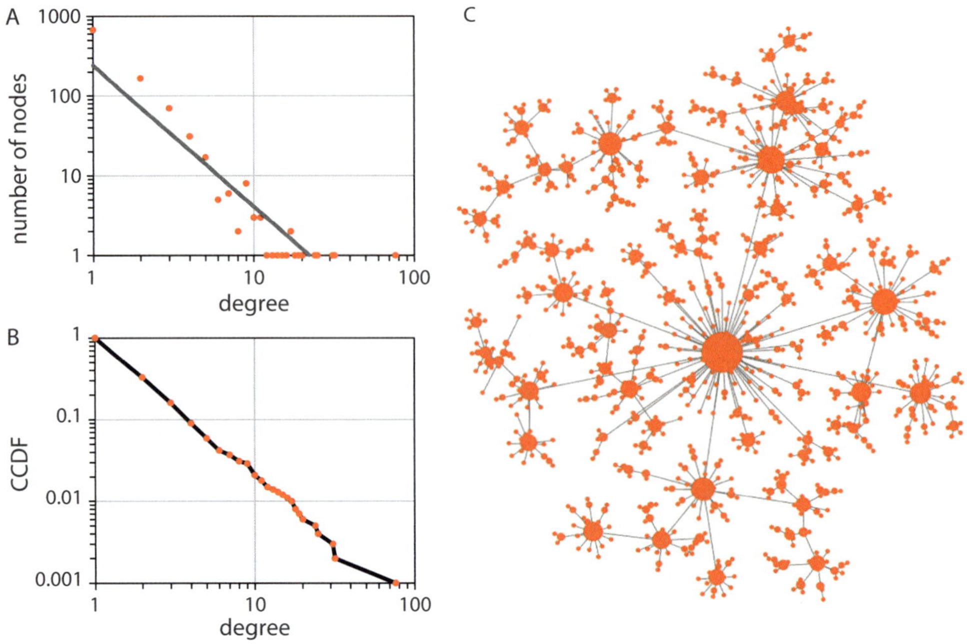

Figure 9.14 A 1000-node network created using the preferential attachment algorithm. (A) A log-log plot of the degree distribution with the best-fit power-law curve. (B) The complementary cumulative distribution function (CCDF) of degree on a log-log scale. (C) Visualization of the model, with the size of the node scaled proportional to the square root of its degree.

in this network. Four of these connect to node 1, two of them connect to nodes 4 and 5, and nodes 2 and 3 have only one link each. Once a link is selected, each of its two nodes have equal probability of being selected, and since this is true for all nodes, it doesn't affect their relative probabilities of being selected, but it does help to normalize the probabilities so they sum to one. The probability that node 2 is selected is $\frac{1}{2} \times \frac{1}{5} = \frac{1}{10}$, while the probability that node 1 is selected is $\frac{1}{2} \times \frac{4}{5} = \frac{4}{10}$. In other words, the probability that a given node is chosen is exactly equal to its relative degree. When large networks are created using the preferential attachment algorithm, their degree distributions are heavy-tailed and can be approximated by a power law. Figure 9.14 illustrates a 1000-node network made using the algorithm, with its degree distribution plotted on a log-log scale. It is common to illustrate a heavy-tailed distribution using a **complementary cumulative distribution function** (CCDF), which indicates the proportion of nodes with a degree equal to or greater than k. This provides a somewhat clearer picture of the degree distribution (compare Figures 9.14A and 9.14B).

The preferential attachment algorithm is useful for generating networks when we want to model scenarios in which a network formation process like preferential attachment is at play, or when we want to model *any* social dynamics on networks with heavy-tailed degree distributions. Variants of this algorithm are widely used for modeling social dynamics. Nevertheless, the great simplicity of the model also means that its assumptions are rarely met. The model assumes that the network is continually growing. This assumption might hold for some social networks during their initial growth phase, but is unlikely to be met indefinitely. The model also assumes that the popularity of a node is the sole driver of social connections (other than purely stochastic processes). However, individuals form connections for many reasons other than popularity, including ones related to their social identity, existing relationships, or personality. Some models have arisen to address some of these concerns, and in the next section we'll briefly review some other social drivers of network structure.

9.6 Other Social Drivers of Network Structure

Preferential attachment only models a single driver: individuals are more likely to form connections with nodes with high degree. However, this algorithm ignores many important facets of social life concerning how individuals connect to the relevant people in their lives. Whom one forms a connection with is also influenced by factors like social identity (e.g., individuals may be more likely to form ties with others with whom they share an identity), social inheritance (e.g., individuals may be more likely to connect with others already connected to their family and friends), and personality (e.g., individuals may differ in their propensity to seek new connections). These aspects of social life are important targets for models of social network structures. Most of the work in this area is quite recent, and here I will only briefly summarize a few notable investigations.

9.6.1 Social Identity

Identity is one of the defining features of social life, particular in our modern, interconnected world.[13] In the models we have considered thus far, agents distinguished one another based only on their degree, if at all. However, group identities often influence whether and how someone engages with a new person, shaping the subsequent structure of social networks. One of the simplest ways to incorporate identity into a model of network formation is to assign each agent to a distinct group. Individuals then form ties using a modified random network algorithm: two agents form a tie with probability p_i if they are members of the same group and with probability p_o if they are not, where $p_i \geq p_o$. Such models are typically called either **stochastic block models** (Holland et al., 1983) or **multi-type random network models**. The structure of these models tends to deviate from many real-world social networks (Karrer and Newman, 2011), but they are able to capture some important features of homophily (or positive assortment), in which similar agents tend to connect (Golub and Jackson, 2012).

9.6.2 Social Inheritance

An individual born into a tight-knit community is much more likely to form connections with others in that community than with those outside it, just as new friends can gradually become integrated with an entire friend clique and newly hired faculty become connected with many of those at their new university. Ilany and Akçay (2016b) recently presented a dynamic network generation algorithm that allows for the intergenerational transmission of network connections. This **social inheritance** algorithm starts with a random network, and then iteratively grows by having a randomly selected node "birth" a new node. The

[13]This book has engaged relatively little with identity. This is not because it isn't important, but rather because most models that incorporate identity are extensions of simpler models that do not—it is these simpler models that are our focus here. For more on integrating identity with models of social behavior, see Smaldino (2022).

new node connects to its parent with probability p_b, connects to its parent's connections with probability p_n, and connects to all other nodes with probability p_r. A key feature of the resulting networks is that in addition to realistic levels of clustering, they also produce **assortative** networks, in which nodes with similar heritable properties tend to be connected. This model is particularly attractive because it allows networks to form dynamically but remain bounded in size, and it could be useful for modeling not only the animal social networks it was designed for, but also human networks that vary in the intensity of their kinship institutions. It seems likely that this model may end up being quite influential in future modeling of social processes on networks.

9.6.3 Personality

Individuals often differ in their behavioral tendencies, which can affect how they form network ties. Some agents, for example, may be quite extraverted, willing to make many connections with those outside their initial social networks, while others may be more introverted, making just a few connections and mainly with those close to home. This can be modeled in a number of ways. For example, Muthukrishna and Schaller (2020) created an agent-based model where agents moved around on a square lattice, forming connections with other agents they encountered. Agents' rate of movement was determined by an "extraversion" trait. Their study focused on differences between populations and showed that greater extraversion was associated with denser networks, higher clustering, and shorter average path lengths. In contrast, Ilany and Akçay (2016a) extended their social inheritance model to consider a heterogeneous population where agents varied in their "boldness," which was the propensity to form ties with random agents instead of agents currently connected to their parent node. Their analysis found that, although all agents had similar degree, bolder individuals had higher betweenness centrality (a measure of how many shortest paths require going through a particular node, often used to characterize nodes that form bridges between communities) but lower clustering coefficients, and that agents tended to assort on personality type despite not explicitly using personality in their tie-formation decisions. Characterizing boldness in this way may help to explain the importance of bold individuals in connecting different knowledge domains, and the importance of less bold individuals for maintaining cultural cohesion within groups. This work also highlights the value of formalizing psychological phenomena in terms of their influence on social behavior, which can help us to better understand the origins of social network structure.

9.7 Reflections

I believe that all social scientists should have at least a rudimentary understanding of network theory, some of which I have attempted to provide in this chapter. Network theory has revolutionized the study of social systems, because it provides a framework for understanding how the structure of a population shapes the propagation of communication and influence within that population.

In social science, network edges are metaphors. Edges represent relationships, which are rarely as cut and dry as implied by the common modeling assumption that an edge is either simply present or absent. Rather, each relationship is idiosyncratic, with properties not shared by any other relationship. However, this objection can just as easily be extended to all the models in this book, regardless of whether they use network structures. Each individual is, after all, a unique person with key differences from every other person. We have different personalities, different proclivities, different hopes and dreams. The models obscure some of these differences in order to capture important approximations of reality that lead to explanations, mechanisms, and even predictions.

Still, care is needed. For example, consider the complex contagion models we examined, in which the contagions were halted in small-world networks because they could not spread between weakly-connected communities. This result stems from the hard rule that two or more connections are needed for the contagion to spread. But what if this rule is relaxed? What if two connections is preferred, but contagions still occasionally spread with only one connection. After all, some individuals might be fickle or whimsical. It may turn out that less-clustered networks might not be as much of a problem for the spread of complex contagions as the model we studied suggests (Eckles et al., 2019). Exploring this claim is left as an exercise.

9.8 Going Deeper

In this chapter we kept it simple and limited our discussion to undirected, unweighted, single-layer, unipartite networks. That is, edges were symmetrical and identical (if A is connected to B, then B is connected to A, and an edge is an edge is an edge), and there was only one type of edge and one type of node. And we focused on just a few core network architectures: lattices, random networks, small-world networks, and scale-free networks. We did this because even these simple networks have fairly complex properties, because much of network theory was originally developed around these simple networks, and because many models of social behavior use networks of this type.

However, it is worth mentioning other types of networks, which can be used to model types of relationships not well captured by the simpler networks we have studied here (Figure 9.15). **Weighted networks** allow for the connections between nodes to take on specific characteristics, such that some connections are more important than others, and might even be negative. Artificial neural networks are well-known examples of weighted networks, where negative weights typically reflect inhibitory connections, but social networks can also be weighted, with weights often reflecting the strength of the social tie. **Directed networks** remove the requirement that relationships be symmetrical, so that node A can be connected to node B without node B being connected to node A. In the world of social media, Facebook friendships can be represented by an undirected network, while Twitter followers require a directed network. Most artificial neural networks are both weighted and directed.

Sometimes a network's structure is too complex to be characterized by a single set of nodes and edges. **Bipartite networks** allow two distinct types of nodes, often with the constraint that nodes of one type can only be connected to nodes of the other type. Such networks can model political actors and institutions, or organisms and habitats. Bipartite networks are actually a subcategory of multipartite networks, which can have an arbitrary number of node types. Recently, **multiplex networks**, sometimes called multilayer networks, have grown in popularity due to the emergence of new analytic tools for studying them. These networks are characterized by a set of nodes and multiple sets of edges, with each set of edges known as a layer. Each layer can represent a unique context for social connections, such as work relationships, friendships, and spatial neighborhoods. Multiplex networks can be of particular interest when one considers that relationships in

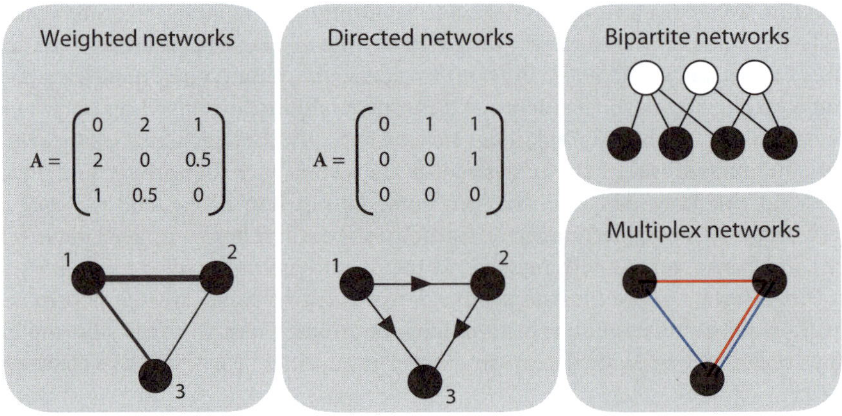

Figure 9.15 Other kinds of networks.

these contexts are not causally independent, and so "spillover" effects may involve influence between the different layers of the network (Boccaletti et al., 2014; Smaldino et al., 2018b; Atkisson et al., 2020).

Pretty much all the topics we have considered throughout this book can be reimagined and modeled using explicit network structures. We already considered how network structure can affect the spread of contagion, and network epidemiology is a robust and growing field (Keeling and Eames, 2005; Bansal et al., 2007; Salathé and Jones, 2010). Researchers have also considered how opinions (Flache and Macy, 2011; Golub and Jackson, 2012; Turner and Smaldino, 2018), cooperative behaviors (Perc et al., 2013; Smaldino and Lubell, 2014), norms (Nakamaru and Levin, 2004; De et al., 2017), and even scientific beliefs (Zollman, 2013; Weatherall et al., 2020) spread and interact on networks. As you might expect, network structure can have serious effects on the dynamics of systems involving social interaction. For example, the addition of long-range ties to an otherwise clustered network can dramatically increase polarization in a model of opinion dynamics with negative influence (Flache and Macy, 2011), and the heterogeneity of social ties in scale-free networks can enhance the stability of cooperation (Santos et al., 2006b). I encourage you to revisit the previous chapters of this book newly armed with network science tools, and to consider how the models and analyses presented there could be extended or enhanced.

Network structures can also change over time. We make new friends and lose touch with old ones. We make new business connections, relocate and make new social connections, go online and connect with people all over the world. In other words, network structures change over time. Although we focused on static or strictly growing networks in this chapter, there is a rich literature on dynamic networks. This literature investigates how the ways in which nodes drop or add ties can influence the emergent dynamics of network structure, information flow, or behavioral change. For example, models for the evolution of cooperation have shown that although changing ties in response to payoffs can enhance cooperation (Santos et al., 2006a), such dynamics can also provide conditions for the future invasion by non-cooperators (Akçay, 2018).

The successful study of networks as mathematical objects has also improved scientists' ability to study empirical systems as networks by taking advantage of the many structural properties we can glean from network analysis. Analyses have tackled sets of connected nodes ranging from computing hubs to genes to neurons to firms to nations. **Social network analysis** is in particular a large and growing area of research (Butts, 2008; Scott and Carrington, 2011; Farine and Whitehead, 2015). One especially useful type of network analysis is the automated detection of communities. **Community detection** algorithms parse network structures to detect communities of nodes that are strongly clustered among themselves but only weakly connected to other clusters. There are several such algorithms, each of which have pros and cons as to what sort of communities they best detect, and which must be calibrated based on how fine-grained a distinction is desired between communities (Lancichinetti and Fortunato, 2009; Yang et al., 2016). It is worth noting that exploratory factor analysis (EFA), a statistical technique used to detect latent variables, is essentially a community detection algorithm, as it divides items into communities which correlate strongly with each other and only weakly with items in other communities.

9.9 Exploration

1. **Network skills.** Consider the 5-node network depicted below. All calculations should be done by hand, without the use of software to compute network statistics.
   (a) What is the node(s) with the lowest degree, and what is that degree? What is the node(s) with the highest degree, and what is that degree? Plot the degree distribution for this network.
   (b) Calculate the average path length for this network.
   (c) Calculate the average clustering for this network.

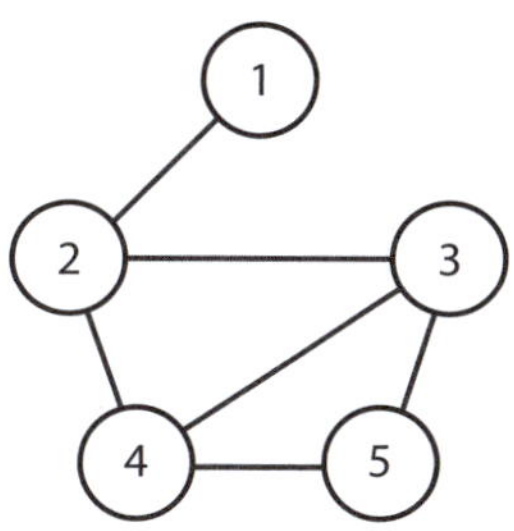

2. **Getting attached, preferentially.** Begin with a model using the preferential attachment algorithm to create scale-free networks. In NetLogo, you can open the Preferential Attachment model (Models Library → Sample Models → Networks → Preferential Attachment.nlogo).
   (a) Run the model until you have a 500-node network. What does the degree distribution of this network look like? Plot the degree distribution on linear and log-log scales. What is the degree of the node with the highest degree?
   (b) In NetLogo, each node's `who` number will tell you the order in which it was created. Nodes with lower `who` numbers will be older. Resize the nodes so that their radius is proportional to the square root of their degree (this can be done by clicking on `resize-nodes`). What relationship do you see between degree and the "age" of the nodes? Plot the degree of each node against its age. How would you explain the resulting relationship in terms of the "rich get richer" narrative?
   (c) Modify your code so that all nodes created after 500 time steps are visualized using a different color (e.g., green) than that used for the first 500 nodes (e.g., red). This

will allow you to plot two distinct degree distributions: one for the newer green nodes and one for older red ones. Perform steps (a) and (b) again, but run the model until you have 1000 nodes—half red, half green. Plot degree distributions for each set of nodes. How do the two degree distributions compare? What is the highest degree among the red nodes and among the green nodes? What is the implication for systems where "the rich get richer" applies?

3. **It's a small world after all.** Open the model for simple and complex contagion on small-world networks (SmallWorldDiffusion.nlogo). For this problem, we are only going to focus on properties of the small-world networks, not on the diffusion dynamics.

   (a) Write code to compute the average path length for the initial lattice (assuming no rewiring), and for the rewired small-world network. The latter quantity divided by the former will yield the normalized path length for small-world networks.
   (b) Similarly, write code to compute the clustering coefficient for the initial lattice (assuming no rewiring), and for the rewired small-world network. The latter quantity divided by the former will yield the normalized clustering coefficient for small-world networks.
   (c) Use BehaviorSpace to run batches of ten iterations each for a 100-node network with `rewiring-probability` varying between 0 and 1 in increments of 0.001 (i.e., $10 \times 100 = 1000$ total simulations), recording the normalized path length and clustering coefficient in each case. Plot the means of the normalized average path length and clustering coefficient against the rewiring probability. You may want to plot the rewiring probability on a log scale.
   (d) How do average path length and clustering change with increasing amounts of rewiring? How do small-world networks compare with ring lattices and random networks? What kinds of real-world networks might share similar properties?

4. **Are complex contagions *really* so hard to spread?** Consider the complex contagion model we developed (SmallWorldDiffusion.nlogo). We assumed that complex contagion meant that individuals would *never* adopt a behavior if fewer than two of their contacts had not also adopted. But what if this is not such a hard and fast rule? After all, individual friends may sometimes be particularly persuasive.

   (a) Create a new parameter called `prob-spread-one`. Rewrite the model code so that complex contagions always generate a new infection if at least two neighbors are infected (as before with $p = 1$), but may also spread if only one neighbor is infected, with probability `prob-spread-one`. Run a few simulations and report the apparent behavior of the model for varying `prob-spread-one`.
   (b) Run an experiment with BehaviorSpace to see how the rewiring probability influences the speed of infection for this modified model. Run at least 10 iterations of the model for each parameter combination, using `rewiring-probability` $\in \{.001, .01, .1, 1\}$ and `prob-spread-one` $\in \{0, .001, .01, .1, 1\}$. Run some sample simulations to assess how long you need to run each simulation for. Plot both the proportion of runs for which the contagion spreads throughout the network as well as the time to fully spread for those runs that did so, as a function of `rewiring-probability` (as in Figure 9.12). Make sure to plot `rewiring-probability` on a logarithmic scale.

(c) Do these new results affect the conclusions we previously drew about complex contagions and network connectivity? If so, how might we rethink what the model tells us about simple and complex contagions in networks?

5. **It's all about networking.** Consider any of the models from chapters 3–7. Recreate this model on a small-world network, and explore the role of rewiring on the subsequent dynamics.

10 Models and Reality

> Metaphor is not only a means of delight but also, and above all, a tool of cognition.
>
> —Umberto Eco, *Semiotics and the Philosophy of Language* (1986)

I hope that, by this point, you are convinced that formalizing theories and hypotheses using mathematical and computational models confers a real benefit to the science of social behavior. Science is ultimately built upon an edifice of empirical facts. But models help us to organize those facts, explain how they fit together, highlight the facts that are more or less important, and point to the unknown facts we most urgently need to discover. Indeed, some of our facts may be about the sorts of outcomes that emerge from particular constraints—these facts derive purely from models! The French mathematician Henri Poincaré (1905) presents an analogy of science as a library that must increase its stores using limited funds. Experimental research is the librarian that makes these purchases; that is, empirical research alone increases our knowledge of the universe. However, without guidance as to which research is likely to be worth investing in, many of these purchases will be frivolous and of little lasting interest. Formal theory acts as the cataloger, enabling those limited resources to be spent judiciously in pursuit of the most important purchases. In other words, models can help us decide what to do with empirical research, and can even indicate which research directions we should be most motivated to pursue.

So far, I have devoted relatively little of this book's space to discussing empirical research, focusing instead on using models to better understand complex social systems. I think this is justifiable. Building a model involves the distillation of meaning, in which the most essential elements of reality (so you hope) are formed into a coherent story that lends itself to both articulation and exploration. Using formal models forces us to lay bare the assumptions we use to make inferences about our study system. When our assumptions are clear, we can better communicate the meaning of our statements; this includes our use of concepts that might otherwise be ambiguous or vague if presented only in the natural language of speech. When we have clear assumptions, we can work through the consequences that must logically follow, and thereby either find support for our theories, by analogy to the real patterns in the world, or highlight the gaps in our understanding, by identifying the mismatches between model and reality. The enterprise of understanding social behavior is a daunting challenge, and formal modeling is an indispensable tool in that endeavor.

Building a model is something of an art form. As with most art forms, a modeler may develop their own unique style, so that different modelers will model a given system in different ways. Another similarity with art forms is that the path to developing your own style should probably involve rigorous training in the classics and in the techniques used by those influential artists who came before. I hope that the previous chapters in this book have helped you to gain some of this training in the fundamentals of modeling social behavior. Putting this training to use is nontrivial. There are many considerations that go into developing and analyzing models of your own. This chapter is about these next steps, about actually doing some modeling in the pursuit of your own questions and ideas. We'll go through things to consider while you design, build, and analyze your model, as well as the often complicated relationship between formal models and empirical data in the social sciences.

10.1 The Mapping Problem

For a model to be useful, its components do not necessarily have to map directly onto things in the real world, as long as *something* about them is meaningfully similar. For example, a network model may be useful not because the nodes are good representations of individual people, but because the structure of the model helps us to understand how information flows through a network under different constraints, which in turn may be useful for making general inferences about information flow in some real-world systems. Even so, most models are models *of* something. This mapping requires some attention.

In particular, I want to call attention to what I term the **mapping problem**, which tends to be particularly pernicious when dealing with social systems. The mapping problem is the potential mismatch between the components of one's models on the one hand and the measurements that one makes of empirical systems on the other. In models of physical systems, the components are typically measurable quantities. Physical theories concern things like mass, position, velocity, and voltage—all things we can measure directly. In the equations used in physical models, the terms have *units*. This means that the outputs of these equations also have units, and the models therefore make predictions about quantities that we can also measure. Because the theoretical models lead directly to precise, quantitative predictions, the value of formal models in the physical sciences is rarely questioned.[1]

In the social sciences, the mapping between measurements and models tends to be quite a bit thornier. Our models often concern concepts like information, cooperation, norms, polarization, and power (the social kind). We can operationalize these concepts in an empirical study, but few would say, for example, that a nation's Gini index accurately and completely captures the social and economic inequality among its citizenry.[2] For example, inequality also relates to social and political power, behavioral affordances, and network structure. In general, the mapping between models of social behavior and the empirical

[1] Even so, measurements don't always capture the folk intuitions associated with their terms. For example, the use of entropy in thermodynamics to capture "disorder" relies on a very particular reading of order and disorder—an evenly distributed gas may appear orderly to some, but it displays maximal disorder to the statistical physicist.

[2] Indeed, the Gini index is a statistical characterization that can apply to *any* distribution of outcomes. This includes topics quite far from those typically studied by social scientists, such as the size and fecundity of plants (Damgaard and Weiner, 2000).

data to which they are related is more qualitative and relies more heavily on analogy than the mappings typically found in the physical sciences.

One might be tempted to use this observation to devalue theoretical models in the social sciences. If they can't make precise predictions, if they're not precisely about the data we can collect in the world, what good are they? I think this devaluation would be a big mistake. I touched on just why it would be a mistake back in chapter 1. But I want to revisit the topic now that you've had a chance to explore some formal models of social phenomena in detail.

The world is dizzyingly complicated. Moreover, our minds are not built to maximize our understanding of how it works. Instead, our minds construct categories that help us to make meaningful predictions for biological and social success. Many of these categories become shared among cultural groups because they help us to communicate ideas and explanations. They are so useful in this enterprise that we often come to treat them as real things in the world. Yet we cannot readily measure things like "happiness," "identity," "agreement," or "inequality" in ways that are objective and not contextually and culturally contingent. When we try to do science that tackles the relationships among these constructed concepts, we bring along an awful lot of baggage. Critically, we risk the imprecision inherent in using colloquial terms rather than measured quantities as explanans and explananda. Models force us to show the assumptions that go into our definitions. And, as I hope I've made clear, studying formal models gives us insight into how complex social systems can work, which scaffolds our ability to reason about real-world complex systems.

On some occasions, models of social systems can be fine-grained, high-dimensionality representations of real-world systems. Some agent-based models in epidemiology, urban studies, and even archaeology are of this type. Each agent corresponds to a real person (or at least to an anonymous member of a community), the physical geography is accurately represented, and movement in model space reflects movement in the corresponding physical space.

Much more often, though, models in the social and behavioral sciences are coarse-grained representations or even very rough abstractions. The output of these models corresponds, at best, to broad, qualitative patterns in the data, and does not reproduce exact measurements. Often modelers are not even interested in reproducing any empirical patterns per se, but only in exploring the consequences of our assumptions. There is value in this. Most people are not nearly as good at mentally simulating the dynamics of complex systems as they think they are. Moreover, most people are not great at mentally accounting for all of the assumptions explicit or implicit in their explanations. Models can therefore help even when they do not make any specific predictions about any particular system. At minimum, models tell us about what we can expect when a particular set of assumptions holds, and that's worth a lot.[3] Now that we've reacquainted ourselves with the value of formal models, we can discuss how you might go about building models of your own.

10.2 Nine Lessons for Turning an Idea into a Model

There are many reasons to build a model. You might build a model because you are trying to explain some empirical pattern in the world. You might build a model in order to articulate some theory or hypothesis, and to see if it would work in principle. You might be interested in a model system just to explore the consequences of some assumptions, or because you

[3] For an extended discussion on the different degrees of coarseness–fineness seen in models of social dynamics, see Bruch and Atwell (2015).

like watching little blobs move pleasantly across a screen. These are all valid reasons to mess around with models! But it is also good to have a sense of the questions you are trying to answer with the model. Below, I lay out a few tips to keep in mind when attempting to turn your idea into a formal model.[4]

10.2.1 Decide on the Parts of the System

To formulate a scientific theory or hypothesis about some system, it is necessary to decompose that system into relevant parts, their properties, and the relationships between them. There is no one right way to do this for a given system; rather, the value of a particular decomposition stems from its ability to answer meaningful questions in useful ways. A model is an instantiation of these parts and relationships. A well-articulated question will constrain the model design in useful ways. Because there are lots of ways to represent any particular system, you need to think carefully about the parts you are going to focus on and the parts you are going to ignore. What questions does your verbal theory address? What parts do you need to include to answer those questions? Is your representation a satisfying analogue to your verbal theory? If not, what is missing? This sort of consideration can also highlight limitations to your verbal theory. Modeling can help us not only test but develop our theories by forcing us to consider each assumption—and each consequence—explicitly.

10.2.2 Choose Your Tools

The formal representation of the components and dynamics of your model will be constrained not only by your research questions, but also by the tools you use to model them. Purely mathematical models are useful because they allow for formal proofs and easy exploration of variables. If you are building a mathematical model, the mathematical tools at your disposal (such as differential equations, probability theory, and linear algebra) will constrain your assumptions. You may end up assuming things like very large populations, minimal population structure (e.g., random mixing), and equal-sized groups. You also may have to rely on things like expected values, which can make it difficult to assess the influence of differential variance. That said, mathematical analysis is often less ambiguous than more stochastic computational approaches, and because these models can be described mostly or even entirely in terms of their equations, they are usually straightforward to communicate.

Agent-based models open up a wide range of possibilities. They can easily represent individual and structural heterogeneity, and can thereby allow you to account for stochastic and rare events. Multiple behaviors and time scales can be accounted for. And there is often an undeniable joy in watching the dynamics unfold in a good visualization. Disadvantages include the sheer number of simulations required to adequately describe the behavior of the model, and the complexity of accurately describing all of its algorithmic details. Poorly thought-out agent-based models may also include arbitrary elements, like spatially constrained interactions that don't map well onto real behavioral affordances.

[4]Much of this section was adapted from my 2020 paper, "How to translate a verbal theory into a formal model."

10.2.3 Get Your Story Straight

As you plan your model design, think hard about what you are trying to accomplish with your model, and what you are trying to show with it. A model is built around an often-implicit story. Consider the stories implied by some of the models we explored in this book. Once there were two groups of people who lived in a city, and each moved when there weren't enough people from their own group living nearby. Or, once there was an infection that arose in some people, and it spread from person to person through physical contact. Or, once there was a land in which there were two norms of behavior, and when people tried to cooperate they faced the dilemma of not knowing which behavior their partner would employ.

Getting your story straight will help you figure out what aspects of reality you are putting into your model, and therefore what aspects you are leaving out. Understanding your story also becomes useful when you write up your model for dissemination. Describing the story of the model is a huge help in interpreting the mathematical or computational details of the model, because the story provides a map that indicates what each component is intended to represent.

10.2.4 Know the Literature

As with any scholarly endeavor, you want to make sure you are neither reinventing the wheel nor failing to give credit to those who blazed the trail ahead of you. For a modeling project, due diligence requires consideration of both the relevant modeling literature and the relevant empirical and theoretical literatures. The modeler is a scientist using a particular tool to ask scientific questions. The modeler therefore needs to engage with the same sort of deep knowledge of their study system that would be expected of any other researcher working on that system. This book was designed to help you start getting familiarized with the relevant modeling literature. However, whatever your topic, it is likely vast. You'll need to do more reading.

This is also a call for a positive attitude toward interdisciplinarity. For example, based on the fact that you are reading this book, you are likely interested in aspects of social behavior like cooperation, coordination, signaling, social influence, and norms. There are already well-developed modeling traditions related to these topics—these include those described in this book as well as others that are not. The models may have been written by researchers and placed in journals marked with different disciplinary labels than your home department, be it psychology, biology, anthropology, sociology, economics, computer science, mathematics, or philosophy. These sciences are deeply interrelated. If you are driven by your research question rather than by your disciplinary identity, it is hard not to come into contact with relevant literatures from neighboring disciplines.

My promotion of an interdisciplinary approach here doesn't stem from some idealistic notion that interdisciplinarity is just morally good (or, at least, not *only* from that notion). An interdisciplinary mindset also yields tangible benefits by providing connections that facilitate better models of complex systems, and hence better theories about their workings. Some examples may help to illustrate this point. Drawing on insights from the physics of magnetic spins, Hopfield (1982) built a model that showed how neural networks can recall stored memories from partial information. Drawing on insights from the study of epistatic gene networks, Lazer and Friedman (2007) built a model that showed how social network structure influences the performance of teams working to solve collective problems. Drawing on insights from ecology and dynamical systems theory, Turchin (2003, 2016) built a model that showed how interconnections between the general population, elites, and government can contribute to cycles of political stability and instability. Drawing on insights from infectious disease modeling, multiple researchers have explored how innovations, products, and behaviors diffuse, showing that social transmission may work in similar—though not identical—ways as disease transmission (e.g., Bass, 1969; Centola, 2018). The interdisciplinary field of cultural evolution, perhaps the best candidate for a unifying framework for understanding human behavior, relies on models that draw from evolutionary ecology, population genetics, social psychology, anthropology, and economics. In sum, disciplines other than your own have developed useful techniques that are likely relevant to your research questions.[5] Don't ignore them. Insights can come from anywhere.

While it is important to acknowledge and learn from what has come before, we should also avoid becoming a slave to the past. The way we model a system constrains the questions that we can ask about that system. Therefore, just because a system tends to be modeled in a particular way doesn't mean that other decompositions aren't valuable. Throughout this book, we considered how different ways of conceptualizing a system allowed us to ask new questions.

10.2.5 Be Constrained by Your Question, Not Your Skills

In Kurt Vonnegut's 1987 novel *Bluebeard*, the protagonist, the abstract expressionist painter Rabo Karabekian, is accused by his wife Dorothy of having chosen his style of painting because he isn't able to paint anything real. He responds by drawing a completely realistic portrait of their sons without even looking at them. Dorothy is flabbergasted, and asks why he didn't simply paint like that all the time. Rabo responds, "It's too fucking *easy*."

When you build a model, it will be tempting to rely on modeling techniques and representations that are familiar and comfortable. And sometimes it is justifiable to yield to this temptation. A commonly used representation may capture the essence of your research question (e.g., representing opportunities for altruistic cooperation as a prisoner's dilemma game). A particular technique may effectively capture the dynamics that you are interested in using your model to study (e.g., discrete-time replicator dynamics). Other times, however, you are trying to do something else. Try describing your questions to your friends—the ones who want to hear about your research, anyway—and consider whether the model you have constructed captures essential features of the world that help address your question.

This all means that the easy way to build a model is not necessarily the right way. It might be, and it's worth remembering that many of the most influential models are simple ones. But even then, simple things can be novel. Building the model you need to address your questions may require developing new skills. If you remain driven by your questions and not by your existing skill set, your research will benefit.

10.2.6 Separate Design from Construction

In his book, *Gödel, Escher, Bach: An Eternal Golden Braid*, the cognitive scientist Douglas Hofstadter (1979) introduced **Hofstadter's Law**: *It always takes longer than you expect, even when you take into account Hofstadter's Law.* This is a good heuristic for almost any complex project, and certainly for modeling projects. The truth is that once you learn some math and

[5] Cailin O'Connor and I have developed a formal model showing how interdisciplinarity can also help to overcome disciplinary biases that prevent better methods from spreading (Smaldino and O'Connor, 2020).

pick up some coding chops, it often doesn't take very long to program a model. This can lead to the mistaken impression that *modeling* doesn't take very long. The error is to conflate the programming with the modeling. The hardest part of modeling is almost always *designing the model*.

The ability to design a model is, more than anything else, what makes a modeler more than just a technician who knows some math and has some programming skills. Modeling social phenomena is a sort of mapmaking where the territory is the hypothesis or theory being modeled. The modeler must decide which aspects of that territory warrant some of the map's limited space, and how best to represent them for the purposes of navigation.

The design process can take ages, with lots of false starts and returns to the drawing board. This is why I usually encourage modelers to write out their entire model before a single line of code is written. In other words, you should separate the process of model *design*, in which you figure out all of your assumptions and how they will be implemented, from the process of model *construction*, in which you actually code the thing so it works. If you are building a house (or a boat, or a cabinet, or a guitar, or whatever DIY project resonates with you), you know you shouldn't start hammering before making sure you've got all your parts, all your tools, and have laid out your plans. If you *do* start before these things are done, you are more likely to make wrong turns, paint yourself into corners, and find yourself committed to unfortunate choices. In practice, I find that the design stage often works best as an iterated process with multiple steps: going from verbal theory to a set of parts and relationships, to a set of parameters, to a set of analyses, to an algorithmic design in pseudocode, and then finally to a coded model. All the while, be willing to reconsider the choices you've made at each step.

10.2.7 Be as Simple as You Can Be, as Complicated as You Need to Be

A model gets its power from its ability to remove complexity from the world, allowing us to focus only on those aspects we think matter. Many of the most powerful models in the cognitive, behavioral, and social sciences are simple models that reveal powerful and counterintuitive dynamics. I have seen extremely complicated models that merely reproduced the insights of simpler models with less transparency and generalizability. Counterintuitively, simple models are not necessarily easier to produce! It's often easier to produce a complicated model that has redundant or unnecessary components than a simple and streamlined one.[6]

I have personally had the experience of starting a modeling project with a very complicated design. As I developed a deeper understanding of the theory I was trying to model, I was able to more readily identify those parts of the system that were critical to the theory as well as those parts that I could, at least for the moment, leave out. It is a common experience among modelers to have nonmodelers ask for more realism in the model design, but giving in to these requests is often a mistake. It is actually by establishing limits that models allow for the creative development of new ideas.

This is not to say that simple models are *always* best. While you should strive to keep your model simple, don't be afraid to get complicated when the situation calls for it. Added complexity creates new affordances for the model's behavior. Heterogeneity and structure create new opportunities for feedback. Once you understand the simple cases, adding complexity to a model can yield important insights that would otherwise be missed. Examples can be

[6]Recall Blaise Pascal's famous postscript: "I have made this letter longer than usual because I have not had time to make it shorter."

found in the fields of **artificial life** and **systems science**, which are known for embracing complicated computational models to explore how the myriad components of a complex system interact to produce emergent outcomes. Complex models are also often necessary when trying to predict outcomes in systems for which rich structural data exist, such as in the case of urban dynamics (Crooks et al., 2019).

10.2.8 Attack Your Design

Once you've designed your model and articulated the relevant parts and relationships, attack it! Seek out its weaknesses and push them until the whole thing falls apart. You need to know its limitations. What must be true for your model assumptions to hold? Does the model convincingly represent your system? Could it just as convincingly represent something else?

It's also worth reevaluating the modeling framework you've employed. If you are working with a mathematical model, are there frustrating simplifications on which an agent-based approach would shed light? If you are using a computational model, are there results that could be expressed more simply and cleanly, perhaps even conclusively proven, using a mathematical model? Mathematical and computational approaches are often complementary and can be usefully employed to explore variations on the same model.

This is perhaps somewhat repetitive with some of the earlier recommendations, but I feel like I can't overstate this point: designing a good model is hard! It's important to continuously strive for ways to improve your design.

10.2.9 Be Open

A model's value comes from its power to show how assumptions lead to consequences. If those assumptions are not transparent, then you have failed to truly offer a formal model. Instead, you have merely added alchemy to your verbal theory. Your model should be described clearly. Model descriptions should be held to this standard: a competent modeler who is otherwise unfamiliar with your project should be able to read your model description and, without looking at your source code, be able to produce their own working version of the model. I sometimes have the impression that vague model descriptions come from modelers who are unsure of the wisdom of their choices. Be confident in what you've done and avoid ambiguity when describing it to others.

While you should be clear in your description, you should also share your code. The code should be well-documented, including in-line comments, and be readily available on a stable repository.[7] There are at least two reasons to do so. First, even the best writers sometimes unintentionally inject ambiguity, and this can be cleared up by referencing the source code. Second, sharing code provides a service to those who wish to extend your work. Some people may be reluctant to simply give away their hard-earned source code. All I can say is that I have never seen anyone's career hurt by their sharing information. Your influence will only increase when others can readily build on your work.

In his book *Steal Like an Artist*, Austin Kleon (2012) writes that the secret to success in the arts is to "do good work and share it with people":

> It's a two-step process. Step one, "do good work," is incredibly hard. There are no shortcuts. Make stuff every day. Know you're going to suck for a while. Fail. Get better. Step

[7]Examples include GitHub, Dryad, and CoMSES. Please don't succumb to the dodgy practice of saying that the code is "available upon request." The success rate of such requests is notoriously low (Minocher et al., 2021).

two, "share it with people," was really hard up until about ten years ago or so. Now, it's very simple: "Put your stuff on the Internet."

I think the parallels to science are pretty clear. A worry some people have is that if you put your work online, someone will steal it, robbing you of opportunity and taking credit for your work. However, that perspective drastically overestimates the likelihood that someone else will take your model and do that awesome analysis that you were planning to do without you, and even more drastically overestimates the likelihood that someone will steal your code entirely and claim it as their own. We're doing science here. Scientists tend to be pretty communitarian. Moreover, have some humility. I've always found a lot of truth to this quip by the physicist Howard Aiken: "Don't worry about people stealing your ideas. If your ideas are any good, you'll have to ram them down people's throats."

10.3 Analyzing Your Model in Light of Itself

When people talk about model analysis, they usually mean one of two things, which are actually quite different from one another. First, you can analyze the model in light of itself, which means that you examine the logical consequences of the model's formal and quantitative assumptions, and explore how those consequences shift when you vary the assumptions (such as particular parameter values). Second, you can analyze the model in light of empirical data, which means that you validate or calibrate the model's assumptions in terms of the goodness of fit between the model outcomes and an empirical data set it is intended to explain. This section focuses on the first type of analysis, while the next section focuses on the second.

The purpose of analyzing a model in light of itself is to characterize just how the model's assumptions lead to particular consequences, and how those consequences change when the assumptions are varied. Sometimes, someone will criticize a modeling study by claiming that its conclusions were "baked in" to the model's design. This is often a baseless critique, because of course it is *always true of every model*. The model's design was chosen by the modeler, and the conclusions from the model analysis are logical, necessary consequences of that design. What is usually meant by the critique is that the fact that a particular set of consequences follow from a particular set of assumptions is so obvious that setting up a formal model in the first place was unnecessary and does nothing to advance our understanding. This *can* be a valid criticism. Not all models are useful, and many models make poor or questionable assumptions that are either unrealistic or else assume the very phenomenon one is ostensibly trying to explain.

On the other hand, what is obvious to one person may not be obvious to someone else, and simple models can help guide people's intuitions. Sometimes a model is useful simply because it allows you to make your assumptions explicit and show exactly what you do or do not mean by your verbal description. A good simple model can later form the basis for more complicated models. Consider the prisoner's dilemma game. It's pretty obvious that mutual cooperation yields higher payoffs than mutual defection, but defection is always preferred in the one-shot game. This insight allows us to reason much more deeply about cooperation than we could without the model, and many complex models of cooperation have used the prisoner's dilemma game as their foundation. Other times, examination of a model can highlight important information that needs to be learned before strong conclusions about your system can be drawn. And other times still, your model generates behavior that is legitimately unexpected, that you would not have thought of when you first started analyzing it. The only way to find out whether something unexpected will emerge is to start analyzing.

10.3.1 Plan Your Analyses

A critical consideration in designing a model is what the outcome measures will be. This forces us to ask again: What are we trying to show? What are the questions we want answered about our system? These can then be reframed to ask how you should structure your model so that it answers your questions. Are you interested in the state of the model at equilibrium? The frequency of a trait or the solution to a problem? Are you more interested in dynamics, the rise and fall of some variable? Will you need to consider the variation among the agents in your model, or the variation of outcomes between simulation runs?

The challenge of coming up with good questions should not be underestimated, and these considerations should be prominent in your mind. As noted, it is often valuable to write out a formal, detailed description of the model *before* any analyses or simulations are performed. That way, you make sure that you are extremely clear on how the model works. The description also provides a record of what you were intending to do, which is important as model designs can otherwise change as decisions are made on the fly while coding. An important part of this documentation is a detailed description of outcome measures, particularly if you are working with a computational model, for which the number of possible analyses are more numerous. What are the parameters and initialization conditions you will consider? If your model is very complicated, how will you deal with the issue of many interacting variables? The more complicated your model, the larger the space of possible parameter combinations. You will need to think carefully about how you will cover this space, given that once your model has grown past a few parameters, a full sweep of all combinations becomes logistically tricky.

For an agent-based model, there are pretty much *always* more analyses one could do—more parameter values one could test, alternative assumptions one could impose. This is where the importance of your motivating questions comes into play. A report of an agent-based model should be designed to address a set of questions. But while you should use this scope to constrain your analyses, you should also make sure you sufficiently explore the sensitivity and robustness of your results so that you may be reasonably confident in your conclusions.

Sensitivity refers to the degree to which your conclusions change in response to variations in your assumptions. Often this is exactly what you want to find out, at least for a particular set of assumptions. But there are other assumptions that might not be essential to the verbal description of your theory that you nevertheless have to make decisions about in your formalization. Things like population size, network structure, initial trait frequencies, or movement strategies (among others) can feel like arbitrary choices. Sensitivity analyses are tests of how much your model outcomes vary in response to changes in the assumptions that are still consistent with your underlying theories or hypotheses.[8] **Robustness** is just the flip side of sensitivity; it refers to the extent to which your conclusions are unchanged by these new arbitrary assumptions.

Sensitivity and robustness analyses are extremely important. It is possible for an apparently compelling result to be an artifact of a very particular set of assumptions. And a lack of robustness can itself be an important result. Either way, understanding the robustness of

[8] The boundary between the sort of analyses that do or do not count as sensitivity analyses is quite fuzzy, partly because the analyses can help shape the theories you are trying to formalize.

one's results is a central duty of the modeler during analysis. It is also advisable to actively *try* to find the breaking point(s) of your model. Figuring out when something stops working is an important part of figuring out why it works the way it does. If this all seems like too much, remember that science is iterative. What you learn from one analysis will help make your next model, experiment, or analysis better.

10.3.2 Rethink Statistics

Suppose you are working with an agent-based model, and you have some nice results from your many simulations. How shall you describe the ways in which your outcome measures interact with your model parameters? If you have been trained in the traditions of most of the behavioral and social sciences, you will likely be tempted to use inferential statistics such as a t-test or regression. Although there are exceptions, this is almost never the right decision.

Why? Surely you want to know if the effects you are seeing are valid? Of course you do. But consider that calculating inferential statistics involves constructing a model of your data-generating process, usually with strong simplifying assumptions about the distributions of parameter values and data points. You don't need to do this, however, because you already have a better model of your data-generating process: it's your actual model. There's no reason to model your model with a lower-quality model.

This is true even if your model is extremely complicated, so that understanding the mechanisms underlying its behavior is difficult. Rather than estimating the likely distributions of data produced by the model by appealing to general statistical properties of large data sets, you can simply run enough simulations to obtain arbitrarily precise estimates. I have heard some modelers complain about this suggestion, because running many simulations can take a long time, particularly if one does not have access to a computing cluster. I sympathize. However, we are doing science here, and our job is to get as close as possible to the truth.[9] Think about your colleagues who spend months or years doing fieldwork or collecting longitudinal data. Surely in the interest of precision you can wait a few extra days while your simulations run.

One obvious exception is when you are simulating data for the explicit purpose of preparing your statistical analyses for empirical use or are otherwise trying to explore processes that generate the sorts of statistical patterns you are likely to see in data—an example is the analysis of newsworthy versus trustworthy science in chapter 8. Another exception is when you are directly comparing your model's simulated data with empirical data and have good reason to believe that your theory is sufficiently fine-grained that a statistical fit can be used to calibrate parameter values. This latter scenario will likely be less common than you might think.

10.4 Analyzing Your Model in Light of Empirical Data

If you publicly identify as a modeler, a question you are likely to get asked is: "What data do you work with?" Sometimes, there is a concrete answer, as when you are building models to explain specific patterns in specific data sets. Other times, there will be no data, as when you are exploring the consequences of particular decompositions of your study system. In

[9] This footnote is a placeholder for your preferred caveat about the nature of truth and its relationship with the human mind.

those latter cases, we ask what happens when we conceptualize the world as a particular set of parts, properties, and relationships. This can be important foundational work in the development of strong theory, and it is therefore a mistake to suppose that all models should be employed in the service of explaining specific empirical data. A key consideration is the extent to which the assumptions of a model reflect reasonable assumptions about the forces shaping our empirical phenomenon. If fit mapping between model and reality is poor and the model is sufficiently complex and/or underspecified, then we may end up with good fits of a bad model that tell us little about underlying generative processes.

There is a recurrent dream among some ambitious social scientists that their discipline will one day become like the physical sciences, with lockstep feedback between high-resolution data and precise predictions and retrodictions from formal models. Part of this dream involves solving the so-called **inverse problem**, in which a model is used to determine the causal processes that led to particular patterns in an empirical data set. In most cases, this dream is unlikely to become reality.

Validating a model with empirical data is very difficult. This is particularly true if you are looking to make precise, quantitative predictions about the world. Consider that predictions about the weather are only accurate for about a week or so. The local weather anywhere on earth is entrenched in a globally connected complex system, in which small effects can become amplified over time. This means that extremely small differences in the parameterization of a model can lead to very large differences in the behavior of the model given enough time. The **butterfly effect** is a term first used to describe the work of the mathematician and climate scientist Edward Lorenz, who discovered that minuscule changes to the starting conditions of his climate simulations resulted in massive differences in the resulting trajectories. The idea behind the phrase is a hallmark of what is now known as **deterministic chaos**: extreme sensitivity to initial conditions. Failure to account for the flapping of a butterfly's wings in Brazil will lead you to erroneously predict rain instead of sun the following week in Texas. Many complex systems exhibit this strong sensitivity to initial conditions. The Lorenz attractor (Figure 10.1) is a simple deterministic system defined by three coupled differential equations that illustrates deterministic chaos: the long-term state of the system will always be on the attractor, but precisely *where* on the attractor cannot be known without knowing the exact value used at initialization.[10] Initialize the system even a teeny tiny smidgen away from the correct values, and your predictions will quickly become useless.

Most real social systems have many more moving parts than these simple weather systems. So is predicting social behavior hopeless? Not entirely, for at least two reasons. First, even chaotic systems can exhibit regularities we can observe. For example, we might be able to delineate the states that a system is or is not likely to be observed in. And although the day-to-day weather may be a difficult prediction to make, more coarse-grained patterns are much easier. For example, in central California, the weather is very likely to be hot and dry in the summer and cooler and wetter in the winter. Second, social systems have evolved to be largely **resilient**, which means that they are described by relatively stable states that are largely insensitive to minor fluctuations and can often recover from shocks.

With few exceptions, models of social systems are typically built to capture coarse-grained patterns, so qualitative resemblances between model output and empirical data are usually the best we can hope for. I want to stress that *there is nothing wrong with this.*

[10]Many excellent books have been written about **chaos theory**, which contains useful lessons for understanding and modeling complex systems. A great introduction is still Gleick (1987). For a more mathematical course on the subject, see Strogatz (2015).

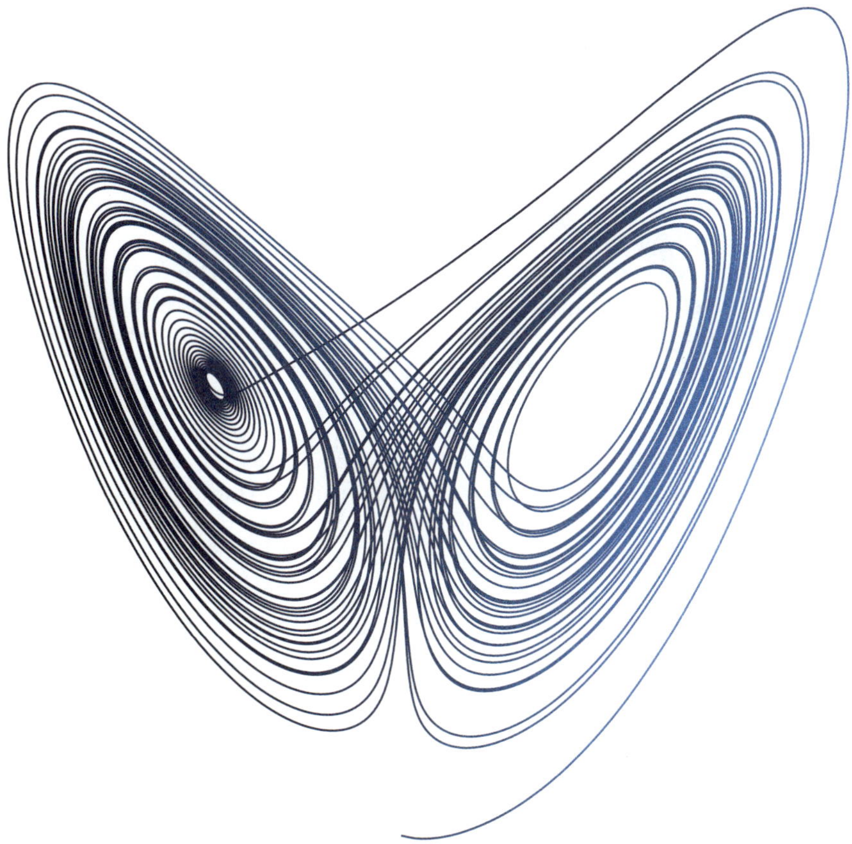

Figure 10.1 Two-dimensional projection of the Lorenz attractor, with color representing the third dimension. The system is initialized at the bottom center and evolves in a complex trajectory. Most social systems are quite a bit more complex than the Lorenz system.

Empirical social science is replete with characterizations like "*A* goes up when *B* goes down," or even "*A* and *B* are not the same!" A formal model can at least help us to specify what we mean by *A* and *B* and to delineate the scope of conditions under which the relationship in question will or will not hold.

Social systems are usually extremely complex. One cannot consume popular science media for long without running into someone claiming that the human brain is the most complex thing in the universe. But human social systems are made up of networked human brains, and those networks are shaped by cultural institutions and technologies that have been evolving for hundreds or thousands of years! Building a model that can account for all the important influences on human social behavior and that yields exact predictions is almost certainly impossible.[11] Moreover, data on human social behavior is usually pretty sparse. We have good aggregate data on economic and cultural indicators, but rarely do we have sufficient data to calibrate the details of human interactions needed for quantitative predictions.

[11]Contrary to the claims of various tech gurus, I think it is very unlikely that we are living in a simulation. The point of a simulation is to simplify reality, not to reproduce it.

This limitation may be changing in the current era of so-called "big data." Our ability to amass very large and fine-grained data sets may allow some amount of prediction, especially in well-defined and constrained systems.[12] Traffic grids, infectious diseases, and communication networks are some of the human domains in which formal models may make precise predictions. Models for the behavior of nonhuman animals are also becoming increasingly precise, since motion capture and radio tagging can yield highly detailed data on animal behavior, which, along with detailed information about climate and ecology, allows researchers to build very rich models.

Most often, however, you will be looking for qualitative patterns and attempting to draw analogies between your formal models and empirical patterns. This can and should be done carefully and deliberately. Models can help you to understand the scope of your theory, and what conditions should or shouldn't be met in order to expect particular qualitative results (e.g., more cooperation, stronger norms, etc.), and empirical data can be used to validate predictions of this sort. But whether you are making qualitative or quantitative predictions with your model, there is a dark cloud you must be on the lookout for. If ignored, it can lead you into the heavy rains of poor and misguided conclusions. This is **equifinality**.

10.4.1 Equifinality

Equifinality is the idea that a particular pattern in data can be generated by multiple, mutually exclusive mechanisms. Concerns about equifinality should be at the front of your mind whenever you evaluate a claim that some phenomenon has been explained by a model. Equifinality is a problem because models are often underspecified by the available data, and so the modeler must make some arbitrary assumptions about those aspects of the system they are unsure about. In his essays on "generative social science," Josh Epstein has touted the value of explanatory agent-based models with the pithy adage, "If you didn't grow it, you didn't explain it" (Epstein, 1999). I tend to agree with this. However, we cannot assume the inverse. If you *did* grow it, you have only generated a candidate explanation in need of further scrutiny. This is because there are often *many* ways to grow it.[13] For example, researchers have shown how Schelling-like patterns of segregation can emerge even if all agents actively prefer to be in the minority (Muldoon et al., 2012), and that similar-looking behavior diffusion curves can be generated by very different social learning strategies (Barrett, 2019).

I'll illustrate the problem of equifinality in greater detail with an example drawn from the literature on human mate choice. There are well-documented regularities in human romantic partnerships. When researchers measure the physical attractiveness of individuals' faces,[14] they reliably find medium-strong correlations between partners, particularly among long-term partners. When researchers look at the distributions of age at first marriage (delightfully known as the marriage hazard rate), they reliably find right-skewed distribution curves, with the probability of first marriage initially increasing rapidly with age and then slowly declining. Modelers interested in understanding the psychology underlying human partner choice have built models that formalized some hypothesized decision rules that individuals might use to select or accept a partner (e.g., Kalick and Hamilton,

[12] This is a footnote reminding you that even when this is possible, it is not always in the interest of society to collect high-precision data on human behavior.

[13] See Frank (2009) for a review of general principles that can generate some commonly observed patterns in data.

[14] Usually by having their research assistants rate them on a linear scale. Blech.

1986; Hills and Todd, 2008). They then used the empirical patterns of attractiveness correlations or marriage hazard rates to support arguments that particular decision rules were especially likely—those decision rules that, when implemented in a simulated interacting population, generated patterns similar to those observed empirically. The patterns seemed to match those seen empirically, so the models' assumptions were supported. Right?

Not so fast. The modelers paid a lot of attention to the decision mechanisms made by individuals in evaluating potential partners, but they appeared to pay less attention to the mechanisms driving how potential partners met one another and created opportunities for evaluation in the first place. For example, the models assumed that populations were well mixed—that is, that individuals met potential partners completely at random. But of course that's not the case in the real world, in which social networks, geographical constraints, and shared interests shape who people tend to meet.

Let's consider a specific example from one of these models. A paper by Kalick and Hamilton (1986) claimed that their model supported the hypothesis that individuals exhibit preferences for maximizing the physical attractiveness of their partner, because when agents employed a decision rule that accepted potential partners with a probability proportional to their attractiveness, the model yielded an intra-pair attractiveness correlation that more closely matched empirical data than did alternative decision rules that incorporated factors like intra-pair similarity. In other words, the paper claimed support for the assumption that people always seek the most attractive possible partner, ignoring all other factors. However, the model also assumed a well-mixed population, which, we just noted, is unrealistic. When the model's dynamics were situated in a spatial setting in which agents moved around their environment and interacted only with other nearby agents, it was revealed that *any* of the evaluated decision rules—including a rule based entirely on maximizing similarity—could generate intra-pair correlations that were arbitrarily close to those observed empirically, merely by tweaking the rate at which agents moved through space (Figure 10.2A). This result doesn't provide support for any particular decision criteria. Rather, it indicates that understanding the network structure of social interactions is absolutely critical for evaluating the empirical validity of individual-level decision rules. Moreover, none of the decision rules tested by Kalick and Hamilton (1986) generated the right-skewed marriage hazard curves seen empirically. Instead, each rule generated very different hazard curves, which varied as a function of individual attractiveness. The decision rule for maximizing similarity implied a negligible effect of attractiveness on the time needed to find a partner, with slightly longer times for the most and least attractive individuals. The decision rule for maximizing attractiveness, on the other hand, implied that the time needed to find a partner should be strongly correlated with attractiveness, so that the least attractive individuals would take many times longer to find a partner than the most attractive individuals—a result that is not observed empirically (Figure 10.2B). All this indicates that the model was never a good candidate for accurately describing human behavior.[15]

10.4.2 Pattern-Oriented Modeling

The critique above contains a clue for dealing with the problem of underspecification and the perils of equifinality. In the partner choice scenario, any model would be much more strongly supported if its output was simultaneously consistent with the empirical data on intra-pair attractiveness correlations *and* marriage hazard curves *and* population network

[15]This analysis is presented in more detail in Smaldino and Schank (2012a).

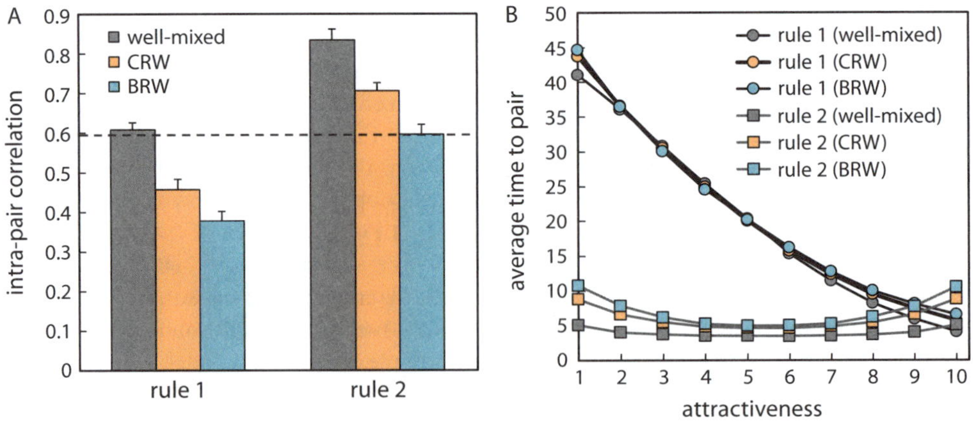

Figure 10.2 Analysis of a spatial mate choice model with two decision rules: one based solely on maximizing partner attractiveness (Rule 1), the other based solely on maximizing partner similarity (Rule 2). Interactions were either well mixed or else assumed a spatial structure in which potential partners could only be found locally and agents moved through space using either a correlated random walk (CRW) that explored widely or a Brownian random walk (BRW) that explored narrowly. (A) Intra-pair correlations matching those seen empirically could be obtained with either decision rule depending on the movement rule. (B) Each decision rule implied very different relationships between attractiveness and average time to find a partner.

structures. Empirical support for a model is strongest when the model's output is consistent with patterns in data across multiple domains, ideally at multiple levels of organization. The model is no longer underspecified by the data when the data specifies numerous constraints on the assumptions the modeler can make in their model design. This approach is sometimes called **pattern-oriented modeling** (Grimm et al., 2005). The key point is this: there is nothing wrong with studying a model in the absence of empirical data, but if you are going to fit your model to data, then you should strive to fit your model to the data in many different ways. There are multiple patterns in the world at different spatial, temporal, and organizational scales. If your model is to represent reality, it should match as many of these patterns as possible. If a model fails to capture some patterns (and it will), those failures can tell you something about the limitations of the model's validity, which may in turn help suggest directions for refining the model.

Pattern-oriented modeling is challenging for a number of reasons. First and foremost, it requires high-quality, high-dimensional data. That's fine if you have access to rich data sets collected over long time periods about multiple facets of a system, but it may not work well in fields with sparser data.[16] Another difficulty is that models are valuable in part because of their ability to simplify complex systems. A model that includes absolutely every detail of the real world is not useful in providing general explanations of phenomena. Thus, good models must hit a sweet spot of complexity. They should be complex enough to generate rich output that makes empirical validation possible, yet simple enough to yield real explanatory value above and beyond the data with which they are being compared. Getting this balance right is as much an art as it is a science.

Pattern-oriented modeling also requires you to have a sufficiently strong theory that you can identify the likely major factors generating your data. If you have a rich set of data and

[16]This fact should ideally encourage more theory-driven data collection in your field.

a strong theory about how the relevant parameters interact, then it is possible to use computational optimization techniques to derive model configurations (i.e., sets of parameter values) that generate those data. This is called **calibrating** the model to the data, which can be accomplished using a variety of computational techniques.[17] The next section briefly reviews some of these.

10.5 Fitting Models to Data on a Rugged Landscape

Suppose you have a model that makes what you think are reasonable assumptions and can generate outcomes seemingly in line with empirical observations, *and* you have data that is sufficiently rich to avoid equifinality. Given the complexity of your model, you likely have many parameters to adjust. How do you go about tuning your model to fit your data?

A useful technique is to conceptualize the space of all possible solutions as a **fitness landscape**. The landscape analogy is very commonly used to represent complex problem spaces, and was first introduced by evolutionary biologists to represent the contributions of different genes to an organism's fitness. Let's start simple and imagine a model with just two free parameters. The landscape is a two-dimensional surface where a position can be described by two coordinates (x, y). Each of these coordinates represents the value of one of the two parameters in your model, so each position on the landscape can be considered a **solution** to the problem you are trying to solve. Crucially, the surface isn't flat. Instead, each position is defined by an elevation, which represents the value, or **fitness**, of the solution. The goal is to fit the model to best match the data, so the fitness of a solution is the quality of the model's fit to the target data under the parameters corresponding to the solution.

If we're very lucky, our two parameters each influence the value of the solution independently and don't interact. In this case the landscape is "smooth" (Figure 10.3, left), and we can use simple **hill climbing** (also called gradient ascent) to optimize our solution. To do this, we adjust each parameter one at a time, settling on the values that maximize the solution's fitness. Search is typically restricted to *neighboring* solutions, which are those solutions involving only small changes to a single parameter, relative to one's current solution. The term "hill climbing" comes from the fact that the process can be conceptualized as moving along the landscape in any direction in solution space that takes us uphill (that is, increases the value of the solution). Once we've done this for each parameter, we should have arrived at the **global maximum**—the best possible solution. When the fitness landscape is smooth, the order in which we adjust the parameters doesn't matter, as all uphill paths will lead to the global maximum.

Unfortunately, we will rarely be so lucky. In complex systems, parameters usually interact in nonadditive ways, and the contribution from one parameter may be strongly influenced by the value of another. For example, recall from chapter 9 that the speed and extent of a contagion's spread on a small-world network increased with the probability of rewiring for simple contagions, but this relationship was reversed for complex contagions. When the fitness contributions of a model's parameters involve interdependencies with other parameters, hill climbing is likely to get us stuck on a **local maximum**, defined as a solution that has

[17] One of the most ambitious attempts to use optimization and machine learning techniques to parameterize agent-based models of social systems has recently been dubbed "inverse generative social science" (See Vu et al., 2019). Other promising techniques not covered here include the use of deep learning algorithms (Dandekar et al., 2020) or approximate Bayesian computation (Kandler and Powell, 2020) to fit model parameters to data.

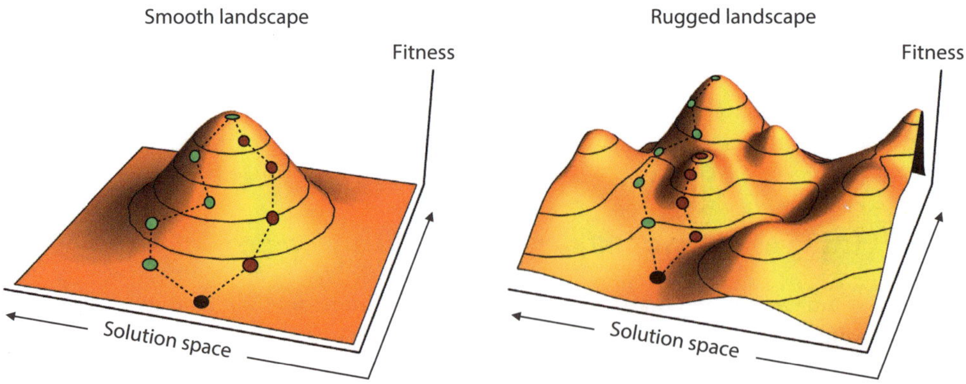

Figure 10.3 Search for high-performing solutions on a smooth (left) and rugged (right) fitness landscape. Left: different search trajectories lead to the same global maximum solution. Right: Different search trajectories can reach solutions of differing quality even when they start from the same initial solution.

a lower value than the global maximum but for which all neighboring solutions have lower value than the local maximum. A **rugged landscape** is characterized by many local maxima (Figure 10.3, right). Rugged landscapes exhibit strong path dependency, such that hill climbing will lead to different local maxima depending on the order in which parameters are explored.

The problem becomes substantially thornier in high-dimensionality fitness landscapes, when there are many parameters that must be simultaneously optimized. Hill climbing will almost certainly lead to a suboptimal solution. And a brute-force search of all possible solutions is usually infeasible. If you have k parameters that can each take n possible values, then the number of unique locations in your solution space is k^n. This number can escalate very rapidly, especially if parameters can be defined by many significant digits. A model with 10 parameters that can each take 100 possible values has more possible configurations than there are atoms in the observable universe.

Because we cannot possibly test each solution for its fit to data, we must strategically explore the space so that a reasonably good solution can be found (even if it is not necessarily the global maximum) in a reasonably short amount of time. The problem of search on high-dimensionality rugged fitness landscapes is a well-studied one in computer science, machine learning, and data science. As such, a number of algorithms have been proposed for this sort of search, most of which have been inspired by natural processes. I will briefly introduce two of the most widely used optimization algorithms: simulated annealing and genetic algorithms.

10.5.1 Simulated Annealing

When smiths and jewelers shape a piece of metal into its desired form (such as a ring, ornament, or tool), the process of hammering and bending the metal compresses and distorts the metal's natural crystalline structure. The molecules in the metal become disaligned in places and assorted into grains of varying sizes, making the metal brittle and easily broken. What the metalworker wants is for the molecules to reach a state of alignment in a nearly uniform crystal. This is the lowest energy state of the metal, and the one in which the metal will bend rather than break in response to stress. In order to achieve this state, the shaped metal is heated to a high temperature in which the individual molecules can move and rotate, then slowly cooled. The heat is necessary to allow the molecules to move, with more uniform crystals being more stable and hence more likely to persist. Repeating this process, known as *annealing*, yields strong, flexible metals.

Recall that the problem with using hill climbing to search on a rugged fitness landscape is that the system can only go to toward a local maximum. **Simulated annealing** increases the "temperature" of the system, allowing search in directions that do not immediately decrease the "energy" of the system—that is, movement *away* from local maxima is allowed. If the metaphorical temperature remains high, the resulting search will be overly random. However, by gradually decreasing this temperature (and therefore gradually increasing the resemblance to simple hill climbing), the system will settle onto a solution that is superior to what could have been found by hill climbing alone.[18]

A commonly used algorithm[19] for simulated annealing works as follows. We start at some initial position in solution space, s, with a known value, $V(s)$. We then consider some neighboring position, s', and assess its value, $V(s')$, where $\Delta V = V(s') - V(s)$. If $\Delta V > 0$, we accept the new position as our current solution, just as with simple hill climbing. However, a new position may also be accepted even if it is worse than the current solution as long as the temperature T is nonzero. More specifically, the probability of accepting a new solution s' over a previous solution s is

$$\Pr(\text{Accept } s') = \begin{cases} 1 & \text{if } (\Delta V + T) > 0 \\ 0, & \text{otherwise.} \end{cases} \qquad (10.1)$$

The temperature T decreases monotonically with the number of solutions tested, k, so that new positions that decrease the solution's fitness are accepted with decreasing probability. As $T(k) \to 0$, the algorithm reduces to simple hill climbing, and so will eventually stop. The rate at which T decreases trades solution value against the time needed to reach that solution.

Simulated annealing is fairly simple to implement and will usually achieve much better solutions than simple hill climbing. However, because it involves a single search that, by definition, must start at some location in solution space and only compares new solutions to the most recently observed solution, it can still often fail to find global maxima. If one's available computational resources permit the necessary flexibility, it might be better to simultaneously compare an entire population of searchers on the fitness landscape, and to allow those searchers who have found better solutions to preferentially recruit more help in their search. Enter the genetic algorithm.

10.5.2 Genetic Algorithms

Evolution by natural selection works on the principles of variation, heredity, and selection. Individuals within a population vary in their traits, and their offspring inherit those traits (with mutation as well as crossover in sexually reproducing species). Certain traits confer differential fitness benefits and costs, which leads some traits to increase in frequency

[18]Some psychologists have proposed that simulated annealing serves as a useful analogy for the development of exploration–exploitation tradeoffs over the course of child development (e.g., Gopnik et al., 2015).

[19]There are several variations on this algorithm. The version shown here uses a threshold temperature, which has been shown to outperform a stochastic alternative. See Moscato and Fontanari (1990) and Franz et al. (2001).

while other traits decrease. These dynamics can be implemented computationally. In fact, we have already seen evolutionary algorithms in some of this book's previous chapters, where traits related to cooperation, coordination, and scientific practices evolved. It is a convenient modeling simplification to assume that traits are transmitted whole cloth from parent to offspring.[20] In reality, genetically transmitted traits typically emerge from the interactions of *many* genes. A single gene may contribute to many traits, while a single trait may arise from the interactions of many genes (evolutionary biologists call the former effect *pleiotropy* and the latter effect *epistasis*). These interactions can be complex and can lead to unintuitive outcomes. For example, the genes that increase the risk of sickle cell anemia are common in some sub-Saharan African populations because they also confer some immunity to malaria, an otherwise deadly disease. Natural selection changes the distributions of genotypes whose combined effects maximize the relative fitness of individuals in a population.

Model parameters are like genes that combine to influence the fitness of a model instantiation. **Genetic algorithms** work by formalizing this analogy. Each parameter of a model can be assigned a number of different values, just like actual genes can have multiple alleles. Instead of starting with one solution (that is, one model instantiation with some set of parameter values) as in simple hill climbing or simulated annealing, a genetic algorithm starts with a *population* of solutions, distributed throughout the space of possible solutions. Just as before, we evaluate each of these solutions by their fit to our data, which we can quantify as a fitness score. Then, we let evolution get to work. The solutions with the worst fit are eliminated from the populations. The top performers (say, the top 10%) remain in the population and also produce *offspring* solutions by **crossover mating** and **mutation**. Mutation is easy: with a small but nonzero probability, the value of a parameter is changed at random. Crossover is slightly more complicated, but is analogous to genetic crossover in sexual reproduction. If you picture each solution as a string of values (each corresponding to one parameter value), mating occurs when two of the solutions pair up, pick a location in their strings (usually at random) as a crossover point that divides each solution into two sections, and then produce new solutions that take one section from each of the two "parent" solutions (Figure 10.4). This allows solutions to benefit from traits evolved in multiple lineages.

Genetic algorithms work well because they recruit more searchers to the areas of solution space where higher-valued solutions tend to be found, making the algorithms more likely to find solutions approaching the global maximum. They have been used to evolve complex models to fit rich data. For example, Schank (2008) studied the movement of young rat pups moving in an arena both alone and in groups, using motion capture to track their changing positions over time. He then modeled each pup as an embodied agent that moved in virtual space, using a transition matrix to represent its contingent movement rules (e.g., the probability of turning left when its right side was touching a wall). Schank then used genetic algorithms to evolve these matrices so that the resulting agent movement patterns best approximated the real pups' movements. By comparing the transition matrices evolved based on the solo and social conditions, he was able to show that rat pups effectively treated other pups no differently from walls at seven days old but responded to pups and walls very differently at ten days old, supporting hypotheses about the developmental timing of social cognition.

[20] This is usually called the **behavioral** or **phenotypic gambit**. See Grafen (1984).

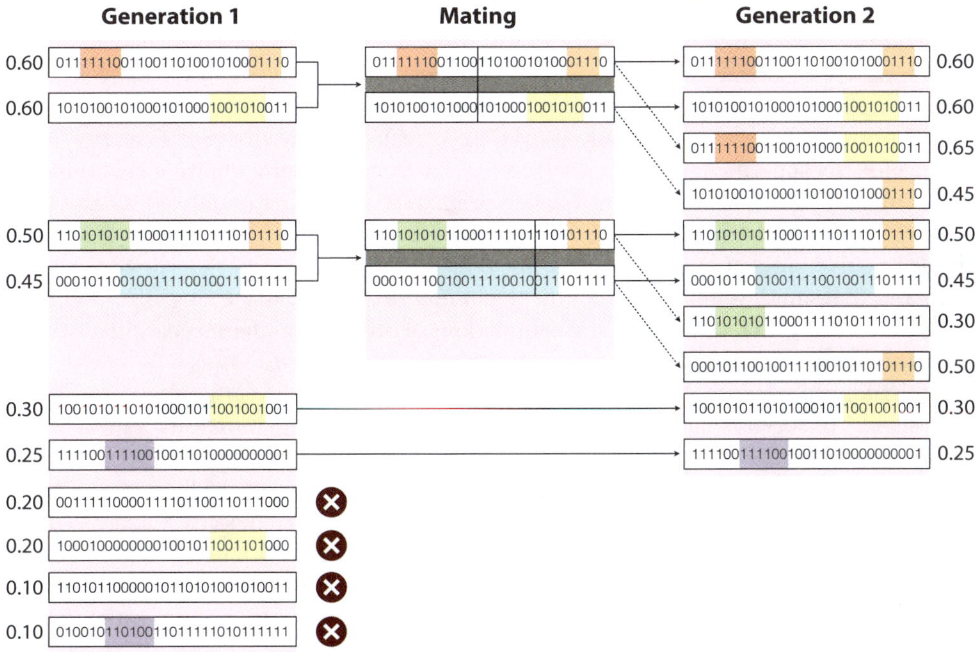

Figure 10.4 Dynamics of a genetic algorithm over one generation. Each solution is represented as a 29-bit string. The highest-performers carry over into the next generation, and some also "mate" to create new solutions. If particular clusters of parameters (indicated by the colored areas) improve the fitness of solutions, cycles of selection and mating will cause these patterns to increase in the population. Over many generations, the population will consist of increasingly high-valued solutions. Adapted from a similar diagram in Holland (1992).

Genetic algorithms can also be used to evolve complex models that exhibit interesting behaviors specified by the programmer, without the need to fit the outcomes to empirical data. For example, Axelrod and colleagues used genetic algorithms to evolve strategies in the iterated prisoner's dilemma game, wondering if something like tit-for-tat would emerge (Axelrod, 1997a). Instead, an even more savvy strategy emerged that could exploit naïve cooperators when possible but otherwise reverted to TFT. Other researchers have shown how complicated movement behaviors of realistically embodied agents can be created by evolving agents with artificial brains (i.e., simulated neural networks) whose fitness is based on their ability to produce user-specified movement goals (Sims, 1994; Reil and Massey, 2001).

Genetic algorithms tend to match or outperform simulated annealing in most complicated optimization tasks. The tradeoff is that they are also more difficult to program and require more computational power. However, genetic algorithms are also often parallelizable, since the individual solutions do not need to exchange messages while the simulations are running, making them convenient for use with computing clusters. An important caveat to using genetic algorithms is that they will produce solutions that optimize the fitness function by whatever means possible, constrained only by the constraints explicitly specified by the programmer. See Lehman et al. (2020) for an entertaining collection of anecdotes about genetic algorithms gone wrong.

10.6 Reflections

There's no one right way to analyze a model, just like there's no one right way to do science. The most important part is not to fool yourself into thinking you know more than you do. Usually, the purpose of model analysis is not so much to exactly re-create the processes that generate your data as it is to identify key relationships among the important factors. Thus, attempting "model validation" is often premature. Recall that multiple processes can generate similar patterns, but also that pattern-oriented modeling can help to surmount the equifinality problem. Even when data analysis is possible, models of social processes are still likely to be most useful as tools for the *generation*, rather than the testing, of hypotheses. More mature models, built on the firm foundations of previous model analyses, can later be used for the refinement and articulation of a hypothesis or theory.

Although integration of modeling and empirical data analysis is the ultimate goal for the sciences of social behavior, we have yet to achieve that integration on a regular basis. This is because social science is very hard. It may be that an exact science of social behavior is impossible, due to constraints on the computational power of the human mind, the feasibility of collecting the necessary data, or both. In 1960, the physicist Eugene Wigner published a thoughtful essay provocatively titled "The unreasonable effectiveness of mathematics in the natural sciences." In it, Wigner notes that it is something of a miracle that humans have discovered mathematical laws of physics that appear to be invariant in important ways. He argues that there is simply no good reason that invariants of nature should be amenable to general mathematical principles discoverable by humans. Perhaps so. But perhaps the existence of mathematical physics is an indication not so much of the majesty of physics, but of the fact that the patterns in nature that were most amenable to mathematization were those first discovered. Social science may not be amenable in the same way. This is at least partly because social scientists are interested in processes that, by definition, are *subjective*. Our explananda are typically creations of human minds, which became widespread due to their utility in colloquially explaining, predicting, and communicating the behaviors of others on scales that were local and observable. The search for formal explanations of these phenomena is challenging because reasonable people often disagree on what the important patterns even *are*.

One could, I suppose, conclude that this subjectivity implies that formal models are mostly useless for understanding social behavior, but I think it implies that formal models are absolutely essential. In explaining large-scale social phenomena that unfurl over long time spans and across disparate communities, we need scaffolds to help us think through possibilities and to help us identify the empirical questions we should even be asking. Ultimately, all knowledge must be supported by empirical evidence. But without models to guide us in our collection and interpretation of that evidence, all we have are competing narratives that enable us to talk past each other without confirming whether we even agree or disagree.

Maps and Territories

11

> The price of metaphor is eternal vigilance.
>
> —Arturo Rosenblueth and Norbert Wiener

The world is complicated, too complicated for our minds to hold all the details thereof. We make sense by simplifying. When we formalize our simplifications, we gain a prosthesis for our imaginations, an engine that churns our assumptions into consequences. But just as models illuminate the darkness of the unknown or overly complex, they also restrict our vision by shining a light in only one direction at a time. When we conceptualize a system with a model, we gain new insights by exploring how our assumptions interact, and new metaphors for explaining new scenarios. The flip side is that, unless we are careful, we may be misled by scenarios in which our model's assumptions do not hold.

Making a map is essentially an act of data compression. We ignore some details of the available data about a physical space so that we can arrange the most relevant and meaningful details in an easy-to-read format. Some models work in exactly the same way—consider the physical model of the San Francisco Bay discussed in chapter 1. For other models, we cannot be quite as sure about the relationship—the mapping—between our assumptions and the real world. Instead, we make educated guesses and postulations. Our task then is to examine the consequences of those assumptions, see whether our mapping is justified, and identify new regions of the world that must be mapped in order to get where we'd like to go.

11.1 The Map Is Not The Territory

If you are in pursuit of explanations for phenomena in the world, then you are a modeler, whether or not you like the moniker. If you are a modeler, then you will be a better modeler by being deliberate about your approach to modeling. Models partition the world into systems of interest, and arrange each particular system into a set of parts, properties, and relations that form the basis of analysis. Models, in other words, make assumptions. Analysis of a model reveals only the consequences of those assumptions, rather than any deep truth about reality. What we learn about reality from a model is dependent on the quality of the mapping between the elements of a model and the elements of reality in which we are interested.

No map is perfectly accurate. I want to highlight a couple of problems that stem from that inaccuracy. First, you may wish to ask questions that require the consideration of elements not present in your model, in which case you must revise your model in light of your new question. We can call this the problem of *inclusion*. Second, your assumptions may not represent key aspects of reality with sufficient precision, so that the conclusions you draw from your model apply to systems too different from those in the real world to be of any actionable use. We can call this the problem of *representation*. Rigorous modeling requires continual scrutiny of one's assumptions in terms of both inclusion and representation. Issues that arise from a failure to consider these problems can be illustrated by analogy to similar problems with less formalized models.

Human existence is inherently relational, and there is perhaps no relationship more central to the lives of adult humans than romantic partnerships, including marriages. Fittingly, there are also few relationships that can give rise to quite the same level of conflict. The Indian philosopher J. Krishnamurti insightfully pegs one important source of relational conflict to a failure of individuals to keep their mental models of their partners updated. In one of his dialogues, he addresses a question about people in relationships being psychologically defensive thusly:

> During the period in which I have lived with my wife. . . I have built an image about her in myself and she has built an image about me. Now these two images have a relationship—*not I with her*. So there is no direct relationship—I see this taking place, all my life it has gone on, the image building and the defending of that image, and I see that as long as I have that image about her, there must be a contradiction, though I may have a relationship with her as a wife, there is a battle going on, and if I want to live without battle, I must first be free of all images. Now, is that possible? (Krishnamurti, 1968, p. 33)

We very often base our expectations concerning how someone will behave or react on our mental representations of those people. Krishnamurti acknowledges that it is an impossible ideal to always base our expectations on the person as they are in the moment, rather than on the models of them we have built up over time. It would also be foolish to ignore all the information we have built up about a person over months or years of knowing them. However, problems arise if we do not allow our models to evolve and update with new information, or if we attempt to apply them in contexts for which they were not calibrated. Healthy relationships are built on communication, in part to give individuals the opportunity to address flaws in their mental models of their partners and to update them accordingly.

Acting with insufficient caution on the insights of flawed models can have larger and more disastrous consequences. Many researchers attribute at least some of the fallout from the 2008 global financial crisis to the limited models of human behavior used by economists, many of which assumed that financial risk was randomly distributed and that each financial event was independent of any other (Rickards, 2008; Colander et al., 2009). This has to do with the computational models used to assign risk to particular financial decisions, and how the mapping between the assumptions of the formal models and reality involved devastating mismatch. Model formalizations were inadequate to capture many factors that economists knew to be important, such as the fact that risk events are often causally linked. Models like these can easily underplay the attention one should give to rare but significant events,

Figure 11.1 Brasília.

particularly when models rely on expected values rather than sufficiently heavy-tailed outcome distributions. These flaws are nuanced and require attention to detail to discover and correct. However, sometimes a model is flawed simply because it contains assumptions that are, at their core, deeply inconsistent with human behavior.

The first half of the twentieth century saw the development of a school of architecture sometimes known as "high modernism," based on a rejection of cultural traditions in favor of scientific insights and technological advancements as an approach to urban design. One of the most influential architects of this school was the Swiss-French Charles-Édouard Jeanneret (1887–1965), also known as Le Corbusier. In his critique of high modernism, the anthropologist James Scott writes that Le Corbusier produced his designs "by stipulating an abstract, simplified human subject with certain material and physical requirements. This schematic subject required so many square meters of living space, so much fresh air, so much sunlight, so much open space, so many essential services" (Scott, 1998, p. 115). Though Le Corbusier produced plans for many urban renewal projects, he was never able to directly implement his ideas on a massive scale. This dubious honor would fall to his student, Lúcio Costa, whose designs were chosen in 1957 for the basis of the new Brazilian city of Brasília (Figure 11.1).

The city was built from scratch in the interior of the country to serve as the new seat of the federal government, and was intended to primarily house government authorities and staff. However, many of the construction staff brought their families and stayed, and many others moved there in search of jobs. The city is characterized by a sharp separation of private and public spheres. In particular, there are essentially no public squares, small parks, sidewalk cafés, or shopping corridors that serve as flexible spaces for meetings or spontaneous social interaction—the city lacks the bustle of street life that characterizes many urban centers.

Residents report feeling isolated from each other and from commerce. Visitors report finding the city difficult to navigate because of the uniformity of its architecture and lack of distinguishing landmarks. Moreover, the city plan did not account for future growth. This, along with other factors, led to the development of an "unplanned Brasília"—a set of satellite communities on the periphery of the planned city. Some of these satellites are home to wealthy residents who have shaped their environments to include some of the more convivial elements missing from the planned city. Real humans need more than an apartment and a job. We are social creatures, and we need space to live, love, work, and congregate. A model of human needs as purely physiological and professional may provide insights and may serve to guide intuitions in useful ways. However, as a basis for a city where humans live out their full lives, such a model falls flat.

Are all models necessarily flawed? By this point it should be clear that they are. All models omit. Does the adoption of models therefore doom us to make bad decisions? No. Recall that *all* decisions are made on the basis of some model, usually a verbal or mental model, because no mind contains an accurate and complete description of reality. So we all use models, it's just that only some of us can write ours down. Reliance on models for decision-making can get us into trouble when we are uncritical of the model's limitations, and when we aren't sufficiently attentive to the mapping between the model's assumptions and the aspects of our system that are likely to be important for our research or policy questions. Model building requires continual feedback, both from direct evaluation of the model's assumptions and from checks against the real-world system it is intended to represent. Models can and should guide interpretations and decisions, but doing so requires an appreciation of the nature and extent of our uncertainty, and sufficient humility to change our models as needed. And although the appropriateness of a model depends on our questions, there may not even be one right model for a given question. This is because most of the questions we can articulate about human behavior can be reframed from multiple perspectives and multiple levels of organization. To answer our question adequately, we sometimes need to account for several of these perspectives simultaneously. In other words, we need many maps to navigate the world successfully.

11.2 We Need Many Maps

Suppose you are a first-time visitor to New York City. You've lost your cell phone, but you have a subway map that shows the routes of all the train lines within the city (Figure 11.2). With this map, you can establish the general layout of the city and you can figure out which trains will get you from Washington Square in downtown Manhattan out to Montrose Avenue in Bushwick, Brooklyn. It's a very handy map, and when I lived in New York I made use of it constantly. However, if you want to go swimming in the pool at Brooklyn's McCarren Park, the map won't help you, because the park isn't on it (it's near the boundary between Williamsburg and Greenpoint). And if you want to drive a car rather than take the train, you'll have little information regarding what roads to take and which routes involve tolls. You'll be completely out of luck if you want to leave the city for a jaunt out to Long Island, because almost all of that region is omitted. Even within the boundaries of the city, the subway map can't tell you where to eat, where to buy clothes, or where the cool underground film festivals happen. That's fine, of course. The purpose of the subway map is to show you the train routes. Really living in New York City, however, requires multiple maps, whether they are printed on paper, illuminated on a smartphone screen, or stored in the memories of individual minds.

Figure 11.2 New York City subway map.

It's easier to navigate the world with a map—or better yet, with many maps—than without one, even if your map is imperfect. Formal models of the sort discussed in this book are mappings that help us understand the world of human social dynamics. Of course, none of these models presents a complete picture of social behavior. Just as a map is made with a particular purpose in mind, a model is designed to answer certain questions and not others. And much like a satisfying outing to the city requires knowing both which subway trains to take and knowledge of just where you'll go when you get off the train, understanding social dynamics involves understanding multiple aspects of your system. An industrious tourist could construct a detailed map with all the information they need for their trip's itinerary, and so a diligent modeler can incorporate aspects of multiple models into a new model designed to address their particular question. The modeling approaches I have covered in this book are intended to get you started on the task of map collecting and map building, but every domain requires additional expertise that could not practically be included here. You'll need to do some additional exploring to chart the maps you need for your own particular journey.

11.3 The Journey Continues

There is an art to modeling, and like any art form, getting good requires regular practice. There are myriad technical skills involved in doing it well, and you must continually hone those skills to build expertise. The good news is that modeling can—perhaps *should*—be fun. The modeler is a storyteller, an animator, a tinkerer, and sometimes even a magician, pulling unanticipated results out of a hat.[1] I invite you to continue your practice of modeling and to have fun doing so.

The only real alternative to formal models is to rely on ambiguous verbal models. Verbal models are often useful, particularly when a few words capture complex situations and relationships whole cloth. Scenarios involving human behaviors and institutions can be difficult to formalize, and verbal models often articulate important nuance that is lost in the formal translation. However, nuance can also stymie progress by preventing us from carefully examining all of the assumptions—or nonassumptions—implicit in our arguments. Models create common frameworks and vocabulary for working on otherwise intractable problems. All mature sciences have integrated formal modeling as key components of the scientist's toolkit. If the social, behavioral, and cognitive sciences are to reach a similar maturity, models will have to be developed that sufficiently capture their core phenomena to a satisfying degree. I am hopeful that the models in this book help move us closer to this goal, even if they don't get us all the way there.

If you feel ready to start building models of your own, the best way to get started is just to start somewhere understanding that you are going to make a lot of mistakes along the way. As you forge ahead, try to make the best models you can, but don't let the perfect be the enemy of the good. Models are analogies, and all analogies are imperfect. They apply to only some aspects of a system and not to others. Understanding a complex system often requires a family of models that cover different aspects of the system. Make a model, keeping your research questions in mind. It's OK if the model doesn't immediately match reality as you know it. If you are diligent, the model's failures to match reality can tell you as much about your system as the model's successes. Keep at it, and the next model you build will be even better. Get crackin'.

[1] Or, more commonly, a computer.

Image Credits

Figure 1.1: PA Images / Alamy Stock Photo. Figure 1.4: Army Corps of Engineers, public domain. Figure 2.2a: Pierre de Latil, La Pensee Artificielle (Gallimard, 1953). Figure 2.3: Wikimedia user Adith George 1840427, CC-BY-SA 4.0 International. Figure 3.1: Erica Fischer, CC-BY-SA 2.0. Figure 9.1: Linnea Mallette, public domain. Figure 9.7, left: Bogdan Giuşcă, CC-BY-SA 3.0 Unported. Figure 9.7, center: Wikimedia user Chris-martin, CC-BY-SA 3.0 Unported. Figure 9.7, right: Wikimedia user Booyabazooka, CC-BY-SA 3.0 Unported. Figure 10.1: Paul Smaldino. Figure 11.1: Westend61 GmbH / Alamy Stock Photo. Figure 11.2: NYC Subway Map ©Metropolitan Transportation Authority. Used with permission.

References

Acerbi, A., Mesoudi, A., and Smolla, M. (2022). *Individual-based models of cultural evolution: A step-by-step guide using R*. Routledge.

Akçay, E. (2018). Collapse and rescue of cooperation in evolving dynamic networks. *Nature Communications*, 9(1):1–9.

Akerlof, G. A. and Michaillat, P. (2018). Persistence of false paradigms in low-power sciences. *Proceedings of the National Academy of Sciences*, 115(52):13228–13233.

Aktipis, C. A. (2004). Know when to walk away: Contingent movement and the evolution of cooperation. *Journal of Theoretical Biology*, 231(2):249–260.

Apicella, C. L. and Silk, J. B. (2019). The evolution of human cooperation. *Current Biology*, 29(11):R447–R450.

Aragonès, E. and Neeman, Z. (2000). Strategic ambiguity in electoral competition. *Journal of Theoretical Politics*, 12(2):183–204.

Atkisson, C., Górski, P. J., Jackson, M. O., Hołyst, J. A., and D'Souza, R. M. (2020). Why understanding multiplex social network structuring processes will help us better understand the evolution of human behavior. *Evolutionary Anthropology*, 29(3): 102–107.

Axelrod, R. (1984). *The evolution of cooperation*. Basic Books.

Axelrod, R. (1997a). *The complexity of cooperation*. Princeton University Press.

Axelrod, R. (1997b). The dissemination of culture: A model with local convergence and global polarization. *Journal of Conflict Resolution*, 41(2):203–226.

Axelrod, R. and Hamilton, W. D. (1981). The evolution of cooperation. *Science*, 211(4489):1390–1396.

Axtell, R., Epstein, J., and Young, H. (2001). The emergence of classes in a multi-agent bargaining model. In Durlauf, S. and Young, H., editors, *Social dynamics*, pages 191–212. MIT Press.

Bakhshandeh, R., Samadi, M., Azimifar, Z., and Schaeffer, J. (2011). Degrees of separation in social networks. In *Fourth Annual Symposium on Combinatorial Search*, 2(1):18–23.

Ball, P. (2004). *Critical mass: How one thing leads to another*. Macmillan.

Bansal, S., Grenfell, B. T., and Meyers, L. A. (2007). When individual behaviour matters: Homogeneous and network models in epidemiology. *Journal of the Royal Society Interface*, 4(16):879–891.

Barabási, A.-L. (2016). *Network science*. Cambridge University Press.

Barabási, A.-L. and Albert, R. (1999). Emergence of scaling in random networks. *Science*, 286(5439):509–512.

Barnett, A. G., Zardo, P., and Graves, N. (2018). Randomly auditing research labs could be an affordable way to improve research quality: A simulation study. *PLOS ONE*, 13(4):e0195613.

Barrett, B. J. (2019). Equifinality in empirical studies of cultural transmission. *Behavioural Processes*, 161:129–138.

Bass, F. M. (1969). A new product growth for model consumer durables. *Management Science*, 15(5):215–227.

Battiston, F., Nicosia, V., Latora, V., and San Miguel, M. (2017). Layered social influence promotes multiculturality in the Axelrod model. *Scientific Reports*, 7(1):1–9.

Bedau, M. A. (1999). Can unrealistic computer models illuminate theoretical biology? In *Proceedings of the 1999 Genetic and Evolutionary Computation Conference Workshop Program*, pages 20–23.

Bedson, J., Skrip, L. A., Pedi, D., Abramowitz, S., Carter, S., Jalloh, M. F., Funk, S., Gobat, N., Giles-Vernick, T., Chowell, G., et al. (2021). A review and agenda for integrated disease models including social and behavioural factors. *Nature Human Behaviour*, 5:834–846.

Benjamin, D. J., Laibson, D., Mischel, W., Peake, P. K., Shoda, Y., Wellsjo, A. S., and Wilson, N. L. (2020). Predicting mid-life capital formation with pre-school delay of gratification and life-course measures of self-regulation. *Journal of Economic Behavior & Organization*, 179:743–756.

Berger, J. and Heath, C. (2007). Where consumers diverge from others: Identity signaling and product domains. *Journal of Consumer Research*, 34(2):121–134.

Berger, J. and Heath, C. (2008). Who drives divergence? Identity signaling, outgroup dissimilarity, and the abandonment of cultural tastes. *Journal of Personality and Social Psychology*, 95(3):593.

Bergstrom, C. T., Foster, J. G., and Song, Y. (2016). Why scientists chase big problems: Individual strategy and social optimality. *arXiv preprint arXiv:1605.05822*.

Bicchieri, C. (2005). *The grammar of society: The nature and dynamics of social norms*. Cambridge University Press.

Bishop, B. (2009). *The big sort: Why the clustering of like-minded America is tearing us apart*. Houghton Mifflin Harcourt.

Bjørnstad, O. N. (2018). *Epidemics: Models and data using R*. Springer.

Boccaletti, S., Bianconi, G., Criado, R., Del Genio, C. I., Gómez-Gardenes, J., Romance, M., Sendina-Nadal, I., Wang, Z., and Zanin, M. (2014). The structure and dynamics of multilayer networks. *Physics Reports*, 544(1):1–122.

Boyd, R. and Lorberbaum, J. P. (1987). No pure strategy is evolutionarily stable in the repeated Prisoner's Dilemma game. *Nature*, 327:58–59.

Boyd, R. and Richerson, P. J. (1985). *Culture and the evolutionary process*. University of Chicago Press.

Boyd, R. and Richerson, P. J. (1992). Punishment allows the evolution of cooperation (or anything else) in sizable groups. *Ethology and Sociobiology*, 13:171–195.

Boyd, R. and Richerson, P. J. (2002). Group beneficial norms can spread rapidly in a structured population. *Journal of Theoretical Biology*, 215(3):287–296.

Boyd, R. and Richerson, P. J. (2009). Voting with your feet: Payoff biased migration and the evolution of group beneficial behavior. *Journal of Theoretical Biology*, 257(2): 331–339.

Bramson, A., Grim, P., Singer, D. J., Berger, W. J., Sack, G., Fisher, S., Flocken, C., and Holman, B. (2017). Understanding polarization: Meanings, measures, and model evaluation. *Philosophy of Science*, 84(1):115–159.

Brand, C. O., Mesoudi, A., and Smaldino, P. E. (2021). Analogy as a catalyst for cumulative cultural evolution. *Trends in Cognitive Sciences*, 25(6):450–461.

Brantingham, P. J. (2003). A neutral model of stone raw material procurement. *American Antiquity*, 68(3):487–509.

Brembs, B. (2018). Prestigious science journals struggle to reach even average reliability. *Frontiers in Human Neuroscience*, 12:37.

Brewer, M. B. (1991). The social self: On being the same and different at the same time. *Personality and Social Psychology Bulletin*, 17(5):475–482.

Broido, A. D. and Clauset, A. (2019). Scale-free networks are rare. *Nature Communications*, 10(1):1–10.

Brooks, J., Reyes-García, V., and Burnside, W. (2018). Re-examining Balinese subaks through the lens of cultural multilevel selection. *Sustainability Science*, 13(1):35–47.

Bruch, E. and Atwell, J. (2015). Agent-based models in empirical social research. *Sociological Methods & Research*, 44(2):186–221.

Bruch, E. E. and Mare, R. D. (2006). Neighborhood choice and neighborhood change. *American Journal of Sociology*, 112(3):667–709.

Bshary, R. and Raihani, N. J. (2017). Helping in humans and other animals: A fruitful interdisciplinary dialogue. *Proceedings of the Royal Society B*, 284(1863):20170929.

Bunce, J. A. (2021). Cultural diversity in unequal societies sustained through cross-cultural competence and identity valuation. *Humanities and Social Sciences Communications*, 8:238.

Burt, R. S. (1992). *Structural holes*. Harvard University Press.

Butts, C. T. (2008). Social network analysis: A methodological introduction. *Asian Journal of Social Psychology*, 11(1):13–41.

Calcott, B. (2008). The other cooperation problem: Generating benefit. *Biology and Philosophy*, 23(2):179–203.

Camerer, C. F. (2003). *Behavioral game theory: Experiments in strategic interaction*. Princeton University Press.

Campbell, D. T. (1965). Variation and selective retention in socio-cultural evolution. In H. Barringer, G. Blanksten, and R. Mack, editors, *Social change in developing areas: A reinterpretation of evolutionary theory*, pages 19–49. Schenkman.

Campbell, D. T. (1976). Assessing the impact of planned social change. Hanover, NH: The Public Affairs Center.

Cangelosi, A. and Parisi, D. (1998). The emergence of a "language" in an evolving population of neural networks. *Connection Science*, 10(2):83–97.

Carley, K. (1991). A theory of group stability. *American Sociological Review*, 56(3):331–354.

Castellano, C., Fortunato, S., and Loreto, V. (2009). Statistical physics of social dynamics. *Reviews of Modern Physics*, 81:591–646.

Centola, D. (2010). The spread of behavior in an online social network experiment. *Science*, 329(5996):1194–1197.

Centola, D. (2018). *How behavior spreads*. Princeton University Press.

Centola, D. and Macy, M. (2007). Complex contagions and the weakness of long ties. *American Journal of Sociology*, 113(3):702–734.

Choi, J.-K. and Bowles, S. (2007). The coevolution of parochial altruism and war. *Science*, 318(5850):636–640.
Chudek, M. and Henrich, J. (2011). Culture–gene coevolution, norm-psychology and the emergence of human prosociality. *Trends in Cognitive Sciences*, 15(5):218–226.
Cialdini, R. B., Reno, R. R., and Kallgren, C. A. (1990). A focus theory of normative conduct: Recycling the concept of norms to reduce littering in public places. *Journal of Personality and Social Psychology*, 58(6):1015.
Clark, W. A. and Fossett, M. (2008). Understanding the social context of the Schelling segregation model. *Proceedings of the National Academy of Sciences*, 105(11):4109–4114.
Cohen, J. (1962). The statistical power of abnormal-social psychological research: A review. *Journal of Abnormal and Social Psychology*, 65(3):145.
Colander, D., Goldberg, M., Haas, A., Juselius, K., Kirman, A., Lux, T., and Sloth, B. (2009). The financial crisis and the systemic failure of the economics profession. *Critical Review*, 21(2–3):249–267.
Coleman, J. S. (1988). Social capital in the creation of human capital. *American Journal of Sociology*, 94:S95–S120.
Contreras Kallens, P. A., Dale, R., and Smaldino, P. E. (2018). Cultural evolution of categorization. *Cognitive Systems Research*, 52:765–774.
Couzin, I. D. (2009). Collective cognition in animal groups. *Trends in Cognitive Sciences*, 13(1):36–43.
Crooks, A., Malleson, N., E. Manley, and Heppenstall, A. (2019). *Agent-based modelling and geographical information systems: A practical primer*. Sage.
Cully, A., Clune, J., Tarapore, D., and Mouret, J.-B. (2015). Robots that can adapt like animals. *Nature*, 521(7553):503–507.
Daley, B. (2016). When baby is due, genetic counselors seen downplaying false alarms. *Boston Globe*. Accessed September 17, 2019.
Damgaard, C. and Weiner, J. (2000). Describing inequality in plant size or fecundity. *Ecology*, 81(4):1139–1142.
Dandekar, P., Goel, A., and Lee, D. T. (2013). Biased assimilation, homophily, and the dynamics of polarization. *Proceedings of the National Academy of Sciences*, 110(15):5791–5796.
Dandekar, R., Rackauckas, C., and Barbastathis, G. (2020). A machine learning-aided global diagnostic and comparative tool to assess effect of quarantine control in COVID-19 spread. *Patterns*, 1(9):100145.
Danon, L., Ford, A. P., House, T., Jewell, C. P., Keeling, M. J., Roberts, G. O., Ross, J. V., and Vernon, M. C. (2011). Networks and the epidemiology of infectious disease. *Interdisciplinary Perspectives on Infectious Diseases*, 2011.
Darwin, C. (1871). *The descent of man, and selection in relation to sex*. John Murray.
Datseris, G., Vahdati, A. R., and DuBois, T. C. (2022). Agents.jl: A performant and feature-full agent based modelling software of minimal code complexity. *Simulation*. doi:10.1177/00375497211068820.
Dawkins, R. (1976). *The selfish gene*. Oxford University Press.
De, S., Nau, D. S., and Gelfand, M. J. (2017). Understanding norm change: An evolutionary game-theoretic approach. In *Proceedings of the 16th Conference on Autonomous Agents and MultiAgent Systems*, pages 1433–1441.
de Solla Price, D. (1976). A general theory of bibliometric and other cumulative advantage processes. *Journal of the American Society for Information Science*, 27(5):292–306.
Deffuant, G., Neau, D., Amblard, F., and Weisbuch, G. (2000). Mixing beliefs among interacting agents. *Advances in Complex Systems*, 3(01n04):87–98.

DeGroot, M. H. (1974). Reaching a consensus. *Journal of the American Statistical Association*, 69(345):118–121.

Dennett, D. C. (2006). *Breaking the spell: Religion as a natural phenomenon*. Penguin.

Doebeli, M. and Hauert, C. (2005). Models of cooperation based on the Prisoner's Dilemma and the Snowdrift game. *Ecology Letters*, 8:748–766.

Drăgulescu, A. and Yakovenko, V. M. (2000). Statistical mechanics of money. *European Physical Journal B-Condensed Matter and Complex Systems*, 17(4):723–729.

Dugatkin, L. A. and Wilson, D. S. (1991). Rover: A strategy for exploiting cooperators in a patchy environment. *American Naturalist*, 138(3):687–701.

Duggins, P. (2017). A psychologically-motivated model of opinion change with applications to American politics. *Journal of Artificial Societies and Social Simulation*, 20(1):13.

Easley, D. and Kleinberg, J. (2010). *Networks, crowds, and markets*, volume 8. Cambridge University Press.

Eckles, D., Mossel, E., Rahimian, M. A., and Sen, S. (2019). Long ties accelerate noisy threshold-based contagions. *Available at SSRN 3262749*.

Efferson, C., Vogt, S., and Fehr, E. (2020). The promise and the peril of using social influence to reverse harmful traditions. *Nature Human Behaviour*, 4(1):55–68.

Eisenberg, E. M. (1984). Ambiguity as strategy in organizational communication. *Communication Monographs*, 51(3):227–242.

Enquist, M. and Leimar, O. (1993). The evolution of cooperation in mobile organisms. *Animal Behaviour*, 45(4):747–757.

Epstein, J. M. (1999). Agent-based computational models and generative social science. *Complexity*, 4(5):41–60.

Epstein, J. M. (2008). Why model? *Journal of Artificial Societies and Social Simulation*, 11(4):12.

Erdös, P. and Rényi, A. (1959). On random graphs I. *Publicationes Mathematicae*, 6:290–297.

Falandays, J. B. and Smaldino, P. E. (2022). The emergence of cultural attractors: How dynamic populations of learners achieve collective cognitive alignment. *Cognitive Science* 46:e13183.

Farine, D. R. and Whitehead, H. (2015). Constructing, conducting and interpreting animal social network analysis. *Journal of Animal Ecology*, 84(5):1144–1163.

Feynman, R. P. (1974). Cargo cult science. *Engineering and Science*, 37(7):10–13.

Fitch, W. T. (2012). *The evolution of language*. Cambridge University Press.

Flache, A. and Macy, M. W. (2011). Small worlds and cultural polarization. *Journal of Mathematical Sociology*, 35(1–3):146–176.

Flache, A., Mäs, M., Feliciani, T., Chattoe-Brown, E., Deffuant, G., Huet, S., and Lorenz, J. (2017). Models of social influence: Towards the next frontiers. *Journal of Artificial Societies and Social Simulation*, 20(4):2.

Flamson, T. J. and Bryant, G. A. (2013). Signals of humor: Encryption and laughter in social interaction. In Dynel, M., editor, *Developments in linguistic humour theory*, volume 1, pages 49–73. John Benjamins Publishing.

Foramitti, J. (2021). Agentpy: A package for agent-based modeling in Python. *Journal of Open Source Software*, 6(62):3065.

Franconeri, S., Padilla, L., Shah, P., Zacks, J., and Hullman, J. (2021). The science of visual data communication: What works. *Psychological Science in the Public Interest*, 22(3):110–161.

Frank, S. A. (2009). The common patterns of nature. *Journal of Evolutionary Biology*, 22(8):1563–1585.

Franz, A., Hoffmann, K. H., and Salamon, P. (2001). Best possible strategy for finding ground states. *Physical Review Letters*, 86(23):5219.

Friedkin, N. E., Proskurnikov, A. V., Tempo, R., and Parsegov, S. E. (2016). Network science on belief system dynamics under logic constraints. *Science*, 354(6310):321–326.

Funk, S., Salathé, M., and Jansen, V. A. (2010). Modelling the influence of human behaviour on the spread of infectious diseases: A review. *Journal of the Royal Society Interface*, 7(50):1247–1256.

Galesic, M., Olsson, H., Dalege, J., van der Does, T., and Stein, D. L. (2021). Integrating social and cognitive aspects of belief dynamics: Towards a unifying framework. *Journal of the Royal Society Interface*, 18(176):20200857.

Gardner, M. (1970). The fantastic combinations of John Conway's new solitaire game, LIFE. *Scientific American*, October:120–123.

Gavrilets, S. (2020). The dynamics of injunctive social norms. *Evolutionary Human Sciences*, 2:E60.

Gelman, A. and Carlin, J. (2014). Beyond power calculations: Assessing type S (sign) and type M (magnitude) errors. *Perspectives on Psychological Science*, 9(6): 641–651.

Gerdelan, A. P. (2010). Behavioural AI in movies and games: The last 20 years. Technical Report CSTN-105, Computer Science, Massey University, Albany, North Shore 102-904, Auckland, New Zealand.

Gerkey, D. (2013). Cooperation in context: Public goods games and post-Soviet collectives in Kamchatka, Russia. *Current Anthropology*, 54(2):144–176.

Gigerenzer, G. (2018). Statistical rituals: The replication delusion and how we got there. *Advances in Methods and Practices in Psychological Science*, 1(2):198–218.

Gigerenzer, G. and Selten, R. (2002). *Bounded rationality: The adaptive toolbox*. MIT Press.

Gleick, J. (1987). *Chaos: Making a new science*. Penguin.

Gleick, J. (2004). *Isaac Newton*. Vintage.

Goldberg, A. and Stein, S. K. (2018). Beyond social contagion: Associative diffusion and the emergence of cultural variation. *American Sociological Review*, 83(5):897–932.

Golman, R., Bugbee, E. H., Jain, A., and Saraf, S. (2021). Hipsters and the cool: A game theoretic analysis of identity expression, trends, and fads. *Psychological Review*, 129(1): 4–17.

Golub, B. and Jackson, M. O. (2012). How homophily affects the speed of learning and best-response dynamics. *Quarterly Journal of Economics*, 127(3):1287–1338.

Goodman, N. D. and Frank, M. C. (2016). Pragmatic language interpretation as probabilistic inference. *Trends in Cognitive Sciences*, 20(11):818–829.

Gopnik, A., Griffiths, T. L., and Lucas, C. G. (2015). When younger learners can be better (or at least more open-minded) than older ones. *Current Directions in Psychological Science*, 24(2):87–92.

Grafen, A. (1984). Natural selection, kin selection and group selection. In Krebs, J. and Davies, N. B., editors, *Behavioural ecology: An evolutionary approach*, pages 62–84. Blackwell.

Granovetter, M. (1978). Threshold models of collective behavior. *American Journal of Sociology*, 83(6):1420–1443.

Granovetter, M. S. (1973). The strength of weak ties. *American Journal of Sociology*, 78(6):1360–1380.

Grimm, V., Berger, U., DeAngelis, D. L., Polhill, J. G., Giske, J., and Railsback, S. F. (2010). The ODD protocol: A review and first update. *Ecological Modelling*, 221(23):2760–2768.

Grimm, V., Railsback, S. F., Vincenot, C. E., Berger, U., Gallagher, C., DeAngelis, D. L., Edmonds, B., Ge, J., Giske, J., Groeneveld, J., et al. (2020). The ODD protocol for describing agent-based and other simulation models: A second update to improve clarity, replication, and structural realism. *Journal of Artificial Societies and Social Simulation*, 23(2):7.

Grimm, V., Revilla, E., Berger, U., Jeltsch, F., Mooij, W. M., Railsback, S. F., Thulke, H.-H., Weiner, J., Wiegand, T., and DeAngelis, D. L. (2005). Pattern-oriented modeling of agent-based complex systems: Lessons from ecology. *Science*, 310(5750):987–991.

Hamilton, W. D. (1964). The genetical evolution of social behaviour. *Journal of Theoretical Biology*, 7(1):1–52.

Hammond, R. A. and Axelrod, R. (2006). The evolution of ethnocentrism. *Journal of Conflict Resolution*, 50(6):926–936.

Hatna, E. and Benenson, I. (2012). The Schelling model of ethnic residential dynamics: Beyond the integrated-segregated dichotomy of patterns. *Journal of Artificial Societies and Social Simulation*, 15(1):6.

Healy, K. (2017). Fuck nuance. *Sociological Theory*, 35(2):118–127.

Healy, K. (2019). *Data visualization: A practical introduction*. Princeton University Press.

Hegselmann, R. and Krause, U. (2002). Opinion dynamics and bounded confidence models, analysis, and simulation. *Journal of Artificial Societies and Social Simulation*, 5(3):2.

Henrich, J. (2004). Cultural group selection, coevolutionary processes and large-scale cooperation. *Journal of Economic Behavior & Organization*, 53(1):3–35.

Henrich, J. and Boyd, R. (2008). Division of labor, economic specialization, and the evolution of social stratification. *Current Anthropology*, 49(4):715–724.

Henrich, J., Boyd, R., Bowles, S., Camerer, C., Fehr, E., Gintis, H., McElreath, R., Alvard, M., Barr, A., Ensminger, J., et al. (2005). "Economic man" in cross-cultural perspective: Behavioral experiments in 15 small-scale societies. *Behavioral and Brain Sciences*, 28(6):795–815.

Henrich, J. and Gil-White, F. J. (2001). The evolution of prestige: Freely conferred deference as a mechanism for enhancing the benefits of cultural transmission. *Evolution and Human Behavior*, 22(3):165–196.

Herman, E. S. and Chomsky, N. (1988). *Manufacturing consent: The political economy of the mass media*. Random House.

Higginson, A. D. and Munafò, M. R. (2016). Current incentives for scientists lead to underpowered studies with erroneous conclusions. *PLoS Biology*, 14(11):e2000995.

Hills, T. and Todd, P. (2008). Population heterogeneity and individual differences in an assortative agent-based marriage and divorce model (madam) using search with relaxing expectations. *Journal of Artificial Societies and Social Simulation*, 11(4):5.

Hoffman, M., Hilbe, C., and Nowak, M. A. (2018). The signal-burying game can explain why we obscure positive traits and good deeds. *Nature Human Behaviour*, 2(6):397–404.

Hoffman, P. (1998). *The man who loved only numbers: The story of Paul Erdös and the search for mathematical truth*. Hachette Books.

Hofstadter, D. R. (1979). *Gödel, Escher, Bach: An eternal golden braid*. Basic Books.

Holland, J. H. (1992). Genetic algorithms. *Scientific American*, 267:66–73.

Holland, P. W., Laskey, K. B., and Leinhardt, S. (1983). Stochastic blockmodels: First steps. *Social Networks*, 5(2):109–137.

Holman, B. and Bruner, J. (2017). Experimentation by industrial selection. *Philosophy of Science*, 84(5):1008–1019.

Hooper, P. L., Kaplan, J. S., and Boone, J. L. (2010). A theory of leadership in human cooperative groups. *Journal of Theoretical Biology*, 265:633–646.

Hopfield, J. J. (1982). Neural networks and physical systems with emergent collective computational abilities. *Proceedings of the National Academy of Sciences*, 79(8):2554–2558.

Huberman, B. A. and Glance, N. S. (1993). Evolutionary games and computer simulations. *Proceedings of the National Academy of Sciences*, 90(16):7716–7718.

Hull, D. L. (1988). *Science as a process: An evolutionary account of the social and conceptual development of science*. University of Chicago Press.

Iannaccone, L. R. and Makowsky, M. D. (2007). Accidental atheists? Agent-based explanations for the persistence of religious regionalism. *Journal for the Scientific Study of Religion*, 46(1):1–16.

Ilany, A. and Akçay, E. (2016a). Personality and social networks: A generative model approach. *Integrative and Comparative Biology*, 56(6):1197–1205.

Ilany, A. and Akçay, E. (2016b). Social inheritance can explain the structure of animal social networks. *Nature Communications*, 7(1):1–10.

Ioannidis, J. P. (2005). Why most published research findings are false. *PLOS Medicine*, 2(8):e124.

Jones, J. H. and Handcock, M. S. (2003). An assessment of preferential attachment as a mechanism for human sexual network formation. *Proceedings of the Royal Society B*, 270(1520):1123–1128.

Kalick, S. M. and Hamilton, T. E. (1986). The matching hypothesis reexamined. *Journal of Personality and Social Psychology*, 51(4):673.

Karrer, B. and Newman, M. E. (2011). Stochastic blockmodels and community structure in networks. *Physical Review E*, 83(1):016107.

Kauffman, S. A. (1971). Articulation of parts explanation in biology and the rational search for them. In Buck, R. C. and Cohen, R. S., editors, *PSA 1970*, pages 257–272. Philosophy of Science Association.

Kazil, J., Masad, D., and Crooks, A. (2020). Utilizing Python for agent-based modeling: The Mesa framework. In *International Conference on Social Computing, Behavioral-Cultural Modeling and Prediction and Behavior Representation in Modeling and Simulation*, pages 308–317. Springer.

Keeling, M. J. and Eames, K. T. (2005). Networks and epidemic models. *Journal of the Royal Society Interface*, 2(4):295–307.

Keeling, M. J. and Rohani, P. (2008). *Modeling infectious diseases in humans and animals*. Princeton University Press.

Kendal, J., Tehrani, J. J., and Odling-Smee, J. (2011). Human niche construction in interdisciplinary focus. *Philosophical Transactions of the Royal Society B*, 366(1566):785–792.

Kendal, R. L., Boogert, N. J., Rendell, L., Laland, K. N., Webster, M., and Jones, P. L. (2018). Social learning strategies: Bridge-building between fields. *Trends in Cognitive Sciences*, 22(7):651–665.

Kimura, M. and Ohta, T. (1969). The average number of generations until fixation of a mutant gene in a finite population. *Genetics*, 61(3):763.

Kirby, S. (2017). Culture and biology in the origins of linguistic structure. *Psychonomic Bulletin & Review*, 24(1):118–137.

Kish Bar-On, K. and Lamm, E. (2023). The interplay of social identity and norm psychology in the evolution of human groups. *Philosophical Transactions of the Royal Society B*, 378: 2021–412.

Kleon, A. (2012). *Steal like an artist: 10 things nobody told you about being creative*. Workman Publishing.

Koella, J. C. (2000). The spatial spread of altruism versus the evolutionary response of egoists. *Proceedings of the Royal Society of London. Series B: Biological Sciences*, 267(1456):1979–1985.

Krishnamurti, J. (1968). *Talks and dialogues*. Avon Books.
Kuran, T. (1995). *Private truths, public lies*. Harvard University Press.
Laland, K. N. (2004). Social learning strategies. *Learning & Behavior*, 32(1):4–14.
Laland, K. N. (2017). *Darwin's unfinished symphony: How culture made the human mind*. Princeton University Press.
Lancichinetti, A. and Fortunato, S. (2009). Community detection algorithms: A comparative analysis. *Physical Review E*, 80(5):056117.
Lazer, D. and Friedman, A. (2007). The network structure of exploration and exploitation. *Administrative Science Quarterly*, 52(4):667–694.
Legare, C. H. and Nielsen, M. (2015). Imitation and innovation: The dual engines of cultural learning. *Trends in Cognitive Sciences*, 19(11):688–699.
Lehman, J., Clune, J., Misevic, D., Adami, C., Altenberg, L., Beaulieu, J., Bentley, P. J., Bernard, S., Beslon, G., Bryson, D. M., et al. (2020). The surprising creativity of digital evolution: A collection of anecdotes from the evolutionary computation and artificial life research communities. *Artificial Life*, 26(2):274–306.
Lewis, D. (1969). *Convention: A philosophical study*. Harvard University Press.
Ligmann-Zielinska, A., Siebers, P.-O., Magliocca, N., Parker, D. C., Grimm, V., Du, J., Cenek, M., Radchuk, V., Arbab, N. N., Li, S., et al. (2020). 'One size does not fit all': A roadmap of purpose-driven mixed-method pathways for sensitivity analysis of agent-based models. *Journal of Artificial Societies and Social Simulation*, 23(1):6.
Lord, C. G., Ross, L., and Lepper, M. R. (1979). Biased assimilation and attitude polarization: The effects of prior theories on subsequently considered evidence. *Journal of Personality and Social Psychology*, 37(11):2098.
Luce, R. D. and Raiffa, H. (1957). *Games and decisions: Introduction and critical survey*. Courier Corporation.
Luke, S., Cioffi-Revilla, C., Panait, L., Sullivan, K., and Balan, G. (2005). Mason: A multiagent simulation environment. *Simulation*, 81(7):517–527.
Lynn, S. K. and Barrett, L. F. (2014). "Utilizing" signal detection theory. *Psychological Science*, 25(9):1663–1673.
Makowsky, M. D. and Smaldino, P. E. (2016). The evolution of power and the divergence of cooperative norms. *Journal of Economic Behavior & Organization*, 126:75–88.
Mark, N. (1998). Beyond individual differences: Social differentiation from first principles. *American Sociological Review*, 63(3):309–330.
Mason, L. (2018). *Uncivil agreement: How politics became our identity*. University of Chicago Press.
Matsuda, H., Ogita, N., Sasaki, A., and Satō, K. (1992). Statistical mechanics of population: The lattice Lotka-Volterra model. *Progress of Theoretical Physics*, 88(6):1035–1049.
May, C. J., Schank, J. C., Joshi, S., Tran, J., Taylor, R., and Scott, I.-E. (2006). Rat pups and random robots generate similar self-organized and intentional behavior. *Complexity*, 12(1):53–66.
Maynard Smith, J. (1964). Group selection and kin selection. *Nature*, 201(4924):1145–1147.
Maynard Smith, J. (1982). *Evolution and the theory of games*. Cambridge University Press.
Maynard Smith, J. and Price, G. R. (1973). The logic of animal conflict. *Nature*, 246:15–18.
Maynard Smith, J. and Szathmary, E. (1997). *The major transitions in evolution*. Oxford University Press.
McElreath, R. (2020). *Statistical rethinking: A Bayesian course with examples in R and Stan*. Chapman and Hall/CRC.
McElreath, R. and Boyd, R. (2007). *Mathematical models of social evolution: A guide for the perplexed*. University of Chicago Press.

McElreath, R., Boyd, R., and Richerson, P. J. (2003). Shared norms and the evolution of ethnic markers. *Current Anthropology*, 44(1):122–130.

McElreath, R. and Smaldino, P. E. (2015). Replication, communication, and the population dynamics of scientific discovery. *PLOS ONE*, 10(8):e0136088.

Meehl, P. E. (1967). Theory-testing in psychology and physics: A methodological paradox. *Philosophy of Science*, 34(2):103–115.

Mehr, S. A., Krasnow, M. M., Bryant, G. A., and Hagen, E. H. (2021). Origins of music in credible signaling. *Behavioral and Brain Sciences*, 44:E60.

Mehta, R. S. and Rosenberg, N. A. (2020). Modelling anti-vaccine sentiment as a cultural pathogen. *Evolutionary Human Sciences*, 2:E21.

Menczer, F., Fortunato, S., and Davis, C. A. (2020). *A first course in network science*. Cambridge University Press.

Miller, J. H. and Page, S. (2007). *Complex adaptive systems: An introduction to computational models of social life*. Princeton University Press.

Minocher, R., Atmaca, S., Bavero, C., McElreath, R., and Beheim, B. (2021). Estimating the reproducibility of social learning research published between 1955 and 2018. *Royal Society Open Science*, 8(9):210450.

Mischel, W. and Ebbesen, E. B. (1970). Attention in delay of gratification. *Journal of Personality and Social Psychology*, 16(2):329–337.

Moscato, P. and Fontanari, J. F. (1990). Stochastic versus deterministic update in simulated annealing. *Physics Letters A*, 146(4):204–208.

Moser, C., Ackerman, J., Dayer, A., Proksch, S., and Smaldino, P. E. (2021). Why don't cockatoos have war songs? *Behavioral and Brain Sciences*, 44:E108.

Muldoon, R., Smith, T., and Weisberg, M. (2012). Segregation that no one seeks. *Philosophy of Science*, 79(1):38–62.

Murray, J. D. (1989). *Mathematical biology*. Springer.

Muthukrishna, M. and Henrich, J. (2019). A problem in theory. *Nature Human Behaviour*, 3(3):221–229.

Muthukrishna, M. and Schaller, M. (2020). Are collectivistic cultures more prone to rapid transformation? Computational models of cross-cultural differences, social network structure, dynamic social influence, and cultural change. *Personality and Social Psychology Review*, 24(2):103–120.

Nakamaru, M. and Levin, S. A. (2004). Spread of two linked social norms on complex interaction networks. *Journal of Theoretical Biology*, 230(1):57–64.

Neal, Z. P. (2017). How small is it? Comparing indices of small worldliness. *Network Science*, 5(1):30–44.

Newman, M. (2018). *Networks: An introduction*. Oxford University Press.

Newman, M. E. (2003). The structure and function of complex networks. *SIAM Review*, 45(2):167–256.

Niles, M. T., Schimanski, L. A., McKiernan, E. C., and Alperin, J. P. (2020). Why we publish where we do: Faculty publishing values and their relationship to review, promotion and tenure expectations. *PLOS ONE*, 15(3):e0228914.

Nissen, S. B., Magidson, T., Gross, K., and Bergstrom, C. T. (2016). Publication bias and the canonization of false facts. *eLife*, 5:e21451.

Noë, R. and Hammerstein, P. (1994). Biological markets: Supply and demand determine the effect of partner choice in cooperation, mutualism and mating. *Behavioral Ecology and Sociobiology*, 35(1):1–11.

Norenzayan, A. (2013). *Big gods: How religion transformed cooperation and conflict*. Princeton University Press.

North, D. C. (1990). *Institutions, institutional change and economic performance*. Cambridge University Press.

North, M. J. (2014). A theoretical formalism for analyzing agent-based models. *Complex Adaptive Systems Modeling*, 2(1):1–34.

Nowak, M. and Sigmund, K. (1993). A strategy of win-stay, lose-shift that outperforms tit-for-tat in the Prisoner's Dilemma game. *Nature*, 364(6432):56–58.

Nowak, M. A., Bonhoeffer, S., and May, R. M. (1994). Spatial games and the maintenance of cooperation. *Proceedings of the National Academy of Sciences*, 91(11):4877–4881.

Nowak, M. A. and May, R. M. (1992). Evolutionary games and spatial chaos. *Nature*, 359(6398):826–829.

O'Connor, C. (2019a). The natural selection of conservative science. *Studies in History and Philosophy of Science Part A*, 76:24–29.

O'Connor, C. (2019b). *The origins of unfairness: Social categories and cultural evolution*. Oxford University Press.

O'Connor, C. and Weatherall, J. O. (2019). *The misinformation age: How false beliefs spread*. Yale University Press.

Odling-Smee, F. J., Laland, K. N., and Feldman, M. W. (2003). *Niche construction: The neglected process in evolution*. Princeton University Press.

Okasha, S. (2006). *Evolution and the levels of selection*. Oxford University Press.

Okasha, S. (2016). The relation between kin and multilevel selection: An approach using causal graphs. *British Journal for the Philosophy of Science*, 67(2):435–470.

Oreskes, N. (2019). *Why trust science?* Princeton University Press.

Ostrom, E., Gardner, R., and Walker, J. (1994). *Rules, games, and commonpool resources*. University of Michigan Press.

Page, S. E. (2018). *The model thinker: What you need to know to make data work for you*. Basic Books.

Pennisi, E. (2018). The momentous transition to multicellular life may not have been so hard after all. *Science*.

Pepper, J. W. (2007). Simple models of assortment through environmental feedback. *Artificial Life*, 13(1):1–9.

Perc, M., Gómez-Gardenes, J., Szolnoki, A., Floría, L. M., and Moreno, Y. (2013). Evolutionary dynamics of group interactions on structured populations: A review. *Journal of the Royal Society Interface*, 10(80):20120997.

Perreault, C., Moya, C., and Boyd, R. (2012). A Bayesian approach to the evolution of social learning. *Evolution and Human Behavior*, 33(5):449–459.

Plutynski, A. (2009). The modern synthesis. In *Routledge encyclopedia of philosophy*. Routledge.

Poincaré, H. (1905). *Science and hypothesis*. Walter Scott Publishing Co.

Popper, K. (1963). *Conjectures and refutations: The growth of scientific knowledge*. Routledge.

Popper, K. (1979). *Objective knowledge: An evolutionary approach*. Oxford University Press.

Puglisi, A., Baronchelli, A., and Loreto, V. (2008). Cultural route to the emergence of linguistic categories. *Proceedings of the National Academy of Sciences*, 105(23): 7936–7940.

Rabb, N., Cowen, L., de Ruiter, J. P., and Scheutz, M. (2022). Cognitive cascades: How to model (and potentially counter) the spread of fake news. *PLOS ONE*, 17(1): e0261811.

Raihani, N. (2021). *The social instinct: How cooperation shaped the world*. MacMillan.

Reil, T. and Massey, C. (2001). Biologically inspired control of physically simulated bipeds. *Theory in Biosciences*, 120(3-4):327–339.

Reynolds, C. W. (1987). Flocks, herds and schools: A distributed behavioral model. In *Proceedings of the 14th Annual Conference on Computer Graphics and Interactive Techniques*, pages 25–34.

Richerson, P., Baldini, R., Bell, A. V., Demps, K., Frost, K., Hillis, V., Mathew, S., Newton, E. K., Naar, N., Newson, L., et al. (2016). Cultural group selection plays an essential role in explaining human cooperation: A sketch of the evidence. *Behavioral and Brain Sciences*, 39:E30.

Richerson, P. J. and Boyd, R. (2005). *Not by genes alone: How culture transformed human evolution*. University of Chicago Press.

Rickards, J. G. (2008). A mountain, overlooked: How risk models failed Wall St. and Washington. *Washington Post* October, 2, 2008, page A23.

Ritchie, S. (2020). *Science fictions: How fraud, bias, negligence, and hype undermine the search for truth*. Metropolitan Books.

Roberto, E. (2016). The divergence index: A decomposable measure of segregation and inequality. *arXiv preprint arXiv:1508.01167*.

Roccas, S. and Brewer, M. B. (2002). Social identity complexity. *Personality and Social Psychology Review*, 6(2):88–106.

Rogers, E. M. (2003). *Diffusion of innovations*, 5th ed. Free Press.

Rossi, J. S. (1990). Statistical power of psychological research: What have we gained in 20 years? *Journal of Consulting and Clinical Psychology*, 58(5):646.

Salathé, M. and Jones, J. H. (2010). Dynamics and control of diseases in networks with community structure. *PLOS Computational Biology*, 6(4):e1000736.

Sander, L., Warren, C., Sokolov, I., Simon, C., and Koopman, J. (2002). Percolation on heterogeneous networks as a model for epidemics. *Mathematical Biosciences*, 180: 293–305.

Santos, F. C., Pacheco, J. M., and Lenaerts, T. (2006a). Cooperation prevails when individuals adjust their social ties. *PLOS Computational Biology*, 2(10):e140.

Santos, F. C., Pacheco, J. M., and Lenaerts, T. (2006b). Evolutionary dynamics of social dilemmas in structured heterogeneous populations. *Proceedings of the National Academy of Sciences*, 103(9):3490–3494.

Savage, P. E., Loui, P., Tarr, B., Schachner, A., Glowacki, L., Mithen, S., and Fitch, W. T. (2021). Music as a coevolved system for social bonding. *Behavioral and Brain Sciences*, 44:E59.

Schank, J. C. (2008). The development of locomotor kinematics in neonatal rats: An agent-based modeling analysis in group and individual contexts. *Journal of Theoretical Biology*, 254(4):826–842.

Schelling, T. C. (1971). Dynamic models of segregation. *Journal of Mathematical Sociology*, 1(2):143–186.

Schelling, T. C. (1978). *Micromotives and macrobehavior*. W. W. Norton & Company.

Scott, J. and Carrington, P. J. (2011). *The SAGE handbook of social network analysis*. SAGE Publications.

Scott, J. C. (1998). *Seeing like a state: How certain schemes to improve the human condition have failed*. Yale University Press.

Sedlmeier, P. and Gigerenzer, G. (1989). Do studies of statistical power have an effect on the power of studies? *Psychological Bulletin*, 105(2):309–316.

Serra-Garcia, M. and Gneezy, U. (2021). Nonreplicable publications are cited more than replicable ones. *Science Advances*, 7(21):eabd1705.

Silvetti, M. and Verguts, T. (2012). Reinforcement learning, high-level cognition, and the human brain. In *Neuroimaging: Cognitive and clinical neuroscience*, pages 283–96.

Simon, H. A. (1955). On a class of skew distribution functions. *Biometrika*, 42:425–440.

Simon, H. A. (1963). The architecture of complexity. *Proceedings of the American Philosophical Society*, 106(6):467–482.

Simon, H. A. (1972). Theories of bounded rationality. *Decision and Organization*, 1(1): 161–176.

Sims, K. (1994). Evolving virtual creatures. In *Proceedings of the 21st Annual Conference on Computer Graphics and Interactive Techniques*, pages 15–22.

Skinner, B. F. (1981). Selection by consequences. *Science*, 213(4):501–504.

Skyrms, B. (2003). *The stag hunt and the evolution of social structure*. Cambridge University Press.

Smaldino, P. E. and O'Connor, C. (2020). Interdisciplinarity can aid the spread of better methods between scientific communities. *Collective Intelligence*, 1(2).

Smaldino, P., Pickett, C., Sherman, J., and Schank, J. (2012). An agent-based model of social identity dynamics. *Journal of Artificial Societies and Social Simulation*, 15(4):7.

Smaldino, P. E. (2014). The cultural evolution of emergent group-level traits. *Behavioral and Brain Sciences*, 37:243–295.

Smaldino, P. E. (2017). Models are stupid, and we need more of them. In Vallacher, R. R., Read, S. J., and Nowak, A., editors, *Computational social psychology*, pages 311–331. Routledge.

Smaldino, P. E. (2019). Social identity and cooperation in cultural evolution. *Behavioural Processes*, 161:108–116.

Smaldino, P. E. (2020). How to translate a verbal theory into a formal model. *Social Psychology*, 51(4):207–218.

Smaldino, P. E. (2022). Models of identity signaling. *Current Directions in Psychological Science*, 31(3):231–237.

Smaldino, P. E., Aplin, L. M., and Farine, D. R. (2018a). Sigmoidal acquisition curves are good indicators of conformist transmission. *Scientific Reports*, 8(1):1–10.

Smaldino, P. E., D'Souza, R. M., and Maoz, Z. (2018b). Resilience by structural entrenchment: Dynamics of single-layer and multiplex networks following sudden changes to tie costs. *Network Science*, 6(2):157–175.

Smaldino, P. E. and Epstein, J. M. (2015). Social conformity despite individual preferences for distinctiveness. *Royal Society Open Science*, 2(3):140437.

Smaldino, P. E., Flamson, T. J., and McElreath, R. (2018c). The evolution of covert signaling. *Scientific Reports*, 8(1):1–10.

Smaldino, P. E., Janssen, M. A., Hillis, V., and Bednar, J. (2017). Adoption as a social marker: Innovation diffusion with outgroup aversion. *Journal of Mathematical Sociology*, 41(1):26–45.

Smaldino, P. E. and Jones, J. H. (2021). Coupled dynamics of behaviour and disease contagion among antagonistic groups. *Evolutionary Human Sciences*, 3:E28.

Smaldino, P. E. and Lubell, M. (2011). An institutional mechanism for assortment in an ecology of games. *PLOS ONE*, 6(8):e23019.

Smaldino, P. E. and Lubell, M. (2014). Institutions and cooperation in an ecology of games. *Artificial Life*, 20(2):207–221.

Smaldino, P. E., Lukaszewski, A., von Rueden, C., and Gurven, M. (2019a). Niche diversity can explain cross-cultural differences in personality structure. *Nature Human Behaviour*, 3(12):1276–1283.

Smaldino, P. E. and McElreath, R. (2016). The natural selection of bad science. *Royal Society Open Science*, 3(9):160384.

Smaldino, P. E., Palagi, E., Burghardt, G. M., and Pellis, S. M. (2019b). The evolution of two types of play. *Behavioral Ecology*, 30(5):1388–1397.

Smaldino, P. E. and Schank, J. C. (2012a). Human mate choice is a complex system. *Complexity*, 17(5):11–22.

Smaldino, P. E. and Schank, J. C. (2012b). Movement patterns, social dynamics, and the evolution of cooperation. *Theoretical Population Biology*, 82(1):48–58.

Smaldino, P. E., Schank, J. C., and McElreath, R. (2013). Increased costs of cooperation help cooperators in the long run. *American Naturalist*, 181(4):451–463.

Smaldino, P. E. and Turner, M. A. (2022). Covert signaling is an adaptive communication strategy in diverse populations. *Psychological Review*, 129(4):812–829.

Smaldino, P. E., Turner, M. A., and Contreras Kallens, P. A. (2019c). Open science and modified funding lotteries can impede the natural selection of bad science. *Royal Society Open Science*, 6(7):190194.

Smith, E. R. and Conrey, F. R. (2007). Agent-based modeling: A new approach for theory building in social psychology. *Personality and Social Psychology Review*, 11(1):87–104.

Smith?, R. (2014). *Mathematical modelling of zombies*. University of Ottawa Press.

Solomon, C., Harvey, B., Kahn, K., Lieberman, H., Miller, M. L., Minsky, M., Papert, A., and Silverman, B. (2020). History of logo. *Proc. ACM Program. Lang.*, 4(HOPL):1–66.

Spaniel, W. (2014). *Game theory 101: The complete textbook*. CreateSpace.

Spence, M. (1973). Job market signaling. *Quarterly Journal of Economics*, 87(3):355–374.

Sperber, D. and Wilson, D. (1995). *Relevance: Communication and cognition*. Blackwell.

Stewart, A. J. and Plotkin, J. B. (2021). The natural selection of good science. *Nature Human Behaviour*, 5:1510–1518.

Strogatz, S. H. (2015). *Nonlinear dynamics and chaos*. CRC press.

Stryker, S. and Burke, P. J. (2000). The past, present, and future of an identity theory. *Social Psychology Quarterly*, 63(4):284–297.

Sugden, R. (1986). *The economics of rights, cooperation and welfare*. Blackwell.

Sumpter, D. J. (2006). The principles of collective animal behaviour. *Philosophical Transactions of the Royal Society B: Biological Sciences*, 361(1465):5–22.

Sumpter, D. J. (2010). *Collective animal behavior*. Princeton University Press.

Sutton, R. S. and Barto, A. G. (2018). *Reinforcement learning: An introduction*. MIT Press.

Szucs, D. and Ioannidis, J. P. (2017). Empirical assessment of published effect sizes and power in the recent cognitive neuroscience and psychology literature. *PLOS Biology*, 15(3):e2000797.

Taibbi, M. (2019). *Hate Inc.: Why today's media makes us despise one another*. OR Books.

Tauber, S., Navarro, D. J., Perfors, A., and Steyvers, M. (2017). Bayesian models of cognition revisited: Setting optimality aside and letting data drive psychological theory. *Psychological Review*, 124(4):410–441.

Tiokhin, L., Yan, M., and Morgan, T. J. (2021). Competition for priority harms the reliability of science, but reforms can help. *Nature Human Behaviour*, 5:857–867.

Todd, P. M. and Gigerenzer, G., (eds.) (2012). *Ecological rationality: Intelligence in the world*. Oxford University Press.

Tooby, J. and Cosmides, L. (1996). Friendship and the banker's paradox: Other pathways to the evolution of adaptations for altruism. *Proceedings of the British Academy*, 88:119–143.

Touboul, J. D. (2019). The hipster effect: When anti-conformists all look the same. *Discrete & Continuous Dynamical Systems-B*, 24(8):4379.

Travers, J. and Milgram, S. (1969). An exploratory study of the small world problem. *Sociometry*, 32:425–43.

Trivers, R. L. (1971). The evolution of reciprocal altruism. *Quarterly Review of Biology*, 46(1):35–57.

Turchin, P. (2003). *Historical dynamics*. Princeton University Press.

Turchin, P. (2015). *Ultrasociety: How 10,000 years of war made humans the greatest cooperators on earth*. Beresta Books.

Turchin, P. (2016). *Ages of discord*. Beresta Books.

Turner, M. A. and Smaldino, P. E. (2018). Paths to polarization: How extreme views, miscommunication, and random chance drive opinion dynamics. *Complexity*, 2740959.

Turner, M. A. and Smaldino, P. E. (2020). Stubborn extremism as a potential pathway to group polarization. In *Proceedings of the 42nd Annual Conference of the Cognitive Science Society*, pages 866–872.

von Uexküll, J. (1909). *Umwelt und innenwelt der tiere*. Springer.

Vonnegut, K. (1987). *Bluebeard*. Delacorte Press.

Vu, T. M., Probst, C., Epstein, J. M., Brennan, A., Strong, M., and Purshouse, R. C. (2019). Toward inverse generative social science using multi-objective genetic programming. In *Proceedings of the Genetic and Evolutionary Computation Conference*, pages 1356–1363.

Wagenmakers, E.-J., Wetzels, R., Borsboom, D., and Van Der Maas, H. L. (2011). Why psychologists must change the way they analyze their data: The case of psi: Comment on Bem (2011). *Journal of Personality and Social Psychology*, 100:426–432.

Wasserstein, R. L. and Lazar, N. A. (2016). The ASA statement on p-values: Context, process, and purpose. *American Statistician* 70(2):129–133.

Watts, D. J. (1999). Networks, dynamics, and the small-world phenomenon. *American Journal of Sociology*, 105(2):493–527.

Watts, D. J. and Strogatz, S. H. (1998). Collective dynamics of 'small-world' networks. *Nature*, 393(6684):440–442.

Weatherall, J. O. and O'Connor, C. (2021). Conformity in scientific networks. *Synthese*, 198:7257–7278.

Weatherall, J. O., O'Connor, C., and Bruner, J. P. (2020). How to beat science and influence people: Policymakers and propaganda in epistemic networks. *British Journal for the Philosophy of Science*, 71(4):1157–1186.

Weisberg, M. (2013). *Simulation and similarity: Using models to understand the world*. Oxford University Press.

Weisbuch, G. (2015). From anti-conformism to extremism. *Journal of Artificial Societies and Social Simulation*, 18(3):1.

West, G. B. (2017). *Scale: The universal laws of growth, innovation, sustainability, and the pace of life in organisms, cities, economies, and companies*. Penguin.

West, S. A., Griffin, A. S., and Gardner, A. (2007). Social semantics: Altruism, cooperation, mutualism, strong reciprocity and group selection. *Journal of Evolutionary Biology*, 20(2):415–432.

White, O. (1966). *Parliament of a thousand tribes: A study of New Guinea*. Heinemann.

Wigner, E. P. (1960). The unreasonable effectiveness of mathematics in the natural sciences. *Communications in Pure and Applied Mathematics*, 13:1–14.

Wilensky, U. (1999). NetLogo. http://ccl.northwestern.edu/netlogo/, Center for Connected Learning and Computer-Based Modeling, Northwestern University, Evanston, IL.

Wilke, C. O. (2019). *Fundamentals of data visualization: A primer on making informative and compelling figures*. O'Reilly Media.

Wilson, D. S. (2002). *Darwin's cathedral: Evolution, religion, and the nature of society*. University of Chicago Press.

Wilson, D. S. and Wilson, E. O. (2007). Rethinking the theoretical foundation of sociobiology. *Quarterly Review of Biology*, 82(4):327–348.

Wimsatt, W. C. (1987). False models as means to truer theories. In Nitecki, M. H. and Hoffman, A., editors, *Neutral models in biology*, pages 23–55. Oxford University Press.

Wolfram, S. (1984). Cellular automata as models of complexity. *Nature*, 311(5985):419–424.

Wright, S. (1938). The distribution of gene frequencies under irreversible mutation. *Proceedings of the National Academy of Sciences*, 24(7):253.

Wu, J., O'Connor, C., and Smaldino, P. E. (2023). The cultural evolution of science. In Tehrani, J., Kendal, R., and Kendal, J., editors, *Oxford handbook of cultural evolution*. Oxford University Press.

Yan, P. (2008). Separate roles of the latent and infectious periods in shaping the relation between the basic reproduction number and the intrinsic growth rate of infectious disease outbreaks. *Journal of Theoretical Biology*, 251(2):238–252.

Yang, Z., Algesheimer, R., and Tessone, C. J. (2016). A comparative analysis of community detection algorithms on artificial networks. *Scientific Reports*, 6(1):1–18.

Young, H. P. (1993). An evolutionary model of bargaining. *Journal of Economic Theory*, 59(1):145–168.

Young, H. P. (2006). The diffusion of innovations in social networks. In Blume, L. E. and S. N. Durlauf, editors, *The Economy as an Evolving Complex System III: Current Perspectives and Future Directions*. Oxford University Press, pages 267–281.

Yule, G. U. (1925). II.—A mathematical theory of evolution, based on the conclusions of Dr. JC Willis, FRS. *Philosophical Transactions of the Royal Society of London, Series B*, 213(402-410):21–87.

Zefferman, M. R. and Mathew, S. (2015). An evolutionary theory of large-scale human warfare: Group-structured cultural selection. *Evolutionary Anthropology*, 24(2):50–61.

Zollman, K. J. S. (2013). Network epistemology: Communication in epistemic communities. *Philosophy Compass*, 8(1):15–27.

Zollman, K. J. S. (2018). The credit economy and the economic rationality of science. *Journal of Philosophy*, 115(1):5–33.

Index